CANONICAL QUANTUM GRAVITY

Fundamentals and Recent Developments

CANONICAL QUANTUM GRAVITY

Fundamentals and Recent Developments

Francesco Cianfrani

University of Wrocław, Poland

Orchidea Maria Lecian

"Sapienza", University of Rome, Italy,
Max Planck Institute for Gravitational Physics, Germany &
ICRA International Center for Relativistic Astrophysics, Italy

Matteo Lulli

"Sapienza", University of Rome, Italy

Giovanni Montani

ENEA - C.R. Frascati, UTFUS-MAG & "Sapienza", University of Rome, Italy

NEW JERSEY • LONDON • SINGAPORE • BEIJING • SHANGHAI • HONG KONG • TAIPEI • CHENNAI

Published by

World Scientific Publishing Co. Pte. Ltd.
5 Toh Tuck Link, Singapore 596224
USA office: 27 Warren Street, Suite 401-402, Hackensack, NJ 07601
UK office: 57 Shelton Street, Covent Garden, London WC2H 9HE

British Library Cataloguing-in-Publication Data
A catalogue record for this book is available from the British Library.

Use of images from the Millennium Simulation, courtesy of V. Springel & the Virgo Consortium

Cover Art by Angel Patricio Susanna, http: //www.behance.net/kunturi

CANONICAL QUANTUM GRAVITY
Fundamentals and Recent Developments

ISBN 978-981-4556-64-4

Printed in Singapore

Contents

Preface

The request of a coherent quantization for the gravitational field dynamics emerges as a natural consequence of Einstein's Equations: the energy-momentum of a field is source of the spacetime curvature and therefore its microscopic quantum features must be reflected onto microgravity effects. The use of expectation values as sources is well-grounded as a first approximation only, and does not fulfill the requirements of a fundamental theory.

However, as is well-known, the achievement of a consistent Quantum Gravity theory remains a complete open task, due to a plethora of different subtleties, which are here summarized in the two main categories: i) General Relativity is a background-independent theory and therefore any analogy with the non-Abelian gauge formalisms must deal with the concept of a dynamic metric field; ii) the implementation in the gravitational sector of standard prescriptions, associated with the quantum mechanics paradigms, appears as a formal procedure, whose real physical content is still elusive.

There exists a qualitative consensus on the idea that a convincing solution to the Quantum Gravity problem will not arise before General Relativity and Quantum Mechanics are both deeply revised in view of a converging picture. Indeed, over the last fifteen years, the three most promising approaches to Quantum Gravity (*i.e.* String Theories, Loop Quantum Gravity and Non-commutative Geometries) revealed common features in defining a "lattice" nature for the microphysics of spacetime. This consideration makes clear that, up to the best of our present understanding, the main effort to improve fundamental formalisms must be the introduction of a "cut-off" physics, able to replace the notion of a spacetime continuum with a consistent discrete scenario. The correctness of such a statement will probably be regarded as the main success of the end of the last century reached in Theoretical Physics. A proper task for the present century is now to constrain the morphology of such a discrete microstructure of spacetime, to get, at least, a phenomenological description for Quantum Gravity effects. From a theoretical point of view, an important aim would consist of a unified picture containing common features of the present approaches, but synthesized into a more powerful mathematical language. On the one hand, it would be important to recognize non-commutative properties in the loop representation of spacetime; on the other hand, transporting the background independence of the "spin networks" into the interaction framework characterizing String Theories would constitute a relevant progress. The viability of these two goals is an intuitive perspective, but it contrasts with the rigidity of the corresponding formalisms, which confirms the request for more general investigation tools.

The simplest approach to the quantization of General Relativity relies on the implementation of the canonical method on the phase-space structure associated with the gravitational degrees of freedom. This attempt is the one originally pursued by B. DeWitt in 1967 and it immediately revealed all the pathologies contained in the canonical quantum geometrodynamics. It was this clear inconsistence of the canonical quantum procedure (referred to the second order formalism) that attracted the attention of a large number of researchers, active on the last four decades. This strong effort of re-analysis, which could give the feeling of an overestimation of the real chances allowed by the canonical method, found its "merry ending" in the developments in Loop Quantum Gravity and its applications. In fact, the formulation of the Hamiltonian problem for General Relativity provided by A. Ashtekar in 1986, allowed for a new paradigm for the canonical method, in close analogy with an $SU(2)$ gauge theory. The main success of Loop Quantum Gravity is recognizing a discrete structure of spacetime, by starting from continuous variables in the phase space of the theory. The origin of such a valuable result consists in the discrete nature acquired by the spectra of areas and volumes, reflecting to some extent the compactness of the $SU(2)$ group emerging in this formulation. The merit of the loop quantization of gravity can be identified in the possibility to deal with non-local geometric variables, like holonomies and fluxes, instead of the simple metric analysis faced by DeWitt. In fact, using the physical properties associated with the connection and the so-called electric fields, the background independence of the theory naturally emerges, and the quantum structure of spacetime comes out, as far as the Hilbert space is characterized via the spin-network basis.

Quantum Cosmology is a privileged arena in which different Quantum Gravity approaches can be implemented and it is expected to provide a consistent picture of the Universe birth. In this respect, the standard minisuperspace implementations of both the Wheeler-DeWitt formulation and of Loop Quantum Gravity (Loop Quantum Cosmology) are predictive to some extent.

In particular the Wheeler-DeWitt equation, as restricted to the homogeneous cosmological spaces, presents some pleasant features, such that the Universe dynamics resembles that of a scalar field interacting with a non-trivial and non-perturbative potential. In this formulation, the cosmological singularity is not removed, but a consistent Universe evolution throughout the Planck era can be clearly traced.

For what concerns Loop Quantum Cosmology, a Big-Bounce emerges due to the discreteness of the space volume spectrum, and it offers a viable semiclassical picture of the singularity removal by a phenomenological repulsion of the high curvature gravity. Nonetheless, a critical revision of the standard loop Universe quantization can be given, based on observing how the homogeneity restriction implies a gauge fixing on the $SU(2)$ internal symmetry of the full theory. A recent promising alternative reformulation of the Loop Quantum Gravity dynamics for the primordial Universe is presented together with its basic tools ("in primis" the presence of an intertwiner structure), which makes it closer to the full theory than the Loop Quantum Cosmology paradigm.

The analysis of Quantum Cosmology here presented is probably one of the most original contributions offered in this volume to the comprehension of the impact that quantum physics can have on the dynamics of the gravitational field. It can be regarded as the prize the reader gets out of the study of the preceding chapters concerning the successes and the limits of the full Quantum Gravity theory.

The aim of this book is to provide a detailed account of the physical content emerging from the canonical approach to Quantum Gravity. All the crucial steps in our presentation have a rather pedagogical character, providing the reader with the necessary tools to become involved in the field. Such a pedagogical aspect is then balanced and completed by subtle discussions on specific topics which we regard as relevant for the physical insight they outline on the treated questions. Our analysis is not aimed at convincing the reader about a pre-constituted point of view instead, our principal goal is to review the picture of Canonical Quantum Gravity on the basis of the concrete facts at the ground of its clear successes, and also of its striking shortcomings.

In order to focus our attention on the physical questions affecting the consistence of a canonical quantum model, when extended to the gravitational sector, we provide a critical discussion of all the fundamental concepts of quantum physics, by stressing the peculiarity of the gravitational case in their respect. All the key features of Loop Quantum Gravity are described in some details, presenting the specific technicalities, but privileging their physical interpretation. Finally, all the open questions concerning the faced topics are clearly outlined by a precise criticism on the weak points present in the quantum construction.

The detailed structure of the book is

- the main goal of the first chapter is to provide the basic formalisms and concepts at the ground of Einsteinian theory of gravity. We start by constructing the tensor representation of a manifold differential geometry in terms of the so-called embedding procedure, *i.e.* we consider a parametric representation for a generic hypersurface. The concepts of spacetime curvature and geodesic deviation are then fully addressed. We discuss the fundamental principles of General Relativity and show how they fix the form of Einstein's equations. As a last step we provide the vierbein formalism for further applications throughout the book.
- in the second chapter we provide a brief presentation of the Universe evolution based on the isotropic Robertson-Walker geometry. We first discuss the morphology of a homogeneous and isotropic Universe and its kinematic implications, able to account for observations such as the photons redshift and the Hubble law. Then, we address the dynamic features of the isotropic Universe as described by Einstein's equations. The structure of the dynamic equations is outlined by characterizing the matter source via a perfect fluid endowed with a proper equation of state. This allows one to develop a brief sketch of the Universe thermal history, describing the different stages of its evolution. Finally, we analyze the inflationary paradigm, offering a compact review of the main shortcomings of the standard cosmological model, as well as the basic ideas at the ground of the inflationary scenario.

- in the third chapter, we introduce the main concepts of the standard Hamiltonian formulation for constrained systems, namely the distinction between primary and secondary constraints as well as between first-class and second-class constraints. Moreover the definition of canonical transformations is reviewed and broadened introducing the concepts of weak- and gauged-canonicity. In order to give some practical insight, we use an example based on the free electromagnetic field where all the concepts previously faced are sketched in a rather pedagogical way. In this respect we will present the L. Castellani algorithm for constructing the generators of gauge transformations.
- in the fourth chapter, the Lagrangian formulation of General Relativity is critically discussed giving different but equivalent (under certain hypothesis) Lagrangian densities for the gravitational field. We start by introducing the commonly used Einstein-Hilbert Lagrangian density followed by the study of the action principle when matter is included. We discuss the non-covariant Lagrangian density approach due to Einstein and Dirac. The study of an extended theory of gravity in the form of an $f(R)$ Lagrangian density is developed followed by description of the Palatini formulation. The ADM formulation is presented using the embedding technique developed in the first chapter and a critical discussion on the boundary terms of some of the formulations is given allowing one to perform the transformation linking some of the presented approaches, namely the ADM to Einstein's and Dirac's ones.
- in the fifth chapter, we recall the basic features of Quantum Mechanics. In particular, we review the definition of quantum states as elements of a Hilbert space and of observables as linear operators. The classical to quantum correspondence between Poisson brackets and commutators is discussed as a tool for the quantization of a generic mechanical system. Then, we present the Schrödinger and Heisenberg representations. The Hamilton-Jacobi equation and the development of semiclassical coherent states are analyzed in order to bridge the quantum formulation with the classical world. Furthermore, we introduce Weyl quantization and outline the main steps of the GNS construction. The polymer representation is then presented as an application. Finally, the quantization of some physically-relevant constrained systems and its shortcomings are discussed.

- in the sixth chapter, the standard Hamiltonian formulation of General Relativity is reviewed. Then, the two possible scenarios for the canonical quantization of the gravitation field are illustrated. On one hand, we present the reduction to the canonical form and its main issues are briefly outlined. On the other hand, we focus on the Wheeler-De Witt approach. We emphasize how the implementation of the constraints via the Dirac prescription leads to the Wheeler-DeWitt equation. The main technical and interpretative shortcomings of this formulation (which prevents it from being a well-grounded quantum description for the gravitational field) are discussed: the absence of a real Hilbert space structure, the need for a consistent regularization scheme and the problem of time. Some proposals to identify a time variable are presented: the Brown-Kuchař model, the multitime approach and the Vilenkin proposal.
- in the seventh chapter, Yang-Mills gauge theories are presented. The classical Lagrangian and Hamiltonian formulations are analyzed. Then, we introduce some tools which are useful for lattice quantization (spin networks and spin foams). Furthermore, a formal comparison between Yang-Mills models and General Relativity is performed by outlining the similarities but also the unavoidable differences. Poincaré gauge theory is then discussed as an alternative formulation in which gravity behaves as a Yang-Mills interaction. The Holst formulation for gravity is then revised and the emergence of a SU(2) gauge symmetry is emphasized. The analysis of the associated constrained system in vacuum and in the presence of matter fields is carefully discussed in a generic local Lorenz frame. Finally, the Kodama state is presented.
- the eighth chapter is devoted to Loop Quantum Gravity. The role of holonomies and fluxes in quantum theories is briefly discussed, while the quantization of the corresponding algebra in Loop Quantum Gravity is presented. The main achievements of this framework are emphasized: the definition of a kinematical Hilbert space, the success in the quantization of kinematical constraints, the achievement of discrete spectra for the area and volume operators, the implementation of the scalar constraint on a quantum level. A careful explanation of the main technical and conceptual issues (the study of the dynamics, the semiclassical limit, the constraint algebra, the role of the Immirzi parameter) is given, while the Master Constraint program and the Algebraic Quantum Grav-

ity setup are briefly discussed as attempts towards the solution of some shortcomings. Finally, a comparison between the picture of the quantum spacetime coming out from the Loop Quantum Gravity and the Wheeler-DeWitt frameworks is performed.

- in the ninth chapter, Quantum Cosmology is presented as an arena in which the predictions of Quantum Gravity can be investigated. The minisuperspace approximation is discussed: we point out the simplifications that the restriction to homogeneous models provides for the quantization issue, such that a proper quantum description of the Bianchi type I and IX models in the Wheeler-DeWitt approach can be given, even though this is not generically enough to obtain a consistent quantum model for the Universe (the presence of singularity and the chaos in Bianchi IX). Then, the paradigm of Loop Quantum Cosmology is introduced and the bouncing solution replacing the Big-Bang singularity is explicitly shown to arise. Finally, a critical discussion on the foundation of Loop Quantum Cosmology is given and a new research line for the cosmological sector of Loop Quantum Gravity, namely Quantum Reduced Loop Gravity, is presented.

We would like to thank Abhay Ashtekar for interesting discussions on Loop Quantum Gravity during the First Stueckelberg Workshop (Pescara, June 25-July 1 2006), and which inspired the writing of this book.

We would also like to thank Richard Arnowitt, Stanley Deser and Charles W. Misner for their comments and discussion on the Hamiltonian formulation of General Relativity and for having called our attention on the specific relevance of reference [17].

The work of FC was supported by funds provided by the National Science Center under the agreement DEC- 2011/02/A/ST2/00294.

The work of OML was partially supported by the research grant 'Reflections on the Hyperbolic Plane' from the Albert Einstein Institute for Gravitational Physics - MPI, Potsdam-Golm, and partially by the research grant 'Classical and Quantum Physics of the Primordial Universe' from Sapienza University of Rome- Physics Department, Rome.

List of Figures

List of Notations

{-1,1,1,1}	spacetime signature
M^5	five-dimensional Minkowski space
$\vec{n}_I$	basis vectors in five-dimensonal Minkowski space
$\mathcal{M}$	four-dimensional manifold (spacetime)
$\vec{b}_\mu$	basis vectors in four-dimensional spacetime
$g_{\mu\nu}$	four-dimensional metric tensor
$\Gamma^\mu_{\nu\rho}$	four-dimensional Christoffel symbols
∇_μ	four-dimensional covariant derivative
$R^\mu_{\ \nu\rho\sigma}$	four-dimensional Riemann tensor
e^α_μ	vierbein vectors
$\omega^{\alpha\beta}_\mu$	four-dimensonal spin connections
$T_{\mu\nu}$	stress-energy tensor
ρ	energy density
p	pressure
S	action
L	Lagrangian
$\mathcal{L}$	Lagrangian density
$[.,.]$	Poisson brackets
$\sigma^{\cdot\cdot}$	symplectic form
H	Hamiltonian
$\approx$	weak equality, *i.e.* modulo constraints
Σ_{x^0}	spatial hypersurfaces
η^μ	normal vector to spatial hypersurfaces
N	lapse function
N^i	shift vector

h_{ij}	three-dimensional metric tensor
$\bar{\Gamma}^i_{jk}$	three-dimensional Christoffel symbols
D_i	three-dimensional covariant derivative
K_{ij}	extrinsic curvature
$\bar{R}^i{}_{jkl}$	three-dimensional Riemann tensor
$[\hat{\cdot}, \hat{\cdot}]$	commutators
H	Hilbert space
$\mathcal{H}$	super-Hamiltonian
$\mathcal{H}_i$	supermomentum
A_μ	(group-valued) connection
g	group element
$\rho(g)$	irreducible representation
D_ρ	dimension of ρ
ψ_S	spin-network state
E^i_a	inverse densitized dreibein vectors
G_a	Gauss constraint
$\mathcal{S}$	scalar constraint
$\mathcal{V}^i$	vector constraint
α	graph
e	edge
v	vertex
σ	surface
τ_a	$SU(2)$ generator
$U_\alpha(A)$	holonomy of A_i along α
$E_a(\sigma)$	flux of E^i_a across σ
f_α	cylindrical function over α
G	Newton constant
$\hbar$	Planck constant
c	speed of light
l_P	Planck length
γ	Immirzi parameter
$\mu, \nu, \rho, \sigma, \ldots$	four-dimensional spacetime indices
$\alpha, \beta, \gamma, \delta, \ldots$	four-dimensional vierbein indices

$i, j, k, l, \ldots$	three-dimensional spatial indices
$a, b, c, d, \ldots$	three-dimensional dreibein indices

Chapter 1

Introduction to General Relativity

In this chapter we will analyze the affine and metric properties of a 4-dimensional manifold, stressing the tensor character of all fundamental geometric objects. Then we will introduce the concepts of geodesics, curvature, and geodesic deviation in order to characterize the motion of test-particles in a generic spacetime. By means of the fundamental General Relativity and Equivalence principles we develop the basic ideas at the ground of General Relativity theory, matching them with the previously constructed formalism. Hence, we write down Einstein's equations describing the geometrodynamics of spacetime and we also outline the role played by matter fields.

Finally, we provide an important formalism, commonly called vierbein representation, which allows to determine the existence of a local gauge-like symmetry associated to the Lorentz group which can be identified with an internal symmetry. In fact, the action of the local Lorentz group leaves unchanged both the metric and the Riemann tensors. In Chapter 7 this will be the starting point in order to represent General Relativity in the fashion of a non-Abelian Yang-Mills theory.

1.1 Parametric manifold representation

Let us consider a differentiable 4-dimensional manifold $\mathcal{M}$ embedded in a Minkowskian 5-dimensional space M^5, endowed with the natural scalar product.

It is possible to define a parametric representation of $\mathcal{M}$ via a 5-dimensional vector, $\vec{X} \in M^5$, of which each component is a function of

4 parameters u^μ. We can project $\vec{X}$ on an orthonormal basis $\{\vec{n}_I\}$

$$\vec{X} = X^I(u)\,\vec{n}_I \qquad \vec{n}_I \cdot \vec{n}_J = \eta_{IJ} \qquad I, J = 0, \ldots, 4, \tag{1.1}$$

where the $\cdot$ indicates the common scalar product using the Minkowski metric $\eta_{IJ} = \mathrm{diag}\{-1, 1, 1, 1, 1\}$. The manifold is said to be infinitely differentiable if $X^I(u^\mu) \in C^\infty$, for $I = 0, \ldots, 4$.

On the other hand, it is also possible to define an adapted basis on $\mathcal{M}$, which is given at each point by the set of the vectors which are tangent to the manifold itself (see figure 1.1). The tangent vectors $\{\vec{b}_\mu\}$ read

$$\vec{b}_\mu(u) \equiv \frac{\partial X^I(u)}{\partial u^\mu}\vec{n}_I = b^I_\mu(u)\,\vec{n}_I \qquad I = 0, \ldots, 4 \qquad \mu = 0, \ldots, 3. \tag{1.2}$$

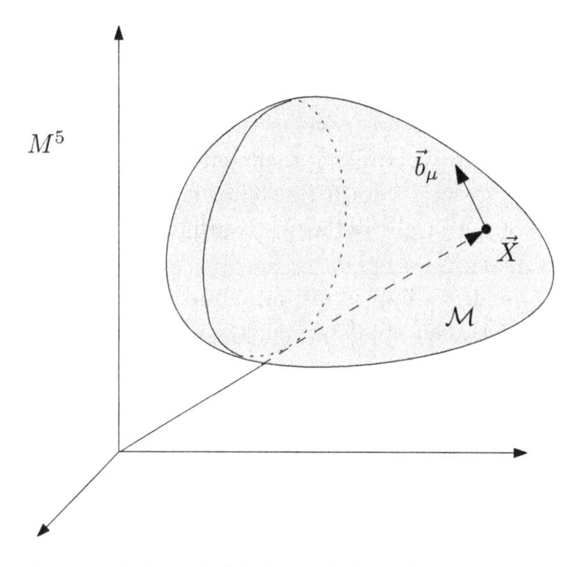

Figure 1.1 A drawing of the embedded manifold $\mathcal{M}$ described by the 5-dimensional vector $\vec{X}$, with the basis vectors $\vec{b}_\mu$.

Then, it follows that any vector $\vec{V} \in \mathcal{M}$ can be projected onto the basis $\{\vec{b}_\mu\}$ as follows

$$\vec{V} = \mathcal{V}^I(u)\,\vec{n}_I = V^\mu(u)\,b^I_\mu(u)\,\vec{n}_I, \tag{1.3}$$

where the $\mathcal{V}^I$ are the components in the 5-dimensional Minkowskian space. In this way we defined the contravariant components of $\vec{V}$ as V^μ. It is possible to define the covariant components via the scalar product with the basis vectors

$$\begin{aligned} V_\rho(u) =& \vec{V}(u)\cdot\vec{b}_\rho(u) = V^\mu(u)\,\vec{b}_\mu(u)\cdot\vec{b}_\rho(u) = V^\mu(u)\,b^I_\mu(u)\,b^J_\rho(u)\,\vec{n}_I\cdot\vec{n}_J \\ =& V^\mu(u)\,b^I_\mu(u)\,b^J_\rho(u)\,\eta_{IJ}. \end{aligned} \tag{1.4}$$

Hence, we generally obtained that $V_\mu \neq V^\mu$, and that the two different kinds of components are linked by the non-trivial relation

$$V^\mu(u)\,b^I_\mu(u)\,b^J_\rho(u)\,\eta_{IJ} = V_\rho(u). \tag{1.5}$$

1.2 Tensor formalism

The set of parameters $\{u^\mu\}$ defines a coordinate system adapted to the tangent hypersurface at every point of $\mathcal{M}$. Any geometrical quantity belonging to the tangent hypersurface can be expressed as a function of $\{u^\mu\}$.

If we choose a different parametrization of the manifold, we obtain a coordinate transformation which we request to be invertible and differentiable, *i.e.* a diffeomorphism, on $\mathcal{M}$

$$z^\mu = z^\mu(u), \tag{1.6}$$

hence, we obtain the transformation of the basis vectors of the tangent hypersurfaces,

$$\vec{b}_{\mu'}(z) = \frac{\partial\vec{X}(z(u))}{\partial z^{\mu'}} = \frac{\partial\vec{X}(u)}{\partial u^\rho}\frac{\partial u^\rho}{\partial z^{\mu'}} = \vec{b}_\rho(u)\,\frac{\partial u^\rho}{\partial z^{\mu'}}. \tag{1.7}$$

Vectors on $\mathcal{M}$ transforming this way, *i.e.* via the derivatives of the old variables with respect to the new ones, are called *covariant*. It then follows

$$\begin{aligned} \vec{V} =& V^\rho(u)\,\vec{b}_\rho(u) = V^\rho(u)\,\frac{\partial z^{\mu'}}{\partial u^\rho}\vec{b}_{\mu'}(z) = V^{\mu'}(z)\,\vec{b}_{\mu'}(z), \\ \Rightarrow\quad & V^{\mu'}(z) = V^\rho(u(z))\,\frac{\partial z^{\mu'}}{\partial u^\rho}, \end{aligned} \tag{1.8}$$

vectors transforming as V^ρ are called *contravariant*, *i.e.* via the derivatives of the new variables with respect to the old ones.

A *scalar* quantity is an object independent of the choice of coordinates. We write it as a contraction of covariant and contravariant quantities

$$\phi = V^\mu(u)\,V_\mu(u) = \frac{\partial u^\mu}{\partial z^{\rho'}}V^{\rho'}(u(z))\,\frac{\partial z^{\rho'}}{\partial u^\mu}V_{\rho'}(u(z)) = V^{\rho'}(u(z))\,V_{\rho'}(u(z)), \tag{1.9}$$

where we used the common property of the derivative chain rule

$$\frac{\partial u^{\mu}}{\partial z^{\rho'}}\frac{\partial z^{\rho'}}{\partial u^{\nu}}=\delta^{\mu}_{\nu}, \qquad \frac{\partial u^{\mu}}{\partial z^{\sigma'}}\frac{\partial z^{\rho'}}{\partial u^{\mu}}=\delta^{\rho'}_{\sigma'}. \tag{1.10}$$

We define a tensor of rank m a geometric object $T_{\rho\dots}{}^{\sigma\dots}$ which transforms as the product of m vectorial components which can be either covariant or contravariant. That is

$$T_{\mu'\dots}{}^{\nu'\dots}(z)=T_{\rho\dots}{}^{\sigma\dots}(u)\,\frac{\partial z^{\nu'}}{\partial u^{\sigma}}\cdots\frac{\partial u^{\rho}}{\partial z^{\mu'}}. \tag{1.11}$$

It follows that if a tensor has all its components equal to zero, in one coordinate system, there is no coordinate transformation able to alter this equality.

A tensor is said to be *symmetric* or *antisymmetric* with respect to the exchange of some pairs of indices of the same kind (*i.e.* both contravariant or both covariant) if, under such an exchange one gets

$$T_{\mu\dots\nu}=T_{\nu\dots\mu}, \qquad T_{\mu\dots\nu}=-T_{\nu\dots\mu}. \tag{1.12}$$

For any tensor we will adopt the following notation in order to indicate its symmetric or antisymmetric part with respect to a pair of indices

$$T_{(\mu\nu)}=\frac{1}{2}\left(T_{\mu\nu}+T_{\nu\mu}\right), \qquad T_{[\mu\nu]}=\frac{1}{2}\left(T_{\mu\nu}-T_{\nu\mu}\right), \tag{1.13}$$

and it is clearly possible to write a generic tensor as the direct sum of its symmetric and antisymmetric parts

$$T_{\mu\nu}=T_{(\mu\nu)}+T_{[\mu\nu]}. \tag{1.14}$$

Another class of important geometrical objects are the so-called tensor/scalar densities: the transformation rules are the usual ones multiplied for some power of the determinant of the transformation itself, *i.e.* the so-called Jacobian. As an example the determinant of any rank two tensor transforms as a scalar density:

$$T_{\mu\nu}(u)=\frac{\partial z^{\rho}}{\partial u^{\mu}}\frac{\partial z^{\sigma}}{\partial u^{\nu}}T_{\rho'\sigma'}(z) \quad\Rightarrow\quad \det(T)=\left[\det\left(\frac{\partial z}{\partial u}\right)\right]^{2}\det(T'). \tag{1.15}$$

1.3 Affine properties of the manifold

1.3.1 *Ordinary derivative*

We now describe the transformation properties of the ordinary derivatives. For the derivative of a vectorial quantity one gets

$$\frac{\partial V_\rho(u)}{\partial u^\sigma} = \frac{\partial z^{\mu'}}{\partial u^\sigma}\frac{\partial}{\partial z^{\mu'}}\left[V_{\nu'}(z)\frac{\partial z^{\nu'}}{\partial u^\rho}\right] = \frac{\partial z^{\mu'}}{\partial u^\sigma}\frac{\partial z^{\nu'}}{\partial u^\rho}\frac{\partial V_{\nu'}(z)}{\partial z^{\mu'}} + V_{\nu'}(z)\frac{\partial^2 z^{\nu'}}{\partial u^\rho \partial u^\sigma}. \tag{1.16}$$

Hence, the ordinary derivative of a vector behaves as a tensor only under linear coordinate transformations for which the second derivative term disappears. This behavior classifies the ordinary derivatives as *pseudotensors*, which are quantities that transform as tensors only under linear transformations.

Otherwise, the ordinary derivative of a scalar behaves as a tensor, in fact one gets

$$\frac{\partial \phi(u)}{\partial u^\rho} = \frac{\partial z^{\mu'}}{\partial u^\rho}\frac{\partial \phi(u(z))}{\partial z^{\mu'}}. \tag{1.17}$$

Hence, from this consideration we see that in order to construct a covariant formulation of some physical phenomenon it is necessary to generalize the concept of ordinary partial derivative. This path leads to the introduction of the so-called covariant derivative.

1.3.2 *Covariant derivative*

Let us consider the derivatives of a vector field $\vec{V}(u) \in \mathcal{M}$ with respect to the set of parameters $\{u^\mu\}$. We obtain

$$\frac{\partial \vec{V}(u)}{\partial u^\rho} = \frac{\partial}{\partial u^\rho}\left(V^\sigma(u)\vec{b}_\sigma(u)\right) = \frac{\partial V^\sigma(u)}{\partial u^\rho}\vec{b}_\sigma(u) + V^\sigma(u)\frac{\partial \vec{b}_\sigma(u)}{\partial u^\rho}. \tag{1.18}$$

We choose a complete basis of M^5 given by the set $\{\vec{b}_\mu,\vec{n}\}$, where $\vec{n}$ is at every point the normal vector to $\mathcal{M}$; it follows that at each point $\vec{b}_\mu \cdot \vec{n} = 0$. We now project the derivative of a basis vector as

$$\frac{\partial \vec{b}_\mu(u)}{\partial u^\nu} = \Gamma^\rho_{\mu\nu}(u)\vec{b}_\rho(u) + \Pi_{\mu\nu}(u)\vec{n}(u), \tag{1.19}$$

and since for the basis vectors one has

$$\frac{\partial \vec{b}_\mu(u)}{\partial u^\nu} = \frac{\partial^2 \vec{X}}{\partial u^\nu \partial u^\mu} = \frac{\partial^2 \vec{X}}{\partial u^\mu \partial u^\nu} = \frac{\partial \vec{b}_\nu(u)}{\partial u^\mu}, \tag{1.20}$$

it follows that $\Gamma^{\rho}_{\mu\nu} = \Gamma^{\rho}_{\nu\mu}$ and $\Pi_{\mu\nu} = \Pi_{\nu\mu}$ by virtue of Schwarz theorem. We then write the projection tangent to the manifold as

$$\begin{aligned}\frac{\partial \vec{V}(u)}{\partial u^{\rho}} \cdot \vec{b}_{\mu}(u) = &\left[\frac{\partial V^{\nu}(u)}{\partial u^{\rho}} + \Gamma^{\nu}_{\rho\sigma}(u) V^{\sigma}(u)\right] \vec{b}_{\nu}(u) \cdot \vec{b}_{\mu}(u) \\ =& \nabla_{\rho} V^{\nu}(u) \vec{b}_{\nu}(u) \cdot \vec{b}_{\mu}(u) .\end{aligned} \tag{1.21}$$

The affine connections $\Gamma^{\nu}_{\rho\sigma}$ contain the information related to the variation of the basis vectors along $\mathcal{M}$. We define the covariant derivative of a contravariant vector as

$$\nabla_{\mu} V^{\rho}(u) = \frac{\partial V^{\rho}(u)}{\partial u^{\mu}} + \Gamma^{\rho}_{\mu\nu}(u) V^{\nu}(u) . \tag{1.22}$$

Starting from the requirement that the covariant derivative of a scalar should be equal to the ordinary derivative and that it should obey Leibniz' rule, we derive the definition for covariant vectors

$$\nabla_{\mu} V_{\nu}(u) = \frac{\partial V_{\nu}(u)}{\partial u^{\mu}} - \Gamma^{\rho}_{\mu\nu}(u) V_{\rho}(u) . \tag{1.23}$$

Hence, we have defined the covariant derivative of a vector as the projection of its ordinary derivative on the manifold the vector belongs to. As we are about to see, by definition, covariant derivatives of tensors transform as tensors.

1.3.3 *Properties of the affine connections*

As shown through (1.20) the affine connections have two symmetric indices, $\Gamma^{\gamma}_{\alpha\beta} = \Gamma^{\gamma}_{\beta\alpha}$. This is not the general case. Indeed, it is possible to introduce an antisymmetric part of the affine connections called Torsion field. This quantity is discussed in section 4.1.6.

With some calculations, it can be shown that the transformation rule of the affine connection is

$$\Gamma^{\mu'}_{\rho'\sigma'}(z) = \left[\frac{\partial u^{\rho}}{\partial z^{\rho'}} \frac{\partial u^{\sigma}}{\partial z^{\sigma'}} \Gamma^{\mu}_{\rho\sigma}(u) + \frac{\partial^2 u^{\mu}}{\partial z^{\rho'} \partial z^{\sigma'}}\right] \frac{\partial z^{\mu'}}{\partial u^{\mu}}, \tag{1.24}$$

taking as a starting point the relation

$$\Gamma^{\rho}_{\mu\nu}(u) \vec{b}_{\rho}(u) \cdot \vec{b}_{\sigma}(u) = \frac{\partial \vec{b}_{\mu}(u)}{\partial u^{\nu}} \cdot \vec{b}_{\sigma}(u) = \frac{\partial^2 \vec{X}}{\partial u^{\mu} \partial u^{\nu}} \cdot \vec{b}_{\sigma}(u) , \tag{1.25}$$

and using the previously defined coordinate transformation rules. The affine connections are pseudotensors because they transform as tensors only under linear coordinate transformations, for which the second derivative term vanishes. Furthermore, it is easy to show that $\Pi_{\mu\nu}$ is a tensor, the so-called *extrinsic curvature*, which we will discuss in detail in section 4.2.1.

However, the transformation law of the affine connection exactly compensates the transformation properties of the ordinary derivative allowing the covariant derivative to transform as a tensor, hence giving

$$\nabla_{\mu'} V_{\nu'}(z) = \frac{\partial z^\rho}{\partial u^{\mu'}} \frac{\partial z^\sigma}{\partial u^{\nu'}} \nabla_\rho V_\sigma(u). \tag{1.26}$$

1.4 Metric properties of the manifold

1.4.1 *Metric tensor*

Considering the scalar product of two vectors $\vec{v}, \vec{w} \in \mathcal{M}$, we can write

$$\begin{aligned} \vec{v}(u) \cdot \vec{w}(u) =& v^\mu(u) w^\nu(u) \vec{b}_\mu(u) \cdot \vec{b}_\nu(u) = v^\mu(u) w^\nu(u) b^I_\mu(u) b^J_\nu(u) \vec{n}_I \cdot \vec{n}_J \\ =& v^\mu(u) w^\nu(u) g_{\mu\nu}(u), \end{aligned} \tag{1.27}$$

defining the rank two covariant tensor

$$g_{\mu\nu} = b^I_\mu(u) b^J_\nu(u) \eta_{IJ} = g_{\nu\mu}, \tag{1.28}$$

which is symmetric, since $\vec{n}_I \cdot \vec{n}_J = \eta_{IJ}$. This object is called *metric tensor.* We can compute the line element of a given metric tensor by contracting it with the contravariant vector du^μ

$$ds^2(u) = g_{\mu\nu}(u) du^\mu du^\nu, \tag{1.29}$$

clearly being a scalar quantity. The contravariant components of the metric tensor, $g^{\mu\nu}$, are uniquely defined by the property

$$g^{\mu\rho}(u) g_{\rho\nu}(u) = \delta^\mu_\nu. \tag{1.30}$$

Another unique property of the metric tensor is that its contraction with a tensor changes the transformation properties of the contracted index, *i.e.*

$$T_{\mu\nu}(u) = g_{\mu\rho}(u) T^\rho_\nu(u) = g_{\mu\rho}(u) g_{\nu\sigma}(u) T^{\rho\sigma}(u), \tag{1.31}$$

as it can be directly argued from (1.5). We now request a property which is called *compatibility* condition of the metric tensor with the affine connection, called Ricci theorem, which translates in the requirement that the covariant derivative of the metric tensor identically vanishes at any point

$$\nabla_\rho g_{\mu\nu}(u) = 0, \qquad \nabla_\rho g^{\mu\nu}(u) = 0. \tag{1.32}$$

We give here a simple demonstration of this theorem: because of (1.19) we can write

$$\nabla_\mu \vec{b}_\nu = \Pi_{\mu\nu} \vec{n}, \tag{1.33}$$

so that for the metric tensor one easily obtains

$$\nabla_\mu g_{\rho\sigma} = \vec{b}_\sigma \cdot \left(\nabla_\mu \vec{b}_\rho\right) + \vec{b}_\rho \cdot \left(\nabla_\mu \vec{b}_\sigma\right) = \Pi_{\mu\rho} \vec{b}_\sigma \cdot \vec{n} + \Pi_{\mu\sigma} \vec{b}_\rho \cdot \vec{n} = 0. \tag{1.34}$$

1.4.2 *Christoffel symbols*

We now use equalities (1.32) in order to obtain a definition of the affine connections $\Gamma^\rho_{\mu\nu}$. One has[1]

$$\nabla_\rho g_{\mu\nu}(u) = \partial_\rho g_{\mu\nu}(u) - \Gamma^\sigma_{\rho\mu}(u)\, g_{\sigma\nu}(u) - \Gamma^\sigma_{\rho\nu}(u)\, g_{\sigma\mu}(u) = 0, \tag{1.35}$$

and, defining $\Gamma_{\mu\nu\,\rho} \equiv g_{\rho\sigma} \Gamma^\sigma_{\mu\nu}$ (which are called Christoffel symbols of the second kind), we get the identity

$$\partial_\rho g_{\mu\nu}(u) = \Gamma_{\rho\mu\,\nu}(u) + \Gamma_{\rho\nu\,\mu}(u). \tag{1.36}$$

Performing now cyclic permutations of the indices, it is easy to find three independent relations which properly combined lead us to

$$\Gamma_{\mu\nu\,\rho}(u) = \frac{1}{2}\left[\partial_\mu g_{\nu\rho}(u) + \partial_\nu g_{\rho\mu}(u) - \partial_\rho g_{\mu\nu}(u)\right], \tag{1.37}$$

and thus to the Christoffel symbols of the first kind

$$\Gamma^\sigma_{\mu\nu}(u) = \frac{1}{2} g^{\sigma\rho}(u)\left[\partial_\mu g_{\nu\rho}(u) + \partial_\nu g_{\rho\mu}(u) - \partial_\rho g_{\mu\nu}(u)\right]. \tag{1.38}$$

We have mentioned above that these quantities are pseudotensors, hence, it is always possible to find a coordinate transformation such that the transformed values equal zero at some point.

We assume that at the origin of both coordinate systems the *old* Christoffel symbols have a non-zero value, $\Gamma^\mu_{\rho\sigma}(u=0) = \Pi^\mu_{\rho\sigma} \neq 0$. In order to have vanishing *new* Christoffel symbol it suffices to write the coordinate transformation as

$$u^\mu = z^{\mu'} - \frac{1}{2} \Pi^\mu_{\rho\sigma} z^{\rho'} z^{\sigma'}, \tag{1.39}$$

as can be checked by direct substitution in (1.24). Since $z = 0$ implies $x = 0$ we have obtained

$$\Gamma^{\mu'}_{\rho'\sigma'}(0) = 0. \tag{1.40}$$

---

[1] From now on, we will use the indices notation of covariant derivatives also for ordinary derivatives: $\partial / \partial x^\mu = \partial_\mu$.

1.5 Geodesic equation and parallel transport

We are now equipped with what we need in order to give a generalization of the concept of *straight line* which is fundamental in Euclidean geometry.

In a Minkowskian space M^n, a straight line is, by construction, a self-parallel curve and it minimizes the distance between two given points. We can write it as a parametric function of the *curvilinear abscissa* s

$$u^\mu(s) = a^\mu s, \qquad a^\mu = \text{const}, \quad \vec{a}\cdot\vec{a} = \eta_{\mu\nu}a^\mu a^\nu = \pm 1, \tag{1.41}$$

so that the tangent vector $t^\mu \equiv du^\mu/ds = a^\mu$ verifies the normalization condition $\vec{t}\cdot\vec{t} = \pm 1$ and the differential relation

$$\frac{d}{ds}t^\mu = \frac{du^\rho}{ds}\partial_\rho t^\mu = t^\rho\partial_\rho t^\mu = 0. \tag{1.42}$$

In particular the sign of the normalization is linked to the space-like ($\vec{t}\cdot\vec{t} = 1$) or time-like ($\vec{t}\cdot\vec{t} = -1$) character of the line. For light-like lines one can define an affine parameter s such that equation (1.42) holds. Now, we want to generalize these properties in order to define a self-parallel curve on the embedded manifold $\mathcal{M}$. This is straightforward as soon as we substitute the Minkowski metric $\eta_{\mu\nu}$ with the manifold metric $g_{\mu\nu}$, and the ordinary derivative with the covariant one. The new normalization condition reads $g_{\mu\nu}t^\mu t^\nu = \pm 1$, and the relation (1.42) becomes

$$\begin{aligned} t^\rho\nabla_\rho t^\mu =& t^\rho\partial_\rho t^\mu + \Gamma^\mu_{\rho\sigma}t^\rho t^\sigma \\ =& \frac{dt^\mu}{ds} + \Gamma^\mu_{\rho\sigma}t^\rho t^\sigma \\ =& \frac{d^2u^\mu}{ds^2} + \Gamma^\mu_{\rho\sigma}\frac{du^\rho}{ds}\frac{du^\sigma}{ds} = 0. \end{aligned} \tag{1.43}$$

This equality is called geodesic equation, which, once solved, gives the equation of a curve that minimizes the distance between any pair of points belonging to the manifold. It is not surprising then, that the same equation can be obtained from the stationarity condition of the distance functional along some path $u^\mu = u^\mu(s)$:

$$\delta D[u(s)] = \delta\int ds\sqrt{\left|g_{\mu\nu}\frac{du^\mu}{ds}\frac{du^\nu}{ds}\right|} = 0. \tag{1.44}$$

It is easy to show that this stationarity condition implies, in the case

$g_{\mu\nu}t^\mu t^\nu = \pm 1$, the much simpler condition

$$\begin{aligned}\delta D\left[u\left(s\right)\right] =&\frac{1}{2}\delta\int ds\; g_{\mu\nu}\frac{dx^\mu}{ds}\frac{dx^\nu}{ds}\\ =&\frac{1}{2}\int ds\left[\partial_\rho g_{\mu\nu}\frac{dx^\mu}{ds}\frac{dx^\nu}{ds}-2\frac{d}{ds}\left(g_{\rho\nu}\frac{dx^\nu}{ds}\right)\right]\delta x^\rho\\ =&-\int ds\left[\frac{d^2x^\nu}{dx^2}-\Gamma^\nu_{\mu\sigma}\frac{dx^\mu}{ds}\frac{dx^\sigma}{ds}\right]g_{\nu\rho}\delta x^\rho=0.\end{aligned} \tag{1.45}$$

Since the variations δx^ρ are arbitrary, but vanishing at the endpoints, the geodesic equation follows.

Moreover, we can generically take the equation

$$A^\mu\nabla_\mu B^\nu = \nabla_{\vec{A}}B^\nu = 0, \tag{1.46}$$

which is the directional derivative of the vector B^μ along the direction of the vector A^μ. Since it equals zero this means that B^μ retains the same direction and the same length along the direction of A^μ. If we substitute the one-parameter vector field $A^\mu = A^\mu\left(s\right)$ the previous equations describe the *parallel transport* of the vector field $B^\mu = B^\mu\left(s\right)$ along the vector field $A^\mu = A^\mu\left(s\right)$. Hence, geodesics are those curves that parallel transport themselves.

1.6 Levi-Civita tensor

Let us now introduce a useful tool: the completely antisymmetric symbol. We examine it in detail for the two-dimensional case

$$\varepsilon_{ij} = \begin{vmatrix}\delta^0_i & \delta^0_j\\ \delta^1_i & \delta^1_j\end{vmatrix} = 2\delta^0_{[i}\delta^1_{j]}, \quad\Rightarrow\quad \varepsilon_{01}=-\varepsilon_{10}=1, \quad \varepsilon_{ii}=0, \tag{1.47}$$

and since we are dealing with algebraic properties we define the contravariant form simply as

$$\varepsilon^{ij} = 2\delta^{[i}_0\delta^{j]}_1. \tag{1.48}$$

Another useful property comes from the product of two completely antisymmetric symbols:

$$\begin{aligned}\varepsilon^{ij}\varepsilon_{rs} =&4\delta^{[i}_0\delta^{j]}_1\delta^0_{[r}\delta^1_{s]}\\ =&\delta^i_0\delta^0_r\delta^j_1\delta^1_s+\delta^i_1\delta^1_r\delta^j_0\delta^0_s-\delta^i_0\delta^0_s\delta^j_1\delta^1_r-\delta^i_1\delta^1_s\delta^j_0\delta^0_r\\ =&\delta^i_k\delta^k_r\delta^j_m\delta^m_s-\delta^i_k\delta^k_s\delta^j_m\delta^m_r\\ =&\delta^i_r\delta^j_s-\delta^i_s\delta^j_r=\begin{vmatrix}\delta^i_r & \delta^i_s\\ \delta^j_r & \delta^j_s\end{vmatrix},\end{aligned} \tag{1.49}$$

from the second to the third line we added and subtracted the symmetric terms $\delta^i_0\delta^0_r\delta^j_0\delta^0_s + \delta^i_1\delta^1_r\delta^j_1\delta^1_s$ which were missing in order to obtain the contracted form in the third line. We can deploy (1.49) and calculate the contraction on one or both indices

$$\begin{aligned}\varepsilon^{ij}\varepsilon_{kj} =&\delta^i_k\delta^j_j - \delta^i_j\delta^j_k = 2\delta^i_k - \delta^i_k = \delta^i_c,\\ \varepsilon^{ij}\varepsilon_{ij} =&\delta^i_i = 2.\end{aligned} \tag{1.50}$$

This quantity can be very handy especially for what concerns determinants of matrices. It is easy to show that for a generic 2×2 matrix M

$$\begin{aligned}\varepsilon^{ij}M_{0i}M_{1j} &= M_{00}M_{11} - M_{01}M_{10} = \det(M),\\ \varepsilon^{ij}M_{1i}M_{0j} &= -\varepsilon^{ij}M_{0i}M_{1j} = -\det(M).\end{aligned} \tag{1.51}$$

We can now calculate the following quantity

$$\varepsilon^{ij}M_{ir}M_{js} = \varepsilon_{rs}\det(M), \tag{1.52}$$

which leads us by virtue of (1.50) to the well-known expression

$$\det(M) = \frac{1}{2}\varepsilon^{ij}\varepsilon^{rs}M_{ir}M_{js}. \tag{1.53}$$

Another interesting algebraic property is at reach: the cofactor matrix

$$\begin{aligned}\varepsilon^{ir}\varepsilon^{js}M_{ij} =&\delta^{ij}\delta^{rs}M_{ij} - \delta^{is}\delta^{rj}M_{ij}\\ =&\mathrm{tr}(M)\,\delta^{rs} - \delta^{is}\delta^{rj}M_{ij},\\ =&\mathrm{cof}(M)^{rs}\end{aligned} \tag{1.54}$$

which serves for the definition of the inverse matrix M^{-1}

$$\left(M^{-1}\right)^{rs} = [\det(M)]^{-1}\,\mathrm{cof}(M)^{rs}. \tag{1.55}$$

Now, showing that all these properties still hold for higher dimensions is just a matter of spare time. We recall them all at once for $d=4$ which is the case we are interested in

$$\varepsilon_{\tau\pi\phi\omega} = \begin{vmatrix}\delta^0_\tau & \delta^0_\pi & \delta^0_\phi & \delta^0_\omega\\ \delta^1_\tau & \delta^1_\pi & \delta^1_\phi & \delta^1_\omega\\ \delta^2_\tau & \delta^2_\pi & \delta^2_\phi & \delta^2_\omega\\ \delta^3_\tau & \delta^3_\pi & \delta^3_\phi & \delta^3_\omega\end{vmatrix}, \qquad \varepsilon_{\tau\pi\phi\omega}\varepsilon^{\mu\nu\rho\sigma} = \begin{vmatrix}\delta^\mu_\tau & \delta^\mu_\pi & \delta^\mu_\phi & \delta^\mu_\omega\\ \delta^\nu_\tau & \delta^\nu_\pi & \delta^\nu_\phi & \delta^\nu_\omega\\ \delta^\rho_\tau & \delta^\rho_\pi & \delta^\rho_\phi & \delta^\rho_\omega\\ \delta^\sigma_\tau & \delta^\sigma_\pi & \delta^\sigma_\phi & \delta^\sigma_\omega\end{vmatrix}, \tag{1.56}$$

$$\varepsilon_{\mu\pi\phi\omega}\varepsilon^{\mu\nu\rho\sigma} = \begin{vmatrix}\delta^\nu_\pi & \delta^\nu_\phi & \delta^\nu_\omega\\ \delta^\rho_\pi & \delta^\rho_\phi & \delta^\rho_\omega\\ \delta^\sigma_\pi & \delta^\sigma_\phi & \delta^\sigma_\omega\end{vmatrix}, \qquad \varepsilon_{\mu\nu\phi\omega}\varepsilon^{\mu\nu\rho\sigma} = 2!\begin{vmatrix}\delta^\rho_\phi & \delta^\rho_\omega\\ \delta^\sigma_\phi & \delta^\sigma_\omega\end{vmatrix}, \tag{1.57}$$

$$\varepsilon_{\mu\nu\rho\omega}\varepsilon^{\mu\nu\rho\sigma} = 3!\delta^\sigma_\omega, \qquad \varepsilon_{\mu\nu\rho\sigma}\varepsilon^{\mu\nu\rho\sigma} = 4!. \tag{1.58}$$

There is also another commonly used symbol used to indicate the product of two non-contracted completely antisymmetric symbols

$$\varepsilon_{\tau\pi\phi\omega}\varepsilon^{\mu\nu\rho\sigma} = 4!\delta^{\mu\nu\rho\sigma}_{\tau\pi\phi\omega}, \tag{1.59}$$

which is clearly antisymmetric for any odd permutation of lower or higher indices. Indeed, the factorial terms come from all the possible permutations of the contracted indices. Hence, we can write the determinant of the covariant metric tensor as

$$g = \frac{1}{4!}\varepsilon^{\tau\pi\phi\omega}\varepsilon^{\mu\nu\rho\sigma}g_{\tau\mu}g_{\pi\nu}g_{\phi\rho}g_{\omega\sigma}. \tag{1.60}$$

The completely antisymmetric symbol has a direct geometrical meaning in Special Relativity: writing a generic Lorentz transformation as Λ we obtain its transformation rule

$$\varepsilon_{\mu\nu\rho\sigma}\Lambda^{\mu}{}_{\mu'}\Lambda^{\nu}{}_{\nu'}\Lambda^{\rho}{}_{\rho'}\Lambda^{\sigma}{}_{\sigma'} = \det(\Lambda)\,\varepsilon_{\mu'\nu'\rho'\sigma'} = \varepsilon_{\mu'\nu'\rho'\sigma'}, \tag{1.61}$$

since transformations belonging to the proper Lorentz group are isometries ($\det(\Lambda) = 1$). Hence, in flat space under Lorentz transformations the completely antisymmetric symbol behaves as a tensor.

This is not true in a generic spacetime in which the transformation looks like

$$\varepsilon_{\mu\nu\rho\sigma}\frac{\partial x^{\mu}}{\partial y^{\mu'}}\frac{\partial x^{\nu}}{\partial y^{\nu'}}\frac{\partial x^{\rho}}{\partial y^{\rho'}}\frac{\partial x^{\sigma}}{\partial y^{\sigma'}} = \det\left(\frac{\partial x}{\partial y}\right)\varepsilon_{\mu'\nu'\rho'\sigma'}, \tag{1.62}$$

so the completely antisymmetric symbol transforms as a tensor density because of the Jacobian of the transformation. In order to obtain a tensor we should multiply the symbol for a quantity that transforms as the inverse of the Jacobian. Indeed, the determinant of any covariant tensor does transform as a power of the inverse of the Jacobian. Usually, this tensor is taken to be the metric tensor, as a natural but not necessary choice. We stress that any rank-two covariant tensor would do the job in the same fashion. Hence, we have

$$g_{\mu\nu} = \frac{\partial y^{\rho'}}{\partial x^{\mu}}\frac{\partial y^{\sigma'}}{\partial x^{\nu}}g_{\rho'\sigma'}, \qquad g = \det(g_{\mu\nu}) = \left[\det\left(\frac{\partial y}{\partial x}\right)\right]^2\det(g'), \tag{1.63}$$

$$\sqrt{-g} = \left|\det\left(\frac{\partial y}{\partial x}\right)\right|\sqrt{-g'}, \tag{1.64}$$

since $\mathrm{sign}(g) = -1$. Thus we can define the covariant Levi-Civita tensor as

$$E_{\mu\nu\rho\sigma} = \sqrt{-g}\,\varepsilon_{\mu\nu\rho\sigma}, \tag{1.65}$$

and its contravariant form is obtained by raising indices via the metric tensor

$$E^{\mu\nu\rho\sigma} = \sqrt{-g}\,g^{\mu\tau}g^{\nu\pi}g^{\rho\phi}g^{\sigma\omega}\varepsilon_{\tau\pi\phi\omega} = \frac{\sqrt{-g}}{g}\varepsilon^{\mu\nu\rho\sigma} = \frac{-1}{\sqrt{-g}}\varepsilon^{\mu\nu\rho\sigma}. \tag{1.66}$$

The results of this section will be of interest in constructing the Lagrangian formalism for General Relativity and in Loop Quantum Gravity.

1.7 Volume element and covariant divergence

We have shown how to perform differentiations in a covariant way. What about integrals? It is already known that when we perform a coordinate transformation in an integral we need to take into account the transformation of the volume element. For one-dimensional integrals this is done via the derivative of the transformation, and for multidimensional integrals via its generalization: the determinant of the Jacobian matrix. Now, we want to perform the integration in such a way that the integral of a tensor still transforms as a tensor, or the integral of a scalar quantity behaves as such.

The common volume element transforms as

$$dx^0 dx^1 dx^2 dx^3 = d^4x = d^4y \left| \det \left(\frac{\partial x}{\partial y} \right) \right|. \tag{1.67}$$

As we have just seen for the Levi-Civita tensor we can use the determinant of the metric tensor to render the volume element an invariant quantity. We can now write the invariant integration measure as

$$d\Omega = d^4x\sqrt{-g} = d^4y\sqrt{-g'}. \tag{1.68}$$

Let us now discuss some interesting properties of the covariant divergence. Indeed, for a vector A^μ we have

$$\nabla_\mu A^\mu = \partial_\mu A^\mu + \Gamma^\mu_{\mu\rho} A^\rho. \tag{1.69}$$

Recalling the definition of the Christoffel symbols it is straightforward to write down the relation

$$\Gamma^\mu_{\mu\rho} = \frac{1}{2} g^{\mu\nu} \partial_\rho g_{\mu\nu} = \frac{1}{2g} \left(g g^{\mu\nu} \partial_\rho g_{\mu\nu} \right), \tag{1.70}$$

this more convoluted form helps us recognize the derivative of $g = \det(g_{\mu\nu})$. By means of (1.60) we can write the determinant variation as

$$\delta g = \left(\frac{1}{3!} \varepsilon^{\tau\pi\phi\omega} \varepsilon^{\mu\nu\rho\sigma} g_{\tau\mu} g_{\pi\nu} g_{\phi\rho} \right) \delta g_{\omega\sigma}. \tag{1.71}$$

Within the parentheses of (1.71) we can see a generalization of the cofactor matrix (1.54) in $d = 4$. A rather simple way to define the cofactor matrix for the metric tensor is

$$g^{\mu\nu} = \mathrm{cof}\,(g)^{\mu\nu}\, g^{-1} \quad \Rightarrow \quad \mathrm{cof}\,(g)^{\mu\nu} = g g^{\mu\nu}. \tag{1.72}$$

Hence, we write

$$\delta g = \left(\frac{1}{3!} \varepsilon^{\tau\pi\phi\omega} \varepsilon^{\mu\nu\rho\sigma} g_{\tau\mu} g_{\pi\nu} g_{\phi\rho} \right) \delta g_{\omega\sigma} = \mathrm{cof}\,(g)^{\mu\nu}\, \delta g_{\mu\nu} = g g^{\mu\nu} \delta g_{\mu\nu}. \tag{1.73}$$

Going back to (1.70) we have

$$\Gamma^{\mu}_{\mu\rho} = \frac{1}{2g}\partial_\rho g = \partial_\rho \log\sqrt{-g}, \tag{1.74}$$

$$\begin{aligned} \nabla_\mu A^\mu =& \partial_\mu A^\mu + A^\rho \frac{1}{\sqrt{-g}}\partial_\rho \sqrt{-g} \\ =& \frac{1}{\sqrt{-g}}\partial_\rho\left(\sqrt{-g}A^\rho\right), \end{aligned} \tag{1.75}$$

which means that the covariant integral of a covariant divergence turns out to be an ordinary divergence of a densitized vector

$$\int d^4x\sqrt{-g}\nabla_\mu A^\mu = \int d^4x \partial_\rho\left(\sqrt{-g}A^\rho\right). \tag{1.76}$$

Generically, all those quantities that transform as products of some power $(-g)^w$ and a tensor are called tensor densities of weight w. Of course these densities have different transformation properties depending on the weight.

The result for the covariant divergence of a vector can be straightforwardly generalized for any antisymmetric tensor $A^{\mu\nu}$, leading to

$$\nabla_\mu A^{\mu\nu} = \frac{1}{\sqrt{-g}}\partial_\mu\left(\sqrt{-g}A^{\mu\nu}\right). \tag{1.77}$$

1.8 Gauss and Stokes theorems

We now briefly discuss how to formulate two fundamental tools of tensor calculus: Stokes and Gauss theorems. Conceptually, these two theorems share a common feature, relating integrals performed on geometric objects of different dimension. In order to formulate them we need, as a first step, to define the other integration measures, other than the four-dimensional one we defined in the previous section. The one-dimensional measure clearly reads as usual, as the differential of a covariant vector dx^μ. Let us now consider the element of surface spanned by two differentials dx^μ and dy^μ: it is well known that the area of a parallelogram, defined by two vectors on a plane, is equal to the determinant of the matrix whose columns are the components of those two vectors. Analogously, in four dimensions, the two-dimensional area element is defined as

$$ds^{\mu\nu} = dx^\mu dy^\nu - dx^\nu dy^\mu = -ds^{\nu\mu}. \tag{1.78}$$

Now, integrations of vectorial fields along surfaces are usually performed by means of a measure which is at each point of the surface orthogonal to it. For two-dimensional surfaces in a four-dimensional space this element is given by the so-called *dual* tensor to $ds^{\mu\nu}$ defined as

$$ds^*_{\mu\nu} = \frac{1}{2}\varepsilon_{\mu\nu\rho\sigma} ds^{\rho\sigma}, \tag{1.79}$$

and it is easy to show that these two elements are orthogonal, *i.e.*

$$ds^*_{\mu\nu} ds^{\mu\nu} = 0. \tag{1.80}$$

For three-dimensional hypersurfaces we follow the same logic and define the rank-three tensor

$$d\Sigma^{\mu\nu\rho} = \begin{vmatrix} dx^\mu & dy^\mu & dz^\mu \\ dx^\nu & dy^\nu & dz^\nu \\ dx^\rho & dy^\rho & dz^\rho \end{vmatrix}, \tag{1.81}$$

as the volume of the parallelepiped formed by the three line elements dx^μ, dy^μ and dz^μ. The dual to (1.81) is a vector defined as

$$d\Sigma_\sigma = -\frac{1}{6}\varepsilon_{\sigma\mu\nu\rho} d\Sigma^{\mu\nu\rho}. \tag{1.82}$$

All the dual measures are naturally covariant under Lorentz transformations and they can be rendered covariant under general coordinate transformations putting $\sqrt{-g}$ as prefactor: $\sqrt{-g}\, ds^*_{\mu\nu}$ and $\sqrt{-g}\, d\Sigma_\mu$. This happens since $\sqrt{-g}\,\varepsilon_{\mu\nu\rho\sigma} = E_{\mu\nu\rho\sigma}$.

Now we can express Gauss theorem as follows: given a four-dimensional integral of a covariant divergence it is possible to write it as an integral over the boundary of the domain as

$$\int_{\mathcal{M}} d^4x \sqrt{-g}\, \nabla_\mu A^\mu = \int_{\mathcal{M}} d^4x \partial_\mu \left(\sqrt{-g} A^\mu\right) = \int_{\partial\mathcal{M}} d\Sigma_\mu \sqrt{-g} A^\mu. \tag{1.83}$$

Stokes theorem can be expressed as follows: the integral of a vector field along a closed-path equals the integral of the derivative of the field performed on the two-dimensional surface bounded by the path

$$\oint_{\partial S} dx^\mu A_\mu = \int_S ds^{\mu\nu} \partial_{[\mu} A_{\nu]}. \tag{1.84}$$

We notice that, since the surface element $ds^{\mu\nu}$ is antisymmetric, we must take the antisymmetric part of $\partial_\mu A_\nu$ which clearly behaves as a tensor, in fact it is straightforward to show that $\partial_{[\mu} A_{\nu]} = \nabla_{[\mu} A_{\nu]}$ (clearly in the absence of torsion).

1.9 The Riemann tensor

We already mentioned that the Christoffel symbols can be made to vanish at a given point with some coordinate transformation (1.39). It is easy to compare equation (1.43) to Newton second law interpreting the Christoffel symbols as force fields. Thus, we can always choose, at each point, a coordinate system in which these pseudoforces vanish (actually, along a generic line). However, since by means of General Relativity we wish to give a geometrical description of the gravitational force, how can we possibly distinguish a "force" given by a coordinate transformation from a real gravitational force[2]? In other words: for a given metric tensor how can we know if the manifold we are studying is intrinsically curved or if the fact that the Christoffel symbols do not vanish is due to some peculiar choice of the coordinate system?

To answer this question we clearly need a tensor quantity which cannot vanish under some coordinate transformation: the Riemann or curvature tensor. In order to construct this quantity, we cannot just use combinations of the Christoffel symbols because these are made up of first order derivatives of the metric tensor and moreover these are not tensors. Second order derivatives are needed. We propose here a construction of this tensor which was firstly given by Levi-Civita [198]. It is important to show that differently from the first order metric derivatives the second ones cannot be all locally reduced to zero by a coordinate transformation.

1.9.1 *Levi-Civita construction*

The idea is the following: given a vector at some point on the manifold, we parallel transport it along a closed geodesic path. If the final vector equals the initial one the manifold is flat, otherwise, it is not (see figure 1.2). This happens because the scalar product between a parallel-transported vector and the tangent vector to a geodesic is conserved. Hence, if the manifold is curved, because of the change of the basis vectors, the final vector will differ from the initial one. This variation for a generic vector V_μ can be written by means of its parallel transport along the union of different geodesics $\{\gamma_i\}$ described by the parametrizations $\{x_i^\mu(s)\}$. For each geodesic the parallel transport (1.46) in differential form reads $d_{(i)}V_\mu = dx_i^\rho \partial_\rho V_\mu = \Gamma^\nu_{\mu\rho} V_\nu dx_i^\mu$, hence

[2]Indeed, this sentence has a very profound physical meaning that we will describe at the end of this chapter.

$$\delta V_\mu = \oint_\gamma d_{(i)} V_\mu = \oint_\gamma \Gamma^\nu_{\mu\rho} V_\nu dx^\mu_i (s) , \tag{1.85}$$

where $\gamma = \bigcup_{i=1}^N \gamma_i$.

It can be shown using Stokes theorem that for infinitesimal closed geodesic paths the last result can be written via the commutator of covariant derivatives acting on V_μ

$$\delta V_\mu = \frac{1}{2} \left[\nabla_\rho , \nabla_\sigma \right] V_\mu \delta s^{\rho\sigma} = -\frac{1}{2} R^\nu{}_{\mu\rho\sigma} V_\nu \delta s^{\rho\sigma} , \tag{1.86}$$

where $\delta s^{\rho\sigma}$ stands for the infinitesimal area element contained by the geodesics loop. Hence, we define the Riemann tensor via the relation[3]

$$\left[\nabla_\mu , \nabla_\nu \right] V_\rho = -R^\sigma{}_{\rho\mu\nu} V_\sigma . \tag{1.87}$$

Indeed, if we consider a flat manifold the covariant derivatives reduce to

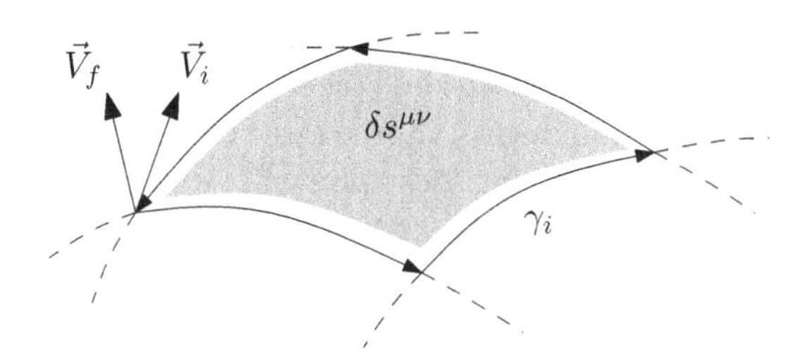

Figure 1.2 Closed loop of geodesics. $\vec{V}_f$ is the result of the parallel transport of $\vec{V}_i$ along the different geodesics γ_i. If they do not differ then the manifold is flat.

the ordinary ones and the commutator is zero because of Schwarz theorem[4]. We give here just one half of the computation

$$\begin{aligned} \nabla_\mu \nabla_\nu V_\rho =& \partial_\mu \left(\nabla_\nu V_\rho \right) - \Gamma^\delta_{\mu\nu} \left(\nabla_\delta V_\rho \right) - \Gamma^\delta_{\mu\rho} \left(\nabla_\nu V_\delta \right) \\ =& - \partial_\mu \Gamma^\sigma_{\nu\rho} V_\sigma + \Gamma^\sigma_{\mu\nu} \Gamma^\tau_{\sigma\rho} V_\tau + \Gamma^\sigma_{\mu\rho} \Gamma^\tau_{\nu\sigma} V_\tau \\ & - \Gamma^\sigma_{\mu\nu} \partial_\sigma V_\rho - 2 \Gamma^\sigma_{\rho(\mu} \partial_{\nu)} V_\sigma + \partial_\mu \partial_\upsilon V_\rho . \end{aligned} \tag{1.88}$$

All symmetric terms in the indices μ, ν will be deleted from those with opposite sign coming from the second term of the commutator leading us to the expression

[3]On a curved manifold covariant derivatives in general do not commute.

[4]This theorem states that for differentiable functions of many variables second order derivatives must commute.

$$R^{\sigma}_{\ \rho\mu\nu} = \partial_\mu \Gamma^{\sigma}_{\nu\rho} - \partial_\nu \Gamma^{\sigma}_{\mu\rho} + \Gamma^{\sigma}_{\mu\tau}\Gamma^{\tau}_{\nu\rho} - \Gamma^{\sigma}_{\nu\tau}\Gamma^{\tau}_{\mu\rho}. \tag{1.89}$$

We notice that the curvature tensor is made up by linear terms in the second derivatives of the metric tensor and quadratic terms of the first order ones. The independent components of the Riemann tensor are 20.

There is a common argument (see for example [238]), which allows us to count the number of linear combinations of second order derivatives of the metric tensor which cannot be made to vanish under diffeomorphisms. We proceed as we did in the case of the Christoffel symbols. Since we aim at cancelling second order derivatives, we need a transformation of the type

$$x^\mu = z^{\mu'} + C^{\mu}_{\ \nu\rho\sigma} z^{\nu'} z^{\rho'} z^{\sigma'}, \tag{1.90}$$

where the constant tensor $C^{\mu}_{\ \nu\rho\sigma}$ has 80 components in total. We can count them by noting that $C^{\mu}_{\ \nu\rho\sigma}$ has to be symmetric under the exchange of ν, ρ and σ for any of the four possible values of μ. For the remaining three indices we can consider three different cases: i) all of them are equal $\nu = \rho = \sigma$ giving four components; ii) just two of them are equal $\nu = \rho, \sigma$, avoiding permutations because of the above mentioned symmetries, producing $2 \cdot 4!/4 = 12$ components; iii) the three indices are different giving $4!/3! = 4$ components. Hence for each value of μ we have 20 components, thus 80 different components for $C^{\mu}_{\ \nu\rho\sigma}$. Since we deal with 100 second order derivatives $\partial_\rho \partial_\sigma g_{\mu\nu}$ (10 components for the derivatives times 10 components for the metric tensor), we are left with 20 linear combinations of the second order derivatives which cannot be made to vanish corresponding to the 20 values of the Riemann tensor.

Given the definition of the Riemann tensor we can state the symmetry and antisymmetry properties of the indices

$$R_{\mu\nu\rho\sigma} = -R_{\nu\mu\rho\sigma}, \qquad R_{\mu\nu\rho\sigma} = -R_{\mu\nu\sigma\rho}, \qquad R_{\mu\nu\rho\sigma} = R_{\rho\sigma\mu\nu}. \tag{1.91}$$

It is possible to consider the same technique applied to a contravariant vector

$$[\nabla_\mu, \nabla_\nu] V^\sigma = [\nabla_\mu, \nabla_\nu] g^{\gamma\sigma} V_\gamma = -g^{\rho\sigma} R^{\tau}_{\ \rho\mu\nu} V_\tau = R^{\sigma}_{\ \rho\mu\nu} V^\rho, \tag{1.92}$$

where we used the symmetry properties of the curvature tensor and the vanishing of the covariant derivative of the metric tensor. Finally, we define the symmetric Ricci tensor

$$R_{\mu\nu} = g^{\rho\sigma} R_{\rho\mu\sigma\nu} = \partial_\rho \Gamma^{\rho}_{\mu\nu} - \partial_\nu \Gamma^{\rho}_{\mu\rho} + \Gamma^{\sigma}_{\mu\nu}\Gamma^{\rho}_{\rho\sigma} - \Gamma^{\sigma}_{\rho\mu}\Gamma^{\rho}_{\sigma\nu}, \tag{1.93}$$

and its contraction, the scalar curvature

$$R = g^{\mu\nu} R_{\mu\nu}. \tag{1.94}$$

1.9.2 *Algebraic properties and Bianchi identities*

The curvature tensor satisfies an algebraic identity

$$R_{\mu\nu\rho\sigma} + R_{\mu\sigma\nu\rho} + R_{\mu\rho\sigma\nu} = 0, \tag{1.95}$$

and a differential one in the metric tensor

$$\nabla_\mu R^\tau{}_{\nu\rho\sigma} + \nabla_\rho R^\tau{}_{\nu\sigma\mu} + \nabla_\sigma R^\tau{}_{\nu\mu\rho} = 0. \tag{1.96}$$

In other words, if we substitute all quantities with their definitions in terms of the metric tensor the last equation becomes an identity, otherwise it represents some constraints to be satisfied either by the Christoffel symbols or by the Riemann tensor itself, depending on which quantities we are considering independent from the others. Equations (1.95) and (1.96) are the so-called algebraic and differential Bianchi identities, respectively.

From equation (1.96) we can derive a tensor of rank two whose covariant divergence is zero; we contract the indices ν, σ and the indices τ, ρ , because the covariant derivative of the metric tensor is zero, obtaining

$$\begin{aligned}
& g^{\nu\sigma}\left(\nabla_\mu R^\rho{}_{\nu\rho\sigma} + \nabla_\rho R^\rho{}_{\nu\sigma\mu} + \nabla_\sigma R^\rho{}_{\nu\mu\rho}\right) \\
=& g^{\nu\sigma}\left(\nabla_\mu R_{\nu\sigma} + \nabla_\rho R^\rho{}_{\nu\sigma\mu} - \nabla_\sigma R_{\nu\mu}\right) \\
=& \left(\nabla_\mu R - 2\nabla_\rho R^\rho_\mu\right) \\
=& -2\nabla_\rho\left(R^\rho_\mu - \frac{1}{2}\delta^\rho_\mu R\right) = 0.
\end{aligned} \tag{1.97}$$

As we will see this is an important property. In fact this combination of the Ricci tensor and of the scalar curvature is called the Einstein tensor

$$G_{\mu\nu} = R_{\mu\nu} - \frac{1}{2}g_{\mu\nu}R, \tag{1.98}$$

which is symmetric and has a vanishing covariant divergence, and represents one side of the gravitational field equations.

1.10 Geodesic deviation

Let us now consider the behavior of a geodesic bundle $x^\mu = x^\mu(s; v)$, where s is the curvilinear abscissa of the single geodesic identified by the parameter v. Our purpose is to determine how curvature influences the behavior of two geodesics close to each other. Indeed, we can always cast one geodesic in the form of a free particle but this is not generally possible considering two geodesics at once because the Christoffel symbols can be made to vanish only at one point or along one single geodesic. Hence, other than the usual tangent vector t^μ we define the infinitesimal vector v^μ linking two geodesics in the bundle

$$t^\mu(s;v) = \frac{\partial x^\mu(s;v)}{\partial s}, \qquad v^\mu(s;v) = \frac{\partial x^\mu(s;v)}{\partial v}. \tag{1.99}$$

In order to describe the dynamics of v^μ, we calculate its second derivative along the geodesic itself

$$t^\rho \nabla_\rho (t^\sigma \nabla_\sigma v^\mu) = t^\rho \nabla_\rho (t^\sigma \partial_\sigma v^\mu + \Gamma^\mu_{\sigma\nu} t^\sigma v^\nu) = t^\rho \nabla_\rho \left(\frac{\partial^2 x^\mu}{\partial s \partial v} + \Gamma^\mu_{\sigma\nu} t^\sigma v^\nu \right). \tag{1.100}$$

Now, it is straightforward to verify that

$$t^\sigma \nabla_\sigma v^\mu = \frac{\partial^2 x^\mu}{\partial s \partial v} + \Gamma^\mu_{\sigma\nu} t^\sigma v^\nu = v^\nu \partial_\nu t^\mu + \Gamma^\mu_{\sigma\nu} t^\sigma v^\nu = v^\nu \nabla_\nu t^\mu, \tag{1.101}$$

hence obtaining

$$t^\sigma \nabla_\sigma v^\mu = v^\nu \nabla_\nu t^\mu, \tag{1.102}$$

leading us to

$$\begin{aligned} t^\rho \nabla_\rho (t^\sigma \nabla_\sigma v^\mu) =& t^\rho \nabla_\rho (v^\nu \nabla_\nu t^\mu) \\ =& (t^\rho \nabla_\rho v^\nu) \nabla_\nu t^\mu + t^\rho v^\nu \nabla_\rho \nabla_\nu t^\mu \\ =& v^\rho \nabla_\rho t^\nu \nabla_\nu t^\mu + t^\rho v^\nu \nabla_\rho \nabla_\nu t^\mu \\ =& v^\rho \nabla_\rho (t^\nu \nabla_\nu t^\mu) - v^\rho t^\nu \nabla_\rho \nabla_\nu t^\mu + t^\rho v^\nu \nabla_\rho \nabla_\nu t^\mu \\ =& t^\rho v^\nu R^\mu_{\ \sigma\rho\nu} t^\sigma. \end{aligned} \tag{1.103}$$

Using the geodesic equation (1.43) and the expression of the Riemann tensor via the commutator of covariant derivatives (1.87) we obtained the so-called geodesic equation

$$\nabla^2_{\vec{t}} v^\sigma = t^\rho \nabla_\rho (t^\sigma \nabla_\sigma v^\sigma) = R^\sigma_{\ \rho\mu\nu} t^\rho t^\mu v^\nu, \tag{1.104}$$

which is equal to zero only for flat manifolds when $v^\mu \neq 0$. Hence, if we are given a determined form for the metric tensor, $g_{\mu\nu} = g_{\mu\nu}(x)$, the relation (1.104) becomes a differential equation for $v^\mu = v^\mu(s,v)$ as a function of s for a given value of v.

The geodesic deviation equation (1.104) is written in a rather unclear way as for its differential equation nature. We then give here its explicit form in terms of ordinary derivatives with respect to the curvilinear abscissa s, *i.e.* $d/ds = t^\sigma \partial_\sigma$. For the left-hand side of (1.104) one gets

$$\begin{aligned} t^\mu \nabla_\mu (t^\nu \nabla_\nu v^\sigma) =& \frac{d^2}{ds^2} v^\sigma + 2 t^\nu \Gamma^\sigma_{\nu\mu} \frac{d}{ds} v^\mu \\ &+ R^\sigma_{\ \rho\mu\nu} t^\rho t^\mu v^\nu + \left(\partial_\nu \Gamma^\sigma_{\mu\rho} + \Gamma^\sigma_{\rho\tau} \Gamma^\tau_{\mu\nu} - \Gamma^\sigma_{\mu\tau} \Gamma^\tau_{\nu\rho} \right) t^\rho t^\mu v^\nu, \end{aligned} \tag{1.105}$$

hence, for the whole expression we get the explicit form

$$\frac{d^2}{ds^2}v^\sigma + 2t^\nu\Gamma^\sigma_{\nu\mu}\frac{d}{ds}v^\mu + \left(\partial_\nu\Gamma^\sigma_{\mu\rho} + \Gamma^\sigma_{\rho\tau}\Gamma^\tau_{\mu\nu} - \Gamma^\sigma_{\mu\tau}\Gamma^\tau_{\nu\rho}\right)t^\rho t^\mu v^\nu = 0, \quad (1.106)$$

which is still covariant, although not explicitly, but clearly shows its differential structure.

1.11 Einstein's equations

In this section we will construct the geometrodynamics associated with the covariance structure of the spacetime manifold. The present derivation will start from fundamental principles and it will have an inductive character, *i.e.* we will construct the gravitational field equations starting from well-motivated criteria.

1.11.1 *Equivalence Principle*

The General Theory of Relativity stems from an experimental evidence which is at the ground of the *equivalence principle*. All bodies, regardless of their weight, freely fall with exactly the same acceleration in earth's gravitational field. Although it could sound as an astonishing result, it becomes even more remarkable when looked from the theoretical perspective. In Newtonian Physics, the equations of motion of a freely falling body can be written as follows

$$m_I a_k = -m_G \partial_k \phi, \quad (1.107)$$

where ϕ is some scalar function describing the gravitational potential, m_I represents the *inertial* mass of the body and m_G represents the gravitational mass of the same body. Now, the gravitational mass has exactly the same logical role of an electric charge for the electromagnetic field in the static case. Indeed we could refer to the gravitational mass as a gravitational charge. As we said, the equivalence principle establishes a universal proportionality relation between these two quantities

$$m_I \propto m_G, \quad (1.108)$$

which can be written as an equality using the proper measure units.

This principle is telling us that these two properties, that would be logically completely unrelated, are in fact numerically equivalent. Let us continue now by noting that because of the special theory of relativity we know that there is an equivalence between inertial mass and energy which then becomes a relation between energy and gravitational mass

$$E = m_I c^2 = m_G c^2. \tag{1.109}$$

Hence, we can always associate a gravitational mass to any energy density which then will interact with the gravitational field: *e.g.* the electromagnetic field will interact with it.

We are thus led to the conclusion that looking at the motion of an accelerated system we would not be able to distinguish if the acceleration is caused by a gravitational field or by the action of inertial forces, *i.e.* a change of spacetime coordinates. This is the core of the *equivalence principle*: locally, an equivalence exists between the formulation of the physics in an inertial frame and a non-inertial one in which a gravitational field produces an opposite acceleration.

This especially means that taking the reference frame of a free falling object any other free falling body, close enough to the first one, will look as if it was steady because the accelerations would be equal. Hence any free falling frame looks, locally, like an inertial frame.

Let us focus on a crucial consequence of the equivalence principle. Since all bodies fall with the same acceleration we can discuss their motion in a gravitational field regardless of their mass. Hence, it is possible to define the theory in a purely geometrical setting where the accelerated particles move along shortest paths on a curved manifold, *i.e.* geodesics, as free particles move along straight lines on a flat manifold.

1.11.2 *Theory requirements*

Indeed, the equivalence principle is an experimental evidence which demands an appropriate physical description which was still lacking at the beginning of the 20th century. Let us now summarize some principles which can be traced in Einstein's work.

- *The Principle of General Relativity*: all physical laws take the same form in any physical reference frame. Any equation of motion, when is verified, can be cast into the form of some function required to be equal to zero. Indeed, if physical laws do not depend on the reference frame, then the clearest way of preserving the validity of the equations of motion is writing them using tensor quantities which if equal to zero in some reference frame will still be zero in any other. This is a consequence of the linearity and homogeneity of the action of general coordinate transformations on tensors.

- *The Equivalence Principle*: as we have seen, experimental evidences combined with the results of the Special Relativity imply that the motion of any particle in a gravitational field does not depend on its energy, and, moreover, that any energy density interacts with the gravitational field as well. This properties suggest that the theory should be written in a geometrical fashion.
- The equivalence between mass and energy implies that the gravitational field must be self-interacting because it must interact with its own self-energy, and therefore the corresponding field theory cannot be linear in the field strength. Since General Relativity has a geometrical structure a notion of energy for the gravitational field can be recovered only in simplifying approximations, like the weak-field limit. Nonetheless, the requirement of a self-interaction of the field emerges in the non-linear structure of the field equations, *i.e.* of the curvature tensor in terms of the metric field and its derivatives.
- The equations of motion of the gravitational field should contain, in order to assure its propagation, second derivatives and in the weak field limit Newton law should be retrieved. It is worth noting that the Einsteinian request to deal with second order derivatives of the metric tensor (the metric components can be regarded as gravitational potential) is not a fundamental requirement, mainly due to historical reasons, and, *de facto*, corresponding to the simplest choice. This requirement can be easily violated by considering field equations containing higher-order derivatives, however preserving all other prescriptions, see for instance the $f(R)$ theories in section 4.1.5.
- The matter[5] coupled to the gravitational field should enter in the equations of motion via its stress-energy tensor. In fact, it is easy to check that when matter is described by a macroscopic fluid, its properties are well-described by a rank two tensor, naturally emerging from the covariance requirements. The stress-energy tensor can be associated to any non-geometrical field and for any physical entity cannot identically vanish. Therefore, in what follows we shall assume that the stress-energy tensor of matter is the proper repre-

[5]By matter we mean non-geometrical physical entities, *e.g.* the electromagnetic field or a macroscopic fluid.

sentation for any (non-geometrical) source.

Indeed, we can now see that the formalism we developed so far has exactly all these required characteristics. In fact the principle of General Relativity can be implemented via the General Covariance of tensors belonging to a curved manifold.

The equivalence principle is implemented by the fact that, locally, there is always a coordinate transformation giving vanishing Christoffel symbols and thus leading to geodesic equations which are just those of a free falling particle, *i.e.* living in a locally inertial frame. The curvature tensor contains second order derivatives of the metric tensor and they all identically vanish for flat manifolds only. This way we can implement the local equivalence between an accelerated frame and a frame in a gravitational field while always being able to distinguish between them: manifolds describing a real gravitational field will always have a non-vanishing curvature, while accelerated frames, having vanishing curvature, might have at most the Christoffel symbols numerically equal, only along a given geodesic, to those of a curved manifold.

The equivalence principle holds, however, only locally. Indeed if we analyze more carefully the equation of geodesic deviation (1.104), it tells us that regardless of the chosen reference frame two close-by geodesics will always experience a relative acceleration. This acceleration gives rise to the so-called *tidal forces.* Hence, if we consider a homogeneous particle system in free fall, it will become more and more non-homogeneous as the curvature increases. Indeed, this feature is captured by the geodesic deviation equation (1.104) since the relative acceleration is proportional to the distance between the two geodesics and to the Riemann tensor.

From an experimental point of view, once assigned a sensibility threshold for the measuring instruments, there will always be some characteristic length scale at which they would no longer measure any tidal force for a given value of the curvature. This is the physical counterpart of the mathematical concept that curvature can be neglected in a sufficiently small neighborhood of a spacetime point.

1.11.3 *Field equations*

In the vacuum case, the simplest equation satisfying all the requirements we just discussed is

$$g^{\rho\sigma}R_{\rho\mu\sigma\nu} = R_{\mu\nu} = \partial_\rho\Gamma^\rho_{\mu\nu} - \partial_\nu\Gamma^\rho_{\mu\rho} + \Gamma^\sigma_{\mu\nu}\Gamma^\rho_{\rho\sigma} - \Gamma^\sigma_{\rho\mu}\Gamma^\rho_{\sigma\nu} = 0, \tag{1.110}$$

which is a set of ten partial non-linear differential equations for the ten independent components of the metric tensor $g_{\mu\nu}$. The Newtonian limit is retrieved in the weak and static field limit

$$g_{00} \simeq -1 + \frac{2\phi}{c^2}, \quad g_{ik} = \delta_{ik}, \tag{1.111}$$

where ϕ represents the Newtonian gravitational potential.

We observe how these vacuum equations can be separated into a linear part in the second derivatives of the metric tensor plus quadratic terms in the first order ones. The latter can be thought as those accounting for the self-interaction of the gravitational field, playing, in some sense, the role of vacuum "gravitational sources". Conceptually, we must regard Einstein's vacuum equations as a Nature prescription for those curved manifolds which could be regarded as describing physical spacetimes.

Since vacuum field equations (1.110) are second order, the Cauchy problem is specified by assigning the value of the metric tensor components and their time derivatives on a space-like (non-singular) initial hypersurface, $x^0 =$ const. It is possible to show that such problem separates into evolutionary and non-evolutionary equations. We will extract this property of the gravitation field when analyzing the Hamiltonian formulation of gravity in section 6.1. The non-evolutionary equations are then constraints to be satisfied either by the initial conditions and by the dynamics itself.

Let us consider the equations in the presence of matter. As discussed above, the gravitational field should couple to matter via its stress-energy tensor $T^{\mu\nu}$. This tensor has by construction, in flat spacetime, a vanishing ordinary divergence which must be promoted to a vanishing covariant divergence in order to preserve its covariance

$$\nabla_\mu T^{\mu\nu} = 0. \tag{1.112}$$

When dealing with a pressureless (dust-like) fluid, equation (1.112) naturally reduces to the geodesic equation.

Looking now at the geometrical side we already encountered a tensor which satisfies the requirements we wrote above and that has a vanishing covariant divergence: the Einstein tensor. We can then write

$$R_{\mu\nu} - \frac{1}{2} g_{\mu\nu} R = \chi T_{\mu\nu}, \tag{1.113}$$

χ being some dimensional constant, known as Einstein constant. If we consider the case of vacuum, *i.e.* where $T_{\mu\nu} = 0$, the last equation reduces to (1.110), in fact we have

$$\begin{aligned} &R_{\mu\nu} - \frac{1}{2} g_{\mu\nu} R = 0 \\ &g^{\mu\nu} R_{\mu\nu} - 2R = 0 \\ &R = g^{\mu\nu} R_{\mu\nu} = 0 \\ &R_{\mu\nu} = 0. \end{aligned} \tag{1.114}$$

This set of equations yields to results that have been experimentally tested, such as the precession of Mercury perihelion which is a phenomenon that only General Relativity can satisfactory account for. In this respect it is worth noting that Einstein's equations, using the trace relation

$$-R = \chi T^{\rho}_{\ \rho}, \tag{1.115}$$

can be equivalently written in the form

$$R_{\mu\nu} = \chi \left(T_{\mu\nu} - \frac{1}{2} g_{\mu\nu} T^{\rho}_{\ \rho} \right), \tag{1.116}$$

which outlines how it is the full stress-energy tensor structure which gives a non-zero Ricci tensor to the manifold. For instance, using only the trace part of $T_{\mu\nu}$ we would immediately violate the equivalence principle because for the electromagnetic field $T^{\rho}_{\ \rho} = 0$. However, this would lead to the paradox of a non-gravitating electromagnetic field. We also notice that the trace tensor governs the behavior of the scalar curvature R which is always vanishing for traceless stress-energy tensors.

When assigning the Cauchy problem in the presence of matter we must specify the spatial configuration and the velocity field of the corresponding matter source. It is worth noting that also in this case constraints appear for the initial data problem, corresponding to the $G^{0}_{\ \mu} - \chi T^{0}_{\ \mu} = 0$.

We conclude by observing how the Newtonian limit in which Einstein's equations reduce to Poisson's ones for the Newtonian potential (in a given static matter distribution) provides us with the following value of Einstein constant

$$\chi = \frac{8\pi G}{c^4}. \tag{1.117}$$

It is worth noting that the field equations (1.113) are affected by an ambiguity consisting in a term of the form $\Lambda g_{\mu\nu}$ where $\Lambda \in \mathbb{R}$ is a generic constant. This term can be equivalently added to the left and right side of (1.113) and it is permitted since it has a vanishing divergence. Such a term is known as the *cosmological constant term* and no experimental evidence of it appeared until the discovery of the Universe accelerated expansion (see Chapter 2) which could be explained by such a term.

1.12 Vierbein representation

A very useful reformulation of General Relativity is the one in which the configuration variables are vierbein vectors e^{α}_{μ} and their inverses e^{μ}_{α}. [6] These objects constitute a set of four orthonormal vectors for each point of spacetime, whose main property is

$$g_{\mu\nu} = \eta_{\alpha\beta} e^{\alpha}_{u} e^{\beta}_{\nu}, \qquad g^{\mu\nu} = \eta^{\alpha\beta} e^{\mu}_{\alpha} e^{\nu}_{\beta}. \tag{1.118}$$

Vierbein vectors can be seen as maps from vectors V^{μ}, in the spacetime manifold, to vectors V^{α} in an internal Minkowski space (local Lorentz frame):

$$V^{\alpha} = e^{\alpha}_{\mu} V^{\mu}. \tag{1.119}$$

This definition can be extended to contravariant vectors and to generic tensors as well. It is worth noting how the indices α can be lowered/raised via the Minkowski metric $\eta_{\alpha\beta}/\eta^{\alpha\beta}$.

One can define the directional derivative in the local Lorentz frame $\partial_{\alpha} = e^{\mu}_{\alpha}\partial_{\mu}$ and construct the derivative which accounts for the variation of vierbein vectors as well, *i.e.*

$$D^{(\gamma)}_{\alpha} V^{\beta} = \partial_{\alpha} V^{\beta} + \gamma^{\beta}{}_{\delta\alpha} V^{\delta} = e^{\mu}_{\alpha} e^{\beta}_{\nu} \nabla_{\mu} V^{\nu}, \tag{1.120}$$

where $\gamma^{\beta}{}_{\delta\alpha}$ live in the local Lorentz frame, whose components are know as Ricci coefficients and they read

$$\gamma^{\alpha}{}_{\beta\gamma} = -e^{\nu}_{\gamma} e^{\mu}_{\beta} \nabla_{\nu} e^{\alpha}_{\mu}. \tag{1.121}$$

From Ricci coefficients one can define the anholonomy coefficients $\lambda^{\alpha}{}_{\beta\delta}$

$$\lambda^{\alpha}{}_{\beta\delta} = \gamma^{\alpha}{}_{\beta\delta} - \gamma^{\alpha}{}_{\delta\beta}, \tag{1.122}$$

whose explicit expression is given by

$$\lambda^{\alpha}{}_{\beta\gamma} = e^{\mu}_{\beta} e^{\nu}_{\gamma} \left(\partial_{\mu} e^{\alpha}_{\nu} - \partial_{\nu} e^{\alpha}_{\mu} \right), \tag{1.123}$$

and thus they do not depend explicitly on Christoffel symbols (this usually simplifies their evaluation with respect to Ricci coefficients).

From Ricci coefficients one can also define a spacetime connection $\omega^{\alpha\beta}_{\mu}$, the *spin connection*, via

$$\omega^{\alpha\beta}_{\mu} = -\eta^{\beta\varepsilon} \gamma^{\alpha}{}_{\varepsilon\delta} e^{\delta}_{\mu} = e^{\beta\nu} \nabla_{\mu} e^{\alpha}_{\nu}. \tag{1.124}$$

[6] The indexes $\alpha, \beta, \gamma, .. = 0, .., 3$ labels the coordinates of the internal Minkowski space, having the metric $\eta_{\alpha\beta} = \mathrm{diag}\{-1, 1, 1, 1\}$.

It is worth noting that $\omega_\mu^{\alpha\beta}$ is antisymmetric under the exchange $\alpha \leftrightarrow \beta$, since

$$\omega_\mu^{\alpha\beta} = e^{\beta\nu}\nabla_\mu e_\nu^\alpha = \nabla_\mu \eta^{\alpha\beta} - \left(\nabla_\mu e^{\beta\nu}\right) e_\nu^\alpha = -e^{\alpha\nu}\nabla_\mu e_\nu^\beta = -\omega_\mu^{\beta\alpha}. \quad (1.125)$$

The covariant derivative of vectors in spacetime (the extension to tensor is straightforward) with internal Lorentz indices reads

$$D_\mu^{(\omega)} V^\alpha = \partial_\mu V^\alpha - \omega^\alpha{}_{\beta\mu} V^\beta, \qquad D_\mu^{(\omega)} V_\alpha = \partial_\mu V_\alpha + \omega_\alpha{}^{\beta\mu} V_\beta. \quad (1.126)$$

There are two invariant tensors in the local Lorentz frame. One is the metric $\eta^{\alpha\beta}$,

$$D_\mu^{(\omega)} \eta^{\alpha\beta} = \partial_\mu \eta^{\alpha\beta} - \omega^\alpha{}_{\gamma\mu}\eta^{\gamma\beta} - \omega^\beta{}_{\gamma\mu}\eta^{\alpha\gamma} = -\omega_\mu^{\alpha\beta} - \omega_\mu^{\beta\alpha} = 0 \quad (1.127)$$

where we used that η is constant and the spin connection is antisymmetric under the exchange of the internal indexes. The other invariant tensor is the projected Levi-Civita tensor $E_{\alpha\beta\gamma\delta}$

$$E_{\alpha\beta\gamma\delta} = E_{\mu\nu\rho\sigma} e_\alpha^\mu e_\beta^\nu e_\gamma^\rho e_\delta^\sigma. \quad (1.128)$$

By using the summation properties of $\varepsilon_{\mu\nu\rho\sigma}$ (1.59), the relationship $\sqrt{-g} = e$, e being the determinant of e_μ^α, and the definition of $E_{\mu\nu\rho\sigma}$ (1.65) it can be shown that

$$\begin{aligned} E_{\alpha\beta\gamma\delta} &= e\, \varepsilon_{\mu\nu\rho\sigma}\; e_\alpha^\mu e_\beta^\nu e_\gamma^\rho e_\delta^\sigma \\ &= \frac{1}{4!}\varepsilon^{\mu'\nu'\rho'\sigma'}\varepsilon_{\alpha'\beta'\gamma'\delta'}\; e_{\mu'}^{\alpha'} e_{\nu'}^{\beta'} e_{\rho'}^{\gamma'} e_{\sigma'}^{\delta'}\; \varepsilon_{\mu\nu\rho\sigma}\; e_\alpha^\mu e_\beta^\nu e_\gamma^\rho e_\delta^\sigma \\ &= \delta^{\mu'\nu'\rho'\sigma'}_{\mu\;\nu\;\rho\;\sigma}\; \varepsilon_{\alpha'\beta'\gamma'\delta'}\; e_{\mu'}^{\alpha'} e_{\nu'}^{\beta'} e_{\rho'}^{\gamma'} e_{\sigma'}^{\delta'} e_\alpha^\mu e_\beta^\nu e_\gamma^\rho e_\delta^\sigma \\ &= \varepsilon_{\alpha\beta\gamma\delta}. \end{aligned}$$

Therefore, the projected Levi-Civita tensor coincides with the completely antisymmetric tensor, thus it is constant. Hence, the derivative reads

$$D_\mu^{(\omega)}\, \varepsilon_{\alpha\beta\gamma\delta} = \omega_\alpha{}^{\alpha'}{}_\mu \varepsilon_{\alpha'\beta\gamma\delta} + \omega_\beta{}^{\alpha'}{}_\mu \varepsilon_{\alpha\alpha'\gamma\delta} + \omega_\gamma{}^{\alpha'}{}_\mu \varepsilon_{\alpha\beta\alpha'\delta} + \omega_\delta{}^{\alpha'}{}_\mu \varepsilon_{\alpha\beta\gamma\alpha'}. \quad (1.129)$$

The expression above vanishes. This can be seen by noting that since the connections and ε are antisymmetric, one gets non-vanishing contributions only when α' differs from α, β, γ and δ, which is possible only if at least two of the free indices coincide. Take for instance, $\alpha = \beta$, it is easy to recognize how the last two terms on the right-hand side of (1.129) vanish, while the first two sum up to zero. Hence, generically we get

$$D_\mu^{(\omega)}\, \varepsilon_{\alpha\beta\gamma\delta} = 0. \quad (1.130)$$

In order to evaluate the scalar curvature in terms of vierbein vectors and spin connections let us note that

$$R^\mu{}_{\nu\rho\sigma} e^{\nu\alpha} = \left[\nabla_\rho, \nabla_\sigma\right] e^{\mu\alpha}, \quad (1.131)$$

which can be rewritten in terms of spin connections as follows

$$R^{\mu}{}_{\nu\rho\sigma}e^{\nu\alpha} = \nabla_{\rho}\left(\omega_{\sigma}^{\alpha\beta}e^{\mu}_{\beta}\right) - \nabla_{\sigma}\left(\omega_{\rho}^{\alpha\beta}e^{\mu}_{\beta}\right)$$
$$= e^{\mu}_{\beta}(\partial_{\rho}\omega_{\sigma}^{\alpha\beta} - \partial_{\sigma}\omega_{\rho}^{\alpha\beta} + \omega^{\alpha}_{\gamma\sigma}\omega_{\rho}^{\gamma\beta} - \omega^{\alpha}_{\gamma\rho}\omega_{\sigma}^{\gamma\beta}),$$

so that one finds

$$R^{\mu}{}_{\nu\rho\sigma}e^{\nu\alpha} = e^{\mu}_{\beta}R^{\alpha\beta}_{\rho\sigma}, \tag{1.132}$$

in which

$$R^{\alpha\beta}_{\mu\nu} = \partial_{\mu}\omega_{\nu}^{\alpha\beta} - \partial_{\nu}\omega_{\mu}^{\alpha\beta} - \omega^{\alpha}{}_{\gamma\mu}\omega_{\nu}^{\gamma\beta} + \omega^{\alpha}{}_{\gamma\mu}\omega_{\nu}^{\gamma\beta}. \tag{1.133}$$

Hence, for the scalar curvature one gets

$$R = R_{\nu\sigma}e^{\nu\alpha}e^{\sigma}_{\alpha} = R^{\mu}{}_{\nu\mu\sigma}e^{\nu\alpha}e^{\sigma}_{\alpha} = -e^{\mu}_{\alpha}e^{\nu}_{\beta}R^{\alpha\beta}_{\mu\nu}. \tag{1.134}$$

The vierbein formalism is, on one hand, a technical tool used to simplify the calculations of geometrical objects like the curvature tensor, while on the other hand it belongs to a paradigm in which one can transport into fully covariant General Relativity formalism local special relativistic properties of the physical fields. As we shall see in section 7.2 this formulation will allow us to rewrite General Relativity with a structure which formally resembles that of a non-Abelian gauge field of the Lorentz group.

Chapter 2

Elements of Cosmology

The birth of a real *Cosmology*, as the science studying the Universe evolution was possible only after the derivation of General Relativity, because the behavior of intense and large scale gravitational field is a necessary framework for developing such a discipline. Indeed, the 19th century physics, essentially based on the Newtonian paradigm of dynamics and gravity and on the Maxwell theory of the electromagnetic interaction, was unable to construct a consistent picture of the Universe, in which such fundamental physics could join the thermodynamic properties of a macroscopic system.

Despite the success of this 19th century paradigm there was a general misunderstanding, to say better a real ignorance, on the morphology of the present Universe. The idea of a homogeneous and isotropic space having a stationary profile appeared a philosophical conjecture more than a well-posed astrophysical notion. These basic assumptions led to well-known difficulties, like the paradox of the night sky for an infinitely extended Universe, or the impossibility of a stable equilibrium when a finite matter distribution was considered.

The cultural relic of this misleading point of view affected also the mind of Einstein, who, addressing the cosmological setting in General Relativity, searched for a stationary Universe, despite the evidence, nowadays very impressive, that geometrodynamics implies the Universe expansion as a natural feature. This surprising morphology of the Universe, predicted by General Relativity, was then experimentally confirmed by the study of Hubble, who demonstrated the existence of a recession among galaxies [171].

The prescription of a homogeneous and isotropic Universe, consistent with the so-called *Cosmological Principle* (say, the Universe appears everywhere isomorphic for any observer), survived across the decades until the notion of modern cosmology emerged. Originally this simplifying request appeared a reasonable assumption, but its real physical validity came out from the discovery of a highly uniform black-body radiation, which could be easily interpreted (to some extent predicted) as the relic photon population coming from the hydrogen recombination age (when the Universe became transparent to the radiation). This fundamental discovery, probably the most impressive of the 20th century, was performed, almost by chance, by Penzias and Wilson in 1965 [245]. Today, the large scale isotropy of the present Universe is reliably confirmed by the galaxy cluster survey and the threshold scale is recognized to be a bit greater than $100Mpc$. Finally, the excellent agreement between the abundance of light elements predicted by the isotropic Big-Bang model, mainly due to Gamow [136], with the observed composition of the Universe provides a further milestone to settle down the high symmetry of cosmological space as a physical evidence (actually since an early instant after the Universe birth).

We trace below the main theoretical steps at the ground of the so-called *Standard Cosmological Model*, including its basic intrinsic shortcomings and the consequently requested inflationary theory.

2.1 The Robertson-Walker geometry

In General Relativity, the choice of the reference frame is merely a "gauge" and the cosmological problem has its natural setting in the synchronous reference ($g_{tt} = -c^2$, $g_{ti} = 0$)[1]. In fact, in a synchronous system it is possible to define a universal notion of cosmological time, that we commonly refer to as the age of the Universe. Furthermore, this frame is a geodesic one (the line of the time variable $x^i = const.$ are geodesics) and, since the present Universe is made up of a *dust* of galaxies (their center of mass behaves as a point-like test particle), we can easily recognize how the synchronous choice is also physically well-motivated (despite the very primordial Universe is no longer properly represented by a dust fluid). Finally, for the case of a homogeneous Universe, the synchronous reference can also be chosen as comoving with the cosmological fluid. This is possible also when a non-zero pressure term is present, because the homogeneity constraint prevents the presence of spatial gradients of the pressure (pressure

[1] In this Chapter and in the next one we use as time-like coordinate the time t instead of $x^0 = ct$, while $\dot{}$ denotes time derivative.

forces) and the comoving and geodesic properties can be reconciled. Therefore, we write down the Universe line element in the following form

$$ds^2 = -c^2dt^2 + dl^2 \,, \tag{2.1}$$

where, the most general homogeneous and isotropic (non-stationary) space line element, reads

$$dl^2 = a^2(t)\left[\frac{dr^2}{1-kr^2} + r^2\left(d\theta^2 + \sin^2\theta d\phi^2\right)\right] , \tag{2.2}$$

θ and ϕ being the usual spherical coordinates, while the radial coordinate r varies in a domain depending on the value of the parameter k. The sign and the amplitude of such a parameter defines the spatial curvature of the considered model. In fact, for $k = 0$, the line element inside the square brackets reduces to the Euclidean line element in spherical coordinates. In this case, the Universe is represented by a non-stationary (expanding or contracting) flat space, whose natural topology is that of a hyperplane $0 \leq r < \infty$ (but also a toroidal closed topology and other nontrivial topological spaces are possible). When k is different from zero, we can rescale it, by redefining the cosmic scale factor $a(t)$. In fact, defining the Universe radius of curvature $a_{curv} \equiv a/\sqrt{\mid k \mid}$ and $r \to r/\sqrt{|k|}$, the case of positive and negative k values admits the following space line element

$$dl^2 = a^2_{curv}(t)\left[\frac{dr^2}{1 \mp r^2} + r^2\left(d\theta^2 + \sin^2\theta d\phi^2\right)\right] . \tag{2.3}$$

The case $k = 1$ correspond to a closed space of constant positive curvature, *i.e.* a hypersphere, while the case $k = -1$ to an open space, having constant negative curvature, *i.e.* a hypersaddle. In what follows, we will refer to the cases $k = 0, \pm 1$, corresponding to a cosmic scale factor coinciding for $k = \pm 1$ with the radius of curvature of the Universe (which clearly diverges for the flat space with $k = 0$).

This spatial geometry is commonly called the Robertson-Walker model [252] and it suitably describes the behavior of a non-stationary isotropic space (the isotropic requirement implies the homogeneity, but not vice versa), in which each spatial distance scales in time with the cosmic factor $a(t)$. Thus, the Universe volume scales like a^3 and if the cosmic scale factor increases with time, the Universe correspondingly expands, just like the present Universe does.

It is important to stress that the galaxy recession is exactly an effect due to the scale of their reciprocal distance with the factor $a(t)$. Nonetheless, this effect has nothing to do with a physical velocity. Indeed, galaxies do not change (apart from their local peculiar motion) the position in space (their space coordinates remain unchanged), but it is the physical distance (a metric notion and hence time dependent in the Robertson-Walker model) which is changing in time. Thus, the velocity of recession for the galaxies, as well as, in general, the Universe expansion (or contraction if a decreases with time) is not a physical velocity and it can exhibit a faster-than-light behavior, without violating the postulates of Special and General Relativity.

2.2 Kinematics of the Universe

The only dynamic degree of freedom of the Robertson-Walker geometry is the cosmic scale factor $a(t)$, whose evolution is prescribed by Einstein's equations. Nonetheless, many important features of the expanding Universe can be fixed by simply studying the kinematics of this model [189], *i.e.* the implications of a non-stationary line element on test particles motion (in what follows we mainly refer to the case of an increasing behavior of a with time, simply because the present Universe is expanding).

If we introduce the notation $dl^2 = a^2\, h_{ij}\, dx^i dx^j$ (where h_{ij} takes different forms in the cases $k = 0, \pm 1$) the 0-component of the geodesic equation (1.43), for the Robertson-Walker geometry, reads

$$u^0 \frac{du^0}{dt} + \frac{1}{a}\frac{da}{dt} u^2 = 0\,, \tag{2.4}$$

$$u^2 \equiv a^2\, h_{ij}\, u^i\, u^j\,, \tag{2.5}$$

where u^μ is the 4-velocity[2] and we made use of the explicit form of the Christoffel symbols in the considered spacetime together with the simple relation $c\,dt = u^0\, ds$ (t being the synchronous time and s the geodesic length, *i.e.* the proper time of the particle times the speed of light). Since $g_{\mu\nu}\, u^\mu\, u^\nu = -(u^0)^2 + u^2 = \mp 1, 0$, by differentiation we easily get $u^0\, du^0 = u\, du$. Substituting this relation in equation (2.4), we arrive at the basic equation for the spatial speed u of a free particle in the Robertson-Walker geometry,

$$\frac{du}{dt} + \frac{u}{a}\frac{da}{dt} = 0\,, \tag{2.6}$$

which admits the solution $u \propto 1/a$. Since the spatial momentum of the particle is related to the speed u via the rest mass m_0, *i.e.* $p \equiv \mid \vec{p} \mid = m_0 u$, we get the fundamental issue that the particle momentum in an expanding

[2] u^μ coincides with t^μ in (1.43).

Universe is "redshifted". In fact, we have $p \propto 1/a$ and as the scale factor increases with the expansion of the Universe, the momentum correspondingly decreases. The notion of *redshift* is commonly used because the same property holds for the photon momentum (our derivation above is independent of the use of the particle proper time or of an affine parameter) hence, we find

$$p_\gamma = \frac{hc}{\lambda} \propto \frac{1}{a}\,, \tag{2.7}$$

which implies that the wavelength λ of the photon is proportional to the cosmic scale factor $a(t)$. Denoting henceforth by a suffix 0 all the quantity evaluated today, we can define the redshift amount $z(t)$, according to the relation

$$1 + z(t) \equiv \frac{a_0}{a(t)} \geq 1\,. \tag{2.8}$$

The quantity z can also be rewritten in terms of the wavelength λ, justifying his denomination, *i.e.* $z = \lambda_0/\lambda - 1$. Thus, since the Universe is expanding, the wavelength of a photon emitted by a remote galaxy is redshifted reaching us, simply because the cosmic scale factor has increased during the period between its emission and observation. This fact results in the simple observational evidence that very far galaxies, rapidly receding from us, appear more red in their emitted light than closer ones, which leads to the determination of the Hubble law

$$z = \frac{v_g}{c} = H_0 \frac{d_L}{c}\,, \tag{2.9}$$

where v_g is the (apparent) galaxy speed of recession, d_L denotes the luminosity distance[3] and H_0 is a constant observationally determined.

Actually, in an expanding Universe, any physical scale increases with the scale factor $a(t)$, summarizing the non-stationary character of the spacetime on which physical phenomena take place.

Since the Universe birth takes place in a given instant, by convention placed in $t = 0$, the maximal distance traveled by a photon is finite and it can be easily calculated by the request $ds^2 = 0$, *i.e.* $dl = cdt/a(t)$. Integrating this differential expression and recalling that to convert the coordinate distance into a physical one, we need to multiply the result by $a(t)$, we eventually get the so-called *physical horizon*, namely

$$d_H(t) = a(t) \int_0^t \frac{c\,dt'}{a(t')}\,, \tag{2.10}$$

[3]The luminosity distance is defined as the square root of the ratio between the source energy power P and the observed flux F, namely $d_L \equiv (P/4\pi F)^{1/2}$.

which, for a power-law scale factor in time $a(t) \sim t^x$, is finite for $x < 1$. The finiteness of the maximal distance that a photon has traveled from the initial singularity, is the main reason for which the night sky is darker than the day one, since we receive light from a finite region of the space (an additional effect is the redshift of the photon energy due to the Universe expansion).

Another important length scale to be considered is the so-called *Hubble length*, *i.e.* the typical scale of the Universe spacetime curvature. This length denoted by L_H is defined as $L_H \equiv cH(t)^{-1}$, where the Hubble function is $H(t) = \dot{a}/a$. This length scale is to be compared with any other scale l of a physical phenomenon, to understand how the latter is influenced by the Universe expansion. In fact, if $l \ll L_H$, that phenomenon will take place in an expanding Universe with the same morphology as in the stationary case, vice versa if $l \sim L_H$, the space expansion must be taken into account.

We observe that for a power-law scale factor the physical horizon (when finite) and the Hubble length are of the same order, being identical for the choice $x = 1/2$ (radiation-dominated Universe, see below). Nonetheless the conceptual difference of these two notions clearly emerges if we observe how, while d_H is a non-local quantity in time, accounting for the whole Universe evolution from the Big-Bang to a time t, the Hubble length is a local notion, characterizing the Universe curvature scale at that stage of evolution.

Finally, we show how the Robertson-Walker geometry is able to reproduce the Hubble law, experimentally determined in 1929. We start by expanding the cosmic scale factor in Taylor series around the present time $t_0 > t$, *i.e.*

$$a(t) = a(t_0) + \dot{a}|_{t=t_0}(t - t_0) + \dots, \tag{2.11}$$

where we truncated the series at the first order of approximation, since we are assuming that $t_0 - t \ll t_0$. Dividing the above relation by $a_0 \equiv a(t_0)$ and recalling the definitions of the redshift (2.8) and the Hubble functions, z and H respectively, we get

$$\frac{1}{1+z} = 1 - H(t_0)(t_0 - t) + \dots. \tag{2.12}$$

Since $a(t) \simeq a(t_0)$, we deal with a redshift $z \ll 1$ and $1/(1+z) \simeq 1 - z$. Furthermore, approximating the scale factor with its present value a_0 within the definition (2.10) of d_H, so estimated at the lowest order, we get the basic relation[4] $d_H \simeq c(t_0 - t)$. Thus, the expression (2.12) can be restated in the following form

$$z = H(t_0)\frac{d_H}{c} + \dots . \tag{2.13}$$

Comparing this expression with the Hubble law discussed above, we are led to the identifications $H_0 = H(t_0)$ and $d_L = d_H$. It is worth observing how the above analysis can be performed at the higher order and then compared with the observational data, which are actually still not accurate to allow a precise determination of the additional parameters emerging other than H_0. The value of H_0 has been measured with a good degree of accuracy by the observation of the Cosmic Microwave Background Radiation (CMBR) [1,169] and it is $\sim 70Km/sMpc$.

We conclude by stressing how the correspondence of the kinematic predictions of the Robertson-Walker geometry to the phenomenology of the observed Universe is a very significant feature which enforced the validity of the Cosmological Principle even before the discovery of the CMBR with its high degree of isotropy on the celestial sphere.

2.3 Isotropic Universe dynamics

The dynamics of the Robertson-Walker geometry is determined by Einstein's equations, once the matter source is specified (in vacuum only a trivial solution exists for $k = -1$ [189]).

The cosmological fluid is microscopically made up of elementary particles at thermal equilibrium, as they appear in the picture of the fundamental interactions, for instance in the Standard Model based on the paradigm of the electroweak and strong interactions. The participation of a given particle species to the thermal equilibrium is ensured by the cross section regulating the relative channels of formation and annihilation of that species. As we shall see, the expansion of the Universe is associated with a corresponding decrease of the temperature and, consequently, the cross sections of specific processes are from a certain instant suppressed, producing the decoupling of that species from the equilibrium.

Despite the complexity of the microscopic features of the primordial Universe thermal bath, its macroscopic behavior is well-described by the

[4] Clearly we are interested here only to the photon path between the instants t and t_0, taken as integration extremes in d_H.

stress-energy tensor of a perfect fluid, *i.e.*

$$T^{pf}_{\mu\nu} = (\rho + p)\, u_\mu\, u_\nu + p\, g_{\mu\nu}\,, \tag{2.14}$$

ρ being the fluid energy density, p its pressure and u_μ its four-velocity field. This macroscopic representation is justified by the observation that, on a macroscopic level, the thermal bath appears as endowed with zero electric charge, zero spin and zero angular momentum. Furthermore, apart from specific phase transitions of the Universe (like when a species decouples or during the re-heating phase after inflation) the expanding fluid is not involved in dissipative (irreversible) processes.

If we choose the comoving reference frame of the perfect cosmological fluid, *i.e.* we fix $u^\mu = (1, \vec{0})$, then the mixed form of the stress-energy tensor (2.14) reduces to the diagonal form

$$T^{pf\,\mu}_{\ \ \nu} = \mathrm{diag}\{-\rho, p, p, p\}\,. \tag{2.15}$$

Calculating by the metric (2.1) the mixed Einstein's tensor G^ν_μ, we can easily set up the dynamic equations of the isotropic Universe. The $0-0$ component of Einstein's equations, known as *Friedmann equation*, takes the form

$$H^2(t) = \left(\frac{\dot{a}}{a}\right)^2 = \frac{8\pi G}{3c^2}\rho - \frac{kc^2}{a^2}\,. \tag{2.16}$$

The $i-i$ components of Einstein's equation provide, due to the isotropy of the model, the same relation for $i = 1, 2, 3$, having the form

$$2\frac{\ddot{a}}{a} + \left(\frac{\dot{a}}{a}\right)^2 + \frac{kc^2}{a^2} = -8\pi G p\,, \tag{2.17}$$

while the $i-j$ components for $i \neq j$ identically vanish. Substituting equation (2.16) into the above spatial component, we arrive at the so-called acceleration equation

$$\frac{\ddot{a}}{a} = -\frac{4\pi G}{3c^2}\left(\rho + 3p\right)\,. \tag{2.18}$$

This equation clearly shows how the Universe has to decelerate as long as $p > -(1/3)\rho$ and this is expected to be the case for the present matter-dominated Universe (the dust of galaxies has an equation of state $p \simeq 0$). From equation (2.16), we see that for the two cases $k = 0, -1$, the Universe expands indefinitely, since there is no finite instant of time when the quantity H^2 vanish. In other words, the flat and negative curved spaces do not admit turning points, where the expansion is converted into collapse (or vice versa). The situation is completely different for the closed Universe, for which a turning point always exists in correspondence to the value of the cosmic scale factor $a_{tp} = (3c^4/8\pi G\rho)^{1/2}$. This turning point is a maximum in the Universe expansion, since in $a = a_{tp}$ the second derivatives $\ddot{a}$ is negative as long as the inequality $p > -(1/3)\rho$ holds.

The two Einstein's equations (2.16) and (2.18) can be completed by the conservation equations of the stress-energy tensor, offering a direct link between the energy density, the pressure and the scale factor. This information can be easily and equivalently recovered by applying the first and second principles of thermodynamics to a generic (coordinate) volume element of the Universe. Since the Universe is an isolated system and all the space points are at the same temperature, the exchanged heat quantity is zero and the evolution results to be mainly isoentropic (apart from brief phase transitions or the re-heating phase after inflation). Then in a generic volume, the variation in the internal energy equates the work made by the pressure, *i.e.*

$$d(\rho a^3) = -pd(a^3)\,, \tag{2.19}$$

which can be easily restated as

$$\frac{d\rho}{dt} = -3\frac{\dot{a}}{a}\left(\rho + p\right)\,. \tag{2.20}$$

This equation, known as the continuity equation, forms, together with equations (2.16) and (2.18), a system in which two equations are independent only, the remaining one being determined by combining the other two. Usually, the adopted couple is formed by equations (2.16) and (2.20). Indeed, as soon as we assign an equation of state for the cosmological fluid, having the form $p = w\,\rho$ (where w is piecewise constant in the Universe thermal history), equation (2.20) provides the expression $\rho(a) \propto a^{-3(w+1)}$. Hence, substituting this relation (with an assigned proportionality constant) into equation (2.16), we can calculate the cosmic scale factor $a(t)$, and eventually $\rho = \rho(t)$.

The limit $a \to 0$ is a physical singularity since for $w > -1$ the Universe energy density diverges. We now observe that for $w > -1/3$, the energy density dominates the curvature term in equation (2.16) as $a \to 0$. This approximation remains indeed valid for the most part of the Universe evolution, since the recent observations of the CMBR confirm that the present value of the spatial curvature is very small (this contribution decreases along the Universe expansion). Then, substituting the expression of $\rho(a)$ into the Friedmann equation, we easily integrate it, via a power-law solution of the form $a(t) \sim t^{2/[3(w+1)]}$. As soon as we insert this expression into the energy density, we arrive at the following form (independent of the value of the proportionality constant)

$$\rho(t) = \frac{c^2}{6\pi G(3w+1)^2 t^2}, \tag{2.21}$$

which provides the value of the energy density at any given instant and confirms the existence of a physical singularity (in the energy density, as well as in the spacetime curvature too) at $a(t=0)=0$. It is just this singular behavior of matter and spacetime geometry in the far cosmological past that is at the ground of the notion of the so-called *Primordial Big-Bang.* In order to characterize the phenomenology associated to this dynamic behavior of the expanding Universe, it is necessary to determine the general features of the Universe thermal history. Indeed, we have to fix the different stages of the evolution, associated to the specific form of the equation of state in order to extract thermodynamic information on the primordial thermal bath.

2.4 Universe thermal history

According to the idea that the younger the Universe the higher its temperature, we can approximate the primordial thermal bath in terms of fundamental particles all having photon-like properties. In fact, all the species are at thermal equilibrium with the photon gas. The high value of the average particle energy, which is of the order of $k_B T$ (k_B denoting the Boltzmann constant and T being the temperature), implies that almost all the species can be treated as ultrarelativistic components, *i.e.* with negligible rest mass. Furthermore, like the photon gas, these species have essentially zero chemical potential. Hence, it is rather natural to speak of the very early Universe as a radiation-dominated system, whose equation of state and energy density behavior versus the scale factor read, respectively

$$p_r = \frac{1}{3}\rho_r, \quad \rho_r \propto \frac{1}{a^4}, \tag{2.22}$$

where the suffix r stands for the radiation-like nature of this contribution. Now we can calculate the same radiation-like energy density component from the kinetic theory (essentially by the equilibrium phase-space distribution function for the particle species) of the expanding Universe and the output of this analysis takes the following form

$$\rho_r = \frac{\pi^2}{30} g^*(T)\, T^4, \tag{2.23}$$

where the Universe temperature T is here identified with the photon gas temperature T_γ, *i.e.* $T \equiv T_\gamma$. The function $g^*(T)$ weights the degrees of freedom and the temperature of the different particle species, according to the expression

$$g^*(T) = \sum_{B_i} g_{B_i} \frac{T_{B_i}^4}{T^4} + \frac{7}{8} \sum_{F_i} g_{F_i} \frac{T_{F_i}^4}{T^4}\,, \tag{2.24}$$

where the labels B_i and F_i refer to the boson and fermion particles species, respectively. This function is piecewise constant along the Universe evolution, since it varies only when a particle species decouples from the thermal equilibrium. Thus, comparing the relations (2.22) and (2.23), we easily get, for the most part of the Universe evolution, an inverse proportionality between the Universe temperature and the cosmic scale factor, *i.e.* $T \propto 1/a$. This relation must remain valid arbitrarily close to the singularity ($a = 0$) since the function g^* reaches a fixed value over a certain temperature (corresponding to the first decoupling of a species). It is just this divergent character of the Universe temperature nearby the instant when the spacetime curvature diverges, that leads to the well-known concept of *hot Big-Bang.* Indeed, from a physical point of view, what is crucial for the construction of the Standard Cosmological Model in terms of an initial hot Big-Bang, is the increasing behavior of the geometrical and physical quantities as the scale factor (*i.e.* the Universe volume) decreases. In particular, in section 9.6.1 we will show that the Loop Quantum dynamics of an isotropic Universe removes the presence of a physical singularity. The presence of an effective backward turning point in the Universe evolution (denominated *Big-Bounce*) is a consequence of the semiclassical features associated to Loop Quantum Cosmology and it is located at the Planckian time $t_P = l_P/c \sim O(10^{-43}s)$ (l_P being the Planck length $l_P \equiv \sqrt{G\hbar/c^3}$). Such a cut-off on the nature of the singularity therefore takes place at a temperature of about $T \sim 10^{19} GeV$, which for the most part of the later Universe phenomenology is indistinguishable from infinity. Hence, no significant modification of the Standard Cosmological Model is predicted by the presence of such removal of the initial singularity.

Since the matter contribution of the Universe, due to non-relativistic particles, has an equation of state $p \simeq 0$, the corresponding energy density ρ_m scales like $1/a^3$. This term decreases simply because the Universe expands, while the mass of the non-relativistic component is conserved. The radiation energy density decreases like $1/a^4$ because, in addition to the volume expansion, the decrease of the energy $k_B T$ of the particles is redshifted by a factor $1/a$.

As result of such a different scaling with the Universe expansion, we see that the ratio between the radiation and matter components of the Universe behaves as $\rho_r/\rho_m \propto 1+z$. Therefore, while the primordial phases of the Universe are radiation-dominated, sooner or later, the matter component prevails. It is possible to verify [189] that, the equivalence age between these two energy contributions takes place for $z \sim 10^4$ (more accurate estimates set this value as $z \sim 3 \cdot 10^3$), corresponding to a Universe age of about $t_{eq} \sim 10^{11}s$. After that phase, the Universe becomes matter-dominated and its cosmic scale factor acquires the power-law behavior $a(t > t_{eq}) \sim t^{2/3}$.

For $t < t_{eq}$ matter and radiation remain in thermal equilibrium with each other till the atomic hydrogen *recombination*. From that instant on photons could (almost) freely travel the Universe, reaching us today. The CMBR is nothing more than the last scattered radiation by the ionized matter, thus it provides a very primordial picture of the matter distribution across the Universe. This background radiation corresponds with surprising agreement to that of a very isotropic black body (more accurate than any black body present in nature or reproduced in laboratory), with temperature fluctuations of about 10^{-5} between couples of points taken on a given angular distance on the sky sphere. The temperature of the CMBR black body is about $T_{CMBR} \sim 2.725K$ [169] and the recombination age is placed at a temperature of about $3 \cdot 10^3 K$, corresponding to a redshift $z_{rec} \sim 1.1 \cdot 10^3$ and to a Universe age $t_{rec} \sim 3.8 \cdot 10^5 y$ at the recombination.

The recombination temperature is much smaller than the hydrogen ionization temperature (about $1.6 \cdot 10^5 K$), because of the high ratio between photons and baryons present in the Universe. Such a ratio is essentially constant during the Universe evolution and it is estimated to be $n_\gamma/n_B \sim 10^9$, n_γ and n_B being the photon and baryon number density, respectively. The large amount of photons with respect to that of the baryon population, implies that also the black body tails at the ionization energy are to be taken into account, at least as soon as such a tail population equates that of the electrons, at a temperature of about $3 \cdot 10^3 K$. Since that time, there are too few photons at ionization energy to maintain the equilibrium between ionized and recombined atoms. However, the last scattering surface has a certain depth, because recombination is not an instantaneous process. Indeed, the large amount of photons allows baryons to remain in equilibrium with radiation up to $z \sim 10^2$ and, *de facto*, only after that time the Universe equation of state approaches the limit $p \simeq 0$, for which the evolution becomes genuinely matter-dominated.

A matter-dominated Universe (as it should be considered the present one) must clearly decelerate, according to equation (2.18). However, 15 years ago, studying the dynamic evolution of the Universe, by means of standard candles (the Super Novae IA), it has been demonstrated, with increasing reliability over the years, that the actual Universe accelerates (see [132] for a review). Estimates of the starting instant of such an accelerated behavior of the Universe fix it at about 7 billions years ago [214], opening the serious question on which effect is responsible for such very unexpected feature. The best fit of the CMBR anisotropies suggests that the origin of this acceleration can be recognized in a non-zero vacuum energy density (*de facto* a cosmological constant term), having equation of state $p = -\rho$. However, the determination of the equation of state, associated to this dark component of the Universe, known as *dark energy*, requires a further refinement of the observational data concerning the present Universe matter distribution. The understanding of this puzzling feature of the present Universe is certainly the most challenging perspective of the cosmological investigations in the near future.

It is worth mentioning that dark energy is not the only dark component of the Universe. In fact, the presence of the so-called *dark matter* has been postulated to match an observational output to some theoretical inconsistencies of the Standard Cosmological model [54, 189].

On one hand, dynamic estimates, like the flatness of the galaxy rotation curves at large distances, indicate the presence of a dark halo component around formed structures, made up of very weakly interacting (non-baryonic) particles, able to create a potential well for the baryonic matter collapse, but unable to produce (via collision processes) electromagnetic emission. On the other hand, the evolution of baryonic matter fluctuations seems to be clearly insufficient to generate the observed structures today populating the Universe, for instance galaxies. In fact, the CMBR temperature anisotropy fixes (via a detailed account for the matter-radiation scattering) the density contrast at the recombination age to be of the order of $\delta\rho_B/\rho_B \sim 10^{-3}$, $\delta\rho_B$ being the baryonic matter fluctuation. From the relativistic perturbation theory, we can determine the behavior of the density fluctuations as the matter-dominated Universe expands and we get $\delta\rho_B/\rho_B \sim a(t)$. In other words, the baryonic density contrast increases by a factor $z_{rec} \simeq 1.1 \cdot 10^3$ from the recombination age up to now. This would imply that today the expected density contrast would be of order one (we would be almost at the end of the linear growth of perturbations). This expectation is in clear contradiction with the observational evidence that the density contrast of galaxies, with respect to the average Universe density, is about 10^5.

A possible solution to this theoretical inconsistency of the standard evolution with the observed structures, can be easily offered by postulating the existence of a dark matter component. Such a component could possess a greater density contrast at the recombination age, since its interaction with the CMBR photons would have been very weak. The weakly interacting particles, constituting dark matter, could have decoupled from the thermal equilibrium much before than the recombination age, when baryons started to decouple. Thus, the density contrast of the dark matter component had much more time to grow with respect to ordinary matter, whose collapse is therefore driven by the non-baryonic gravitational potential well. The crucial point is that the weakly interacting particles leave a very small track on the CMBR. So, the temperature anisotropy of order 10^{-5} can correspond to a non-baryonic density contrast one or two order of magnitude greater than the baryonic 10^{-3} value, determined for the ordinary collisional matter.

The theory of a dark matter component of the Universe is much more settled down than the dark energy hypothesis, which is at the stage of a conjecture. In fact, many possible candidates for weakly interacting particles have been investigated and some of them appear viable proposals (for instance the neutralino particle [54]), but their exotic (still unobserved) nature calls attention for further investigation. In this respect, the dark matter theory is a successful cross-match between the observed halo in clustered structures, the necessity of a weakly interacting Universe component and the predictions, yet to be verified, of high energy particle physics. Another point, still not completely clarified, is the right mixing between *cold dark matter* (which decouples from the equilibrium when it is already a non-relativistic component) and *hot dark matter* (which is a relativistic component at its decoupling) required to reproduce the observed features of the structure formation across the Universe.

2.4.1 *Universe critical parameters*

In order to shed light on how puzzling is the presence of a dark matter and a dark energy component of the Universe in constructing a consistent cosmological picture, we have to estimate their present amount in the Universe, in comparison to the baryonic (visible) matter contribution.

To this end, let us introduce the Universe critical density, defined as

$$\rho_{crit} \equiv \frac{3c^2 H^2}{8\pi G}, \tag{2.25}$$

which corresponds to the Universe density for the flat space $k = 0$. If we introduce the Universe critical parameter as $\Omega \equiv \rho/\rho_{crit}$, then equation (2.16) rewrites as

$$\Omega - 1 = \frac{kc^2}{H^2 a^2} \equiv \Omega_k \,. \tag{2.26}$$

Thus, $\Omega > 1$ corresponds to the closed Universe with $k = 1$, $\Omega < 1$ provides the open Universe with $k = -1$ and, by definition, $\Omega = 1$ defines the flat Universe, associated to $k = 0$.

The determination of the Universe critical parameter Ω from the CMBR data shows clearly that its value is very close to unity and the discrepancy is smaller than one percent. However, the sign of k remains undetermined, $k = 0$ lying within the measurement errors. Thus, if we neglect the quantity Ω_k and separate the critical parameter into its baryonic Ω_B, dark matter Ω_{dm} and dark energy Ω_Λ (vacuum energy) components, we get the basic relation

$$\Omega = \Omega_B + \Omega_{dm} + \Omega_\Lambda \simeq 1 \,. \tag{2.27}$$

The paradox of the Standard Cosmological Model emerges as long as the best fit of the CMBR data is used to determine the relative amounts of such specific critical parameters. In fact, we find that roughly $\Omega_B \sim 0.05$, $\Omega_{dm} \sim 0.23$ and $\Omega_\Lambda \sim 0.72$. Thus, we have a clear understanding and physical control on the five percent of the Universe energy density only, *i.e.* the visible baryonic contribution. Such a peculiar situation is however a measurement of our ignorance on the Universe nature and composition, more than a real physical puzzle to search a solution for.

In this schematic discussion of the Universe thermal history we did not mention important phases of evolution like the baryogenesis or the nucleosynthesis of the light elements (one of the proofs of the Standard Cosmological Model) since they are out of the scope of this volume. We cannot avoid to discuss the very primordial *inflationary era*, since it has a relevant impact on the morphology of the primordial Universe, even with respect to the Planck era, when the quantum effects on the cosmological dynamics of the Universe can no longer be neglected (see Chapter 9).

The inflationary paradigm arises from the necessity to overcome some conceptual paradoxes of the standard cosmological picture as traced by the

Friedmann-Robertson-Walker geometrodynamics, and it provides a deep alteration of the natural Universe evolution, being responsible for a cancellation of the pre-existing initial conditions and the restoring of new morphological (to some extent model dependent) signatures of the Universe.

Before passing in the next subsection to analyze the inflationary theory, by schematically tracing its main features and implications, we stress how the evolution from $z = 10^2$ to $z = 10$, when the galaxy formation takes place, as well as from $z = 10$ up to now is almost under control from the point of view of the basic physical mechanism active in this structure formation phase. The detailed picture of how the observed Universe emerges from the recombination image (determined on the CMBR) remains an open and fascinating question, to be answered by future observational and theoretical tasks.

2.5 Inflationary paradigm

We now trace the general ideas at the ground of the *inflation* theory, starting from the motivations, developing the basic concepts and eventually stressing successes and limits of the paradigm. The presentation will touch only very general points, without entering the subtle technical questions concerning the common ground of cosmology and elementary particle physics the inflationary model relies on. For a complete discussion of this topic we refer the reader to specific monographs or textbooks [189,230,243]. We fix here only those relevant features to better understand how the quantum (Planck) era of the Universe is connected with the later phases of the Universe thermal history, discussed above. It is worth noting how maintaining a general perspective on the inflationary model can also be a good point of view because it allows to better emphasize merits and shortcomings of the presented scenario, avoiding to be too much involved in specific questions which are not yet settled down in many details.

It is just in the general lines here presented that one can find the convincing profile of a real physical theory in the piecewise (to some extent qualitative) picture of the inflationary and re-heating phases of the Universe (the latter one is probably the most heuristic step of the whole conceptual framework).

2.5.1 *Standard Model paradoxes*

The Standard Cosmological Model contains some very subtle paradoxes, whose full understanding opened the way for a new perspective on the primordial Universe cosmology. We are going to briefly discuss these paradoxes, dividing them in the following four points (as in the most common literature on the subject) and ordering them according to their relevance (the order we have chosen relies on how deeply each single paradox affects Standard Cosmological Model beliefs). Before passing to the description of these shortcomings of the Standard Cosmological Model (described by Friedmann-Robertson-Walker geometrodynamics), it is worth noting how almost all of them, especially the first three, arise from an unphysical fine-tuning of the initial conditions.

- i) *The horizon paradox*
  To illustrate this paradox, we have to start by answering the question: how many causal scales are contained in the sky sphere of the CMBR? An estimate of the horizon scale at the recombination age is equivalently offered (in a standard thermal history) by the cosmological horizon d_H, as well as by the Hubble length L_H, both of the order of $ct_{rec} \sim 10^{22}cm$. However, this spatial scale has been enlarged by the Universe expansion (as any other physical scale) and today is a thousand times larger, namely the causal scale on the CMBR sphere is of the order of $l_c \sim 10^{25}cm$. The linear dimension of the CMBR sphere is again of the order of d_H^0 or L_H^0, both $\sim cH_0^{-1} \sim 10^{27}cm$. Thus the number of causal regions we observe when looking at the CMBR is $(cH_0^{-1}/l_c)^2 \sim 10^4$.
  This result naturally leads to a serious physical question (the horizon paradox): why such high number of causally uncorrelated regions have a fine-tuned temperature, with fluctuations of order 10^{-5}?
  The most immediate answer could be searched in the possibility that such regions have been in causal contact in the far past, before the recombination, say, near the singularity. But this is false, simply because, in a radiation-dominated Universe, the scale factor behaves as $\sqrt{t}$, while the causal horizon is linear in t. Thus, if two space points are super-horizon sized at a given instant, they will remain so even in all their past.
  Another more conceptual answer could instead be based on the

opposite point of view, saying that the CMBR sphere is so homogeneous in spite of the causal regions it contains, simply because the Universe was everywhere homogeneous by assumption (as prescribed by the initial conditions). This point of view could appear rather reasonable, but it requires that we calculate how fine tuned would be the initial, say at the Planck era, density contrast. As discussed above, the density contrast at the recombination era is of the order of $\sim 10^{-3}$ (or a bit larger for dark matter). At the equivalence age it loses a factor 10 because the matter density contrast scales with the redshift. Instead, for the radiation energy density we have $\delta\rho_r/\rho_r \sim a^2$ [189, 196]. Hence the density fluctuations in the Planck era is a factor $t_{eq}/t_P \sim 10^{54}$ smaller than the equivalence value. Thus, to have compatibility with the CMBR observations, the initial density fluctuations must be very close to $\sim 10^{-60}$. This very high degree of homogeneity of the Planckian Universe appears rather unphysical, especially in view of the quantum fluctuations of the gravitational and the matter fields. However, even disregarding such a quantum-like argument, the determined initial condition as backward extension of the CMBR anisotropy detection, cannot have any clear physical motivation and it stands as a fine-tuning on the very primordial Universe.

- ii)*Flatness paradox*
Recent observations [169] have shown that the spatial curvature contribution Ω_k is essentially negligible today. Anyway, the simple observation that the light rays coming from far galaxies follow straight lines (instead of positive or negative curved geodesics) suggested, already in the '70s, that the present value of the quantity $\Omega - 1$ had to be in the range of unity (say from 0.1 to 10). Since, going backward to the singularity Ω_k increases slower than the matter or radiation energy density, we can use for estimating $\Omega - 1$, as provided in (2.27), the quantity H^2 in the case $k = 0$. As a consequence, during the matter-dominated era $H^2 \propto 1/a^3$ and $\Omega - 1 \propto a$, while in the radiation-dominated era $H^2 \propto 1/a^4$ and $\Omega - 1 \propto a^2$. Starting with the present value $\Omega_0 - 1 \sim 1$, we easily get, with a calculation very similar to that made above for the energy density fluctuations, that at the Planck era one would have the very fine-tuned initial condition $\Omega_P - 1 \sim 10^{-60}$ ($\Omega_P \equiv \Omega(t = t_P)$). The only

way to escape such unphysical fine-tuning of the Universe critical parameter would be to say that, for some unknown reasons, the Universe emerged from the Planck era, or it was born, as a flat Robertson-Walker model. But, this statement would imply no real physical gain, since the case $k = 0$ is an unstable Universe with respect to the generic case $k \neq 0$ and no convincing explanation for its selection appears. The request of an essentially flat Planckian Universe is the core of the so-called flatness paradox.

- iii)*Entropy paradox*

The Universe evolution, as traced in the previous section, is mainly isoentropic, apart from those phases of non-equilibrium (for instance the decoupling of a particle species), whose contribution to the increase in entropy can be estimated and is not a huge source of irreversibility along the expansion process. The same consideration holds for those local irreversible processes which had taken place across the Universe evolution, *i.e.* star formations and explosions, accretion and emission phenomena around compact astrophysical objects, etc.

The paradox arises from the estimate of the present value of the entropy contained in the Hubble volume $(L_H^0)^3$. The entropy density σ_r associated to the radiation component of the Universe (we recall that the photon to baryon number density is about 10^9) is given by the relation

$$\sigma_r = \frac{\rho_r + p_r}{T} = \frac{4\rho_r}{3T} = \frac{2\pi^2}{45} g^* T^3 \,. \tag{2.28}$$

This expression can be easily estimated, by noting that the radiation and matter energy densities are equal at $z_{eq} \sim 3 \cdot 10^3$ and therefore today $\rho_r \sim 0.3 \cdot 10^{-3} \rho^0_{crit}$ ($\rho^0_{crit}/c^2 \sim 0.9 \times 10^{-29} g/cm^3$ is the actual Universe critical density, *i.e.* its present matter density). A careful estimate provides the following huge value of the actual Universe entropy per Hubble volume

$$\sigma_r^0 (L_H^0)^3 \sim 10^{88} k_B \,. \tag{2.29}$$

Such very large amount of entropy in the Hubble volume cannot be justified, in the Standard Cosmological Model, by physical irreversible processes which concerned the Universe evolution and therefore we are again led to a fine-tuning of the initial condition on the Universe initial entropy, known as the entropy paradox.

- iv) *The unwanted relic paradox*
  Modern theories of fundamental particle physics predict that, at very high energies (say $> 10^{14}GeV$), exotic particles exist, like Planck mass particles or magnetic monopoles. Therefore, when the Universe is close enough to the Planck era ($k_B T_P \sim 10^{19}GeV$), these populations of exotic particles must be produced and they would have to survive in the present Universe too, diluted by the expansion process, but still detectable as any other particle species. The paradox is in the question: why do we not see these exotic species?
  A possible answer could be that they do not exist in nature, but this perspective seems to contradict many reliable general approaches to fundamental particle physics. Therefore the fact that the number of exotic particles in the present Hubble volume is compatible with zero, calls attention for a solution. This paradox is not truly a fine-tuning on the Universe initial conditions. However, it can be set in such a form, simply by requiring that exotic particles species are extremely diluted in the Planckian Universe, so that there are few of them per Hubble volume. Since the redshift of the Planckian era is $z_P \sim 10^{30}$, then, assuming that exotic particles are matter-like species and that their number density is today of a particle per Hubble volume, we get that their Planckian density would be $N_p^{ex} \sim 10^{10}cm^{-3}$. This value is only apparently large, indeed it is so small to generate a paradox, as it clearly stands when we observe that the baryon density extrapolated to the Planckian era would be $n_B(t_P) \sim 10^{85}cm^{-3}$.

Let us now face the analysis of the general features of the inflation mechanism, in order to later focus on how it offers a reliable simultaneous solution to all the paradoxes above.

2.5.2 *Inflation mechanism*

The inflation scenario requires that the Universe performed, in the early stages of its evolution, a phase transition, which in the most accepted models is associated to a spontaneous symmetry breaking (SSB) process. The SSB mechanism takes place when the Lagrangian describing fundamental particle physics is invariant under an internal symmetry, but the same is not true for its vacuum states (identified as the local minima of the potential energy density of the involved fundamental fields) [100, 240].

The SSB is typically associated to that of a Grand Unified Theory (GUT), occurring when electroweak and strong interactions (*i.e.* the interactions of the Standard Model of particle physics) can no longer be unified [295]. Such a phase transition had to take place in a range of Universe temperatures $10^{14} - 10^{15} GeV$ and induced by the Higgs field associated to the GUT (many different models are regarded as viable, with or without Supersymmetry [295]).

During the inflation process the Higgs scalar field ϕ is a source for the Robertson-Walker geometrodynamics and therefore, the homogeneity constraint implies such a field be time dependent only, *i.e.* $\phi = \phi(t)$. Its dynamics will be mainly treated as that of a classical field, according to the idea that it corresponds to a boson wave function at very high occupation numbers (high density of the corresponding boson particles). The situation is at all equivalent to that of a classical electromagnetic field, for which the wave amplitude has a classical behavior in view of the high density of elementary photons it corresponds to. The Higgs field is a self-interacting scalar field, endowed with a potential term, having the typical form

$$V(\phi) = -\frac{1}{2}\mu_0^2\phi^2 + \frac{\lambda}{4!}\phi^4\,, \tag{2.30}$$

where μ_0 and λ are parameters fixed in the GUT. However, we stress how the potential above is associated to a vacuum situation, *i.e.* to a zero-temperature environment. However, the Higgs field is embedded into the primordial thermal bath and the interaction with the finite-temperature environment induces an important modification of the self-interacting potential, which, according to a precise calculation based on the system free energy [189], takes the more general form

$$V_T(\phi) = \frac{1}{2}\mu_T^2\phi^2 + \frac{\lambda}{4!}\phi^4 \quad , \mu_T^2 \equiv \alpha T^2 - \mu_0^2\,, \tag{2.31}$$

where $\alpha > 0$ is a phenomenological parameter. The critical temperature T_c is defined by the massless condition $\mu_{T_c} = \alpha T_c^2 - \mu_0^2 = 0$ and it reads $T_c = \mu_0/\sqrt{\alpha}$. The dependence of the field potential on the temperature must be regarded as a parametric one, which means that at each value of the temperature it takes a precise profile. It is clear that, as long as the temperature is sufficiently high, *i.e.* $T > T_c$, the potential has a minimum in zero, the unique vacuum configuration of the field. When the temperature becomes a bit smaller than such a critical value, T_c, for which a phase transition takes place, $\phi = 0$ becomes a relative maximum (*false* *vacuum*) and two symmetric degenerate minima (*true vacua*) appear in $\phi = \pm\sqrt{6 \mid \mu_T^2 \mid / \lambda}$ (μ_T^2 is now a negative quantity). The gap between the false and the true vacua is given by $\rho^* \equiv 3\mu_T^4/\lambda$ and this is a constant vacuum energy density injected into the Universe expansion, as an effect of the phase transition. This is a SSB process because the two degenerate vacua, symmetric with respect to $\phi = 0$ are clearly not invariant (but mapped into each other) under the discrete symmetry of the considered theory $\phi \to -\phi$. In more general pictures of the inflationary model, not necessarily related to a SSB process, a potential barrier can exist between the true and false vacua, which the Universe has to cross by tunneling effect [230].

Before analyzing the hypotheses under which the inflationary paradigm dynamically influences the Universe evolution, we write the two basic equations describing the dynamics of the two degrees of freedom $a(t)$ and $\phi(t)$, whose behavior is governed by the following Friedmann and generalized Klein-Gordon equation (*i.e.* in the presence of spacetime curvature and of a self-interacting potential), respectively

$$H^2 = \frac{8\pi G}{3c^2}\left[\rho_r + \frac{\dot{\phi}^2}{2c^2} + V_T(\phi)\right] - \frac{kc^2}{a^2} \tag{2.32}$$

$$\ddot{\phi} + 3H\dot{\phi} + c^2\frac{dV_T}{d\phi} = 0\,. \tag{2.33}$$

Now, since for $T \leq T_c \mid \mu_T^2 \mid$ is very small, we can speak of a *slow-rolling phase* of the scalar field toward one of the two minima. In other words, the "generalized coordinate" $\phi(t)$ moves slowly in a quasi-flat profile (the kinetic energy of the scalar degree of freedom is much smaller than the nearly constant potential one). In general, the request that the potential $V_T(\phi)$ has a significantly flat region between the false and true vacua, must be regarded as an *ad hoc* hypothesis of the proposed scenario. A fundamental assumption is that, after the phase transition, the constant potential energy ρ^* dominates the right-hand side of equation (2.32) and, in particular that $\rho^* \gg \rho_r$. This equation then reduces to the equality $H = H^* \equiv \sqrt{8\pi G\rho^*/3c^2}$. This implies that the scale factor takes the exponential time behavior ($a \sim e^{H^*t}$) of a de-Sitter Universe. Hence, the ratio between the scale factor at the end of the slow-rolling phase $t = t_f$ and at its beginning $t = t_i$ takes the form

$$\frac{a_f}{a_i} = \exp\{H^*(t_f - t_i)\} \equiv e^{\tau}\,, \tag{2.34}$$

where the suffix of the scale factors refer to the instant in which they are evaluated and τ is known as the *e-folding* parameter of the de-Sitter phase.

During the slow-rolling the quantity $\ddot{\phi}$ is assumed to be negligible in equation (2.33) and the time evolution of the scalar field is then fixed as soon as we assign the explicit form of $V_T(\phi)$ near the maximum. This potential expression is typically of the form $V_T \sim \rho^* - \beta\phi^4$, where β is a parameter properly fitting the potential behavior. However we do not enter here the details of the scalar field dynamics in the plateau region of the potential, because it is not relevant for the solution of the paradoxes; it is clear that sooner or later the scalar degree of freedom accelerates its motion and falls into one of the two potential well surrounding one of the two minima (which minimum is reached is irrelevant and it depends on the initial conditions, *i.e.* the properties of the field in the pre-existing minimum). We now show that the first two paradoxes are easily simultaneously solved by requiring that the e-folding $\tau \sim 10^2$ (since t_i is fixed by the GUT energy scale and it corresponds to $t_i \sim 10^{-34}s$, we are fixing the end of the de-Sitter phase at $t_f \sim 10^{-32}s$).

First of all we observe that, during the de-Sitter phase, the Hubble microphysical scale L_H remains constant, while the physical horizon d_H exponentially increases.

The matter contained inside a Hubble scale L_H^i (or equivalently a physical horizon d_H^i) is, at the end of the de-Sitter phase, stretched on much greater scales, of the order $L_H^f \sim L_H^i e^{\tau}$, which, if $\tau \sim 10^2$, is much greater than the present Hubble scale (the equality would be obtained for $\tau \simeq 67$).

This result clearly solves the horizon paradox since the matter we are looking at in the CMBR sphere was, before inflation, entirely contained within a single causal and microphysical horizon. As a consequence, there is no longer any surprising feature in the fine-tuned temperature of the CMBR sphere and no unphysical initial perturbation spectrum is then required.

The flatness paradox is solved because of the constant character of $H = H^*$ in equation (2.26), which yields the behavior $\Omega - 1 \propto 1/a^2$. Hence we have that $\Omega - 1$ at t_f is related to the original one at t_i by the relation (we again use obvious notation for the suffixes)

$$\Omega_f - 1 = (\Omega_i - 1)e^{-2\tau} . \tag{2.35}$$

For any physical choice of Ω_i, we get Ω_f essentially equal to unity (with a very high degree of precision). Although, after the de-Sitter phase the quantity $\Omega - 1$ starts to increase again, Ω_0 can naturally take the observed value (nearby one).

During the exponential expansion of the Universe, the temperature of the thermal bath dramatically decays, according to the relation $T_f =$

$T_i\, e^{-\tau}$, and therefore it is very close to the absolute zero. Furthermore, the energy density of all the pre-existing species at $t = t_i$ has been now drastically diluted, for instance, $\rho_r^f = \rho_r^i e^{-4\tau}$. Thus the de-Sitter phase is a process able to cancel the initial conditions of the Universe, *i.e.* any pre-existing structure or information is suppressed as a consequence of the strong stretch of the space geometry. Thus, it is necessary a specific process, able to restore somehow the pre-inflationary thermal bath, solving also the two remaining paradoxes. Such a process, complementary to the de-Sitter phase, is commonly called the *re-heating phase* and it occurs when the scalar field (the inflaton) falls into one of the two well and starts to oscillate around the selected vacuum state.

2.5.3 *Re-heating phase*

As soon as the scalar field begins to oscillate near one of the two minima, it acquires the features of a massive boson, whose effective rest mass m^* is given by the non-zero second derivative of the potential term evaluated in the minimum. The generalized Klein-Gordon equation, which describes the dynamics of such a massive boson, takes the form

$$\ddot{\phi} + 3\,(H + \Gamma)\,\dot{\phi} + c^2{\mu^*}^2\phi = 0\,, \tag{2.36}$$

where $\mu^* \equiv m^*c/\hbar$ is the inverse of the boson Compton length. Here we added the phenomenological parameter $\Gamma = 1/(3\tau_d)$, which accounts for the decay of the field, having an averaged lifetime τ_d. Such a decay process is naturally expected because of the fundamental properties of this Higgs field within the specific GUT model, but it can also be regarded as *a posteriori* requirement for the re-heating process leading to the observed Universe. In fact, the massive bosons, due to the low temperature, form a Bose condensate and their decay re-heats the Universe. The boson field decays into light relativistic particles, whose average temperature is again very high. Indeed, in equation (2.36) both the expansion rate H and the decay rate Γ act as a damping of field oscillations.

This global and huge process of decay that re-heats the Universe is an irreversible phenomenon, able to create a large amount of entropy across the Universe, independently of its initial value at the Planck era. It is just this consideration that justifies the actual huge value of the entropy in the Hubble volume and hence solves the entropy paradox.

The unwanted relic paradox is also solved because all the pre-inflation species are strongly diluted during the de-Sitter phase, but while the other (ordinary) particles are restored by the boson condensate decay, the exotic particles can no longer be produced, since the conditions for their generation are no longer met (or the decay channels are strongly suppressed). Thus, after the whole inflation process, *i.e.* de-Sitter phase plus re-heating, the exotic species remain strongly diluted and *de facto* unobservable.

The temperature at which the Universe is re-heated depends on how relevant the redshift of the expansion rate has been. If the decay process is very efficient ($\Gamma \gg H$), such final temperature T_{rh} must be of order T_i and the re-heating is called *good*, while in the opposite case, T_{rh} results much less than T_i and the re-heating process is said *poor*. Anyway, after such a decay process has re-heated the Universe, the standard thermal history will take again its natural profile as discussed in the previous section.

What about the density perturbation spectrum? The Planckian fluctuations are clearly redshifted and then canceled by the de-Sitter phase. Therefore it is necessary to identify an alternative mechanism for their generation. This can be successfully found in the quantum fluctuations of the inflaton field during the Universe exponential expansion. If we write down the inflaton field as $\phi = \phi_{cl}(t) + \Delta\phi_q(t, x)$, separating it in its classical ϕ_{cl} and quantum $\Delta\phi_q$ parts, respectively, it is possible to show that the perturbation spectrum associated to the Fourier representation of the field $\delta\phi_q$ on a de-Sitter metric, has the very well-known Harrison-Zel'dovich form [189]. Such a spectrum corresponds to having scale-invariant perturbations entering the microphysical horizon L_H. In the inflation case, indeed, the perturbations of cosmological relevant size escape the microphysical horizon during the de-Sitter phase and re-enter it much later, when the Universe is radiation- or matter-dominated. Thus, the quantum fluctuations of the scalar field during the inflation process are the seeds for the structure formation (galaxies, cluster of galaxies, etc.), as soon as they re-enter the microphysics horizon. It is worth noting that when the fluctuations remain out of the microphysics horizon, they are somehow frozen in the form of curvature (geometric and not matter-like) perturbations.

We conclude by observing that the inflation theory contains many non-settled down steps, but its force and compactness rely on the link with reliable fundamental particle physics and overall on its capability to simultaneously solve the fundamental paradoxes of the Standard Cosmological Model. We note that alternative proposals are not able to solve simultaneously all the paradoxes. Furthermore, the predictions of the inflationary model, especially concerning the perturbation spectrum, are in good agreement with the best fit of the CMBR data records [1, 169].

Chapter 3

Constrained Hamiltonian Systems

In this chapter we review the basic concepts related to a special class of theories whose Hamiltonian formulation is characterized by the presence of constraints: relations among the phase-space variables which must be conserved by the dynamic evolution of the system. Clearly, the classical dynamics of the system belongs to a precise region in the phase-space which is called constraints hypersurface. As we will see, some of these constraints play a crucial role being able to generate gauge transformations on any regular function of the phase-space variables. Indeed, the appearance of the constraints in the Hamiltonian formulation of some theory points out the presence of some gauge symmetry even if this symmetry is not clearly visible in the action functional, and some algorithms exist which allow for the construction of the generator of the physically interesting gauge transformations in the phase-space.

We will proceed by recalling all the basic notions for the Hamiltonian formulation of a given theory and then we will introduce the concept of constraints and the difference between primary and secondary constraints. We will give the definition of first-class and second-class constraints. After this we will study how the notion of canonicity of a transformation of variables is changed in this framework. After this general introduction we will study a practical application of many of these concepts using the electromagnetic field as a testing ground.

3.1 Preliminaries

In this section we will use the common notation of mechanic systems since its generalization to continuous systems will be treated later on with the study of the electromagnetic field. Let us consider a physical system described by N generalized configuration coordinates q^i. Given a Lagrangian, function of the coordinates and of their first time derivatives $L = L\left(q^k, \dot{q}^k\right)$, it is possible to obtain the set of N second order differential equations, the so-called equations of motion, which can be integrated once a set of $2N$ initial conditions is imposed. The stationarity condition of the action functional $S = \int dtL\left(q^k, \dot{q}^k\right)$ is a necessary and sufficient condition in order to derive the equations of motion[1]

$$\delta S\left[q^k\right] = \delta \int_{t_1}^{t_2} dtL\left(q^k, \dot{q}^k\right) = \int_{t_1}^{t_2} dt \left(\frac{\partial L}{\partial q^k} - \frac{d}{dt}\frac{\partial L}{\partial \dot{q}^k}\right)\delta q^k + \left[\frac{\partial L}{\partial \dot{q}^k}\delta q^k\right]_{t_1}^{t_2} = 0, \tag{3.1}$$

which are the well-known Euler-Lagrange equations

$$\frac{\partial L}{\partial q^k} - \frac{d}{dt}\frac{\partial L}{\partial \dot{q}^k} = 0. \tag{3.2}$$

In Hamilton's variational principle one of the hypotheses for the arbitrary variations is that

$$\delta q^k\left(t_2\right) = \delta q^k\left(t_1\right) = 0, \tag{3.3}$$

which means that the variables are fixed on the time boundaries. Hence, the last term in (3.1) arises from an integration by part which generates a boundary term that vanishes because of the conditions (3.3). This is the reason why any Lagrangian is always defined up to some total derivatives. One important feature of this approach is that only the q^k are varied, which means that the velocities $\dot{q}^k$ are not considered as independent. Moreover, the stationarity condition of the action determines the same physical trajectory no matter what coordinate system is used: *i.e.* one can use different parametrizations of the configuration space without modifying the physical content of the theory.

There is a way to write the action functional so that the variational principle leads to $2N$ first order differential equations rather than N second order ones. This can be achieved with the Hamiltonian formulation. Two steps are needed to pass from the Lagrangian to the Hamiltonian formulation: the calculation of the conjugate momenta and their inversion with respect to the velocities; the calculation of the Hamiltonian function. Briefly speaking the Hamiltonian is the Legendre transform of the Lagrangian. The conjugate momenta are defined as

$$p_i\left(q^k, \dot{q}^k\right) = \frac{\partial L\left(q^k, \dot{q}^k\right)}{\partial \dot{q}^i}, \tag{3.4}$$

[1] We will keep using the contracted indices convention as much as possible throughout the Chapter.

and for them to be invertible functions of the velocities $\dot{q}^k$ the necessary condition is

$$\det\left[\frac{\partial p_i}{\partial \dot{q}^j}\right] = \det\left[\frac{\partial^2 L}{\partial \dot{q}^i \partial \dot{q}^j}\right] \neq 0. \tag{3.5}$$

Once one has inverted the relation between conjugate momenta and velocities obtaining $\dot{q}^i = \dot{q}^i\left(q^k, p_k\right)$, it is possible to calculate the Hamiltonian defined as

$$H\left(q^k, p_k\right) = p_i \dot{q}^i\left(q^k, p_k\right) - L\left(q^k, p_k\right), \tag{3.6}$$

giving for the action functional

$$S\left[q^k, p_k\right] = \int_{t_1}^{t_2} dt\left[p_i \dot{q}^i\left(q^k, p_k\right) - H\left(q^k, p_k\right)\right]. \tag{3.7}$$

Now, the new set of $2N$ variables $\left\{q^i, p_i\right\}$ belongs to a $2N$-dimensional space which is called phase-space. The coordinates have to be considered independent from the conjugate momenta. Hence, we must require the stationarity of the action functional under arbitrary variations of both the coordinates q^i and the conjugate momenta p_i

$$\begin{aligned} \delta_p S\left[q^k, p_k\right] &= \int_{t_1}^{t_2} dt\left[\dot{q}^i - \frac{\partial H}{\partial p_i}\right]\delta p_i = 0, \\ \delta_q S\left[q^k, p_k\right] &= \int_{t_1}^{t_2} dt\left[-\dot{p}_i - \frac{\partial H}{\partial q^i}\right]\delta q^i + \left[p_i \delta q^i\right]_{t_1}^{t_2} = 0, \end{aligned} \tag{3.8}$$

leading to Hamilton's equations of motion

$$\dot{q}^i = \frac{\partial H}{\partial p_i}, \qquad \dot{p}_i = -\frac{\partial H}{\partial q^i}, \tag{3.9}$$

which are $2N$ first order differential equations. It is worth noticing that the Hamiltonian function (3.6) is not defined up to a boundary term, as it happens for the Lagrangian function, in the sense that this would modify the explicit expression of Hamilton's equations of motion (3.9), even if the stationarity condition has to determine the same trajectory in the configurations space.

Now, let us define the so-called Poisson brackets between two generic smooth functions of the phase-space coordinates, $f = f\left(q^i, p_i\right)$ and $g = g\left(q^i, p_i\right)$, as

$$\left[f\left(q^i, p_i\right), g\left(q^i, p_i\right)\right]_{q,p} = \sum_{k=1}^{N}\left\{\frac{\partial f}{\partial q^k}\frac{\partial g}{\partial p_k} - \frac{\partial g}{\partial q^k}\frac{\partial f}{\partial p_k}\right\}, \tag{3.10}$$

which satisfies, given a third smooth function $h = h(q^i, p_i)$ and two arbitrary coefficients a and b, the following properties

$$\begin{aligned} &[f,g] = -[g,f], \qquad [f, ag + bh] = a[f,g] + b[f,h], \\ &[f, gh] = [f,g]h + g[f,h], \qquad [f,[g,h]] + [g,[h,f]] + [h,[f,g]] = 0, \end{aligned} \tag{3.11}$$

where the last one is called Jacobi's identity. We also write the so-called fundamental Poisson brackets which relate the phase-space variables

$$\left[q^i, q^j\right]_{q,p} = 0, \qquad \left[q^i, p_j\right]_{q,p} = \delta^i_j, \qquad \left[p_i, p_j\right]_{q,p} = 0. \tag{3.12}$$

We can write Hamilton's equations of motion in this formalism and the time derivative of an arbitrary smooth function $f = f(q^i, p_i)$ of the phase-space coordinates as

$$\dot{q}^i = \left[q^i, H\right]_{q,p}, \qquad \dot{p}_i = \left[p_i, H\right]_{q,p}, \tag{3.13}$$

$$\frac{d}{dt} f(q^i, p_i) = \sum_{k=1}^{N} \left\{ \frac{\partial f}{\partial q^k} \dot{q}^k + \frac{\partial f}{\partial p_k} \dot{p}_k \right\} = \sum_{k=1}^{N} \left\{ \frac{\partial f}{\partial q^k} \frac{\partial H}{\partial p_k} - \frac{\partial f}{\partial p_k} \frac{\partial H}{\partial q^k} \right\} = [f, H]_{p,q}. \tag{3.14}$$

It is possible to cast this formalism in a more geometric fashion. Let us call with just one label, x^μ, the whole set of the phase-space coordinates such that $x^k = q^k$ and $x^{k+N} = p_k$ for $k = 1, \ldots, N$. We then define the rank two antisymmetric tensor, which is the so-called *symplectic tensor*

$$\sigma^{\mu\nu} = \begin{pmatrix} 0 & \mathbb{I}_N \\ -\mathbb{I}_N & 0 \end{pmatrix}, \tag{3.15}$$

where $\mathbb{I}_N$ is the N-dimensional identity matrix. In terms of this new quantity we can easily rewrite the Poisson bracket of two generic functions as

$$\left[f(q^i, p_i), g(q^i, p_i)\right]_{q,p} = \sigma^{\mu\nu} \frac{\partial f}{\partial x^\mu} \frac{\partial g}{\partial x^\nu}, \tag{3.16}$$

as well as the whole set of fundamental Poisson brackets and Hamilton's equations of motion

$$\sigma^{\mu\nu} = \sigma^{\rho\sigma} \frac{\partial x^\mu}{\partial x^\rho} \frac{\partial x^\nu}{\partial x^\sigma}, \qquad \dot{x}^\mu = \sigma^{\rho\sigma} \frac{\partial x^\mu}{\partial x^\rho} \frac{\partial H}{\partial x^\sigma}. \tag{3.17}$$

The first equation in (3.17), represents the fundamental Poisson brackets which turn out to be the identical transformation applied to the symplectic tensor in this formalism.

3.2 Constrained systems

In this section we will follow the lines of [153, 167]. For the Hamiltonian formulation of a theory the definition of constraint follows from the condition

$$\det\left[\frac{\partial p_i}{\partial \dot{q}^j}\right] = \det\left[\frac{\partial^2 L}{\partial \dot{q}^i \partial \dot{q}^j}\right] = 0, \tag{3.18}$$

which simply means that not all the velocities $\dot{q}^k$ are invertible in terms of the conjugate momenta p_k, so that some canonical variables will be constrained by some relation $\phi\left(q^k, p_k\right) = 0$. Starting from Euler-Lagrange equations (3.2) we can appreciate the physical meaning of condition (3.18)

$$0 = \frac{\partial L}{\partial q^k} - \frac{d}{dt}\frac{\partial L}{\partial \dot{q}^k} = \frac{\partial L}{\partial q^k} - \frac{\partial^2 L}{\partial \dot{q}^k \partial q^i}\dot{q}^i - \frac{\partial^2 L}{\partial \dot{q}^k \partial \dot{q}^i}\ddot{q}^i,$$

$$\frac{\partial^2 L}{\partial \dot{q}^k \partial \dot{q}^i}\ddot{q}^i = \frac{\partial L}{\partial q^k} - \frac{\partial^2 L}{\partial \dot{q}^k \partial q^i}\dot{q}^i, \tag{3.19}$$

so that if the matrix $\partial^2 L/\partial \dot{q}^i \partial \dot{q}^j$ is not invertible (3.18) some of the accelerations will not be expressible as a function of q^k and $\dot{q}^k$ and thus will be arbitrary. This is the point where the notion of gauge freedom arises: the dynamics of the system is not uniquely defined and arbitrary functions can be used for the undetermined accelerations. For Lagrangian functions defined through polynomials of the configuration variables and their time derivatives, this happens any time there are no quadratic term in the velocities of some variables. This means that we do not have enough variables to define a bijection between the configuration space $\left\{q^k, \dot{q}^k\right\}$ and the phase-space $\left\{q^k, p_k\right\}$. Typically this happens because we are describing the theory with too many variables with respect to its actual number of physical degrees of freedom. Such a situation may occur because we want some symmetry to be manifest, as it happens for the gauge symmetry of the electromagnetic field or the general covariance in General Relativity. Hence, for the electromagnetic field, being two the physical degrees of freedom we would look for a four-dimensional phase-space (clearly, for any given point of the spacetime), while we have to deal with an eight-dimensional one (four degrees of freedom for the vector potential A^μ and four for the conjugate momenta Π_μ). The situation is even worse in General Relativity where the number of the physical degrees of freedom is the same but the dimension of the phase-space is 20 (ten degrees of freedom for the independent components of the symmetric metric tensor $g_{\mu\nu}$ and ten degrees of freedom for the conjugate momenta $p_{\mu\nu}$). It then follows that the dynamics does not cover the whole phase-space but it is limited to some region. In order to recover the invertibility of the transformation we need some new fictitious variables, the Lagrange multipliers, which will help enforcing in the variational principle the existence of such a limited region. We will now classify the constraints in two ways: according to a dynamic or a symmetry criterion. The first one leads to classify the constraints as primary and secondary, while the second one, due to the pioneering work of Dirac [112], classifies the constraints as first- or second-class.

3.2.1 *Primary and secondary constraints*

Given the non-invertibility condition of the Legendre transform (3.18) and the definition of the conjugate momenta (3.4) we define M equations of the kind

$$\phi_k\left(q^i, p_i\right) = 0, \qquad k = 1, \ldots, M, \tag{3.20}$$

which are called *primary constraints*. Let us assume for simplicity that these equations are all independent, it then follows that the hypersurface defined by the primary constraints has dimension $2N - M$. Since these constraints follow from (3.18) they must be enforced through the stationarity condition. In order to obtain this result a simple way is to add M Lagrange multipliers λ^k as new independent variables

$$S\left[q^k, p_k, \lambda\right] = \int_{t_1}^{t_2} dt \left[p_i \dot{q}^i\left(p_k, q^k\right) - H\left(p_k, q^k\right) + \lambda^k \phi_k\left(q^i, p_i\right)\right], \tag{3.21}$$

where the stationarity condition for arbitrary variations of the Lagrange multipliers yields to the enforcement of the primary constraints

$$\delta_\lambda S\left[q^k, p_k, \lambda\right] = \int_{t_1}^{t_2} dt \phi_k\left(q^i, p_i\right) \delta\lambda^k = 0 \quad \Rightarrow \quad \phi_k\left(q^i, p_i\right) = 0. \tag{3.22}$$

Alongside we also get the modified set of Hamilton's equations of motion

$$\begin{aligned} \dot{q}^k &= \left[q^k, H\right]_{q,p} + \lambda^i \left[q^k, \phi_i\right]_{q,p}, \\ \dot{p}_k &= \left[p_k, H\right]_{q,p} + \lambda^i \left[p_k, \phi_i\right]_{q,p}, \end{aligned} \tag{3.23}$$

and for a regular function of the phase-space $f = f\left(q^k, p_k\right)$

$$\frac{d}{dt} f\left(q^k, p_k\right) = \left[f, H\right]_{q,p} + \lambda^i \left[f, \phi_i\right]_{q,p}. \tag{3.24}$$

However, how is the Hamiltonian defined in this case? In fact, since some of the conjugate momenta are invertible as a function of the velocities and some are not it is not clear how to practically act on the expressions. We give here a simple rule: replace each non-invertible velocity with a Lagrange multiplier and follow the same procedure used for unconstrained systems. This is a reasonable choice since the non-invertible velocities truly are arbitrary variables. This way one would automatically obtain the action functional (3.21) where the Hamiltonian H is a function of all invertible conjugate momenta. Since the dynamics is defined on the constraints hypersurface we can add any linear combination of the constraints to the Hamiltonian without changing the physical content of the theory

$$H \to H + c^k \left(q^i, p_i\right) \phi_k . \tag{3.25}$$

Now that we have described the role of the primary constraints it seems rather natural that the dynamics of the system must preserve them. The following relation must be satisfied

$$\dot{\phi}_k = \left[\phi_k, H\right]_{q,p} + \lambda^i \left[\phi_k, \phi_i\right]_{q,p} = 0, \tag{3.26}$$

which could be independent on λ^k, thus only being a function of $\left\{q^i, p_i\right\}$, or it may be a restriction on the Lagrange multipliers λ^k. In the first case, if the new condition cannot be written as a linear combination of the primary constraints, we have obtained a *secondary constraint* $X\left(q^i, p_i\right) = 0$ and, again, we need to request its conservation during the dynamic evolution of the system imposing

$$\dot{X} = \left[X, H\right]_{q,p} + \lambda^i \left[X, \lambda_i\right]_{q,p} = 0, \tag{3.27}$$

which may give rise to 'tertiary' constraints, that, however, are always labelled as secondary. This procedure must be repeated as long as new constraints arise. In the end, we will have a complete set of constraints, M primary and M' secondary, which we label as ϕ_k, $k = 1, \ldots, M + M'$.

We introduce a new notation, which is the symbol of *weak equality* $\approx$ and from now on we will use the notation $\phi_k \approx 0$. In particular any two regular functions f and g will be weakly equal if they differ for an arbitrary linear combination of the constraints

$$f\left(q^k, p_k\right) \approx g\left(q^k, p_k\right) \quad \Rightarrow \quad f\left(q^k, p_k\right) - g\left(q^k, p_k\right) = c^i \phi_i\left(q^k, p_k\right), \tag{3.28}$$

where $c^i = c^i\left(q^k, p_k\right)$ can be generic regular functions of the phase-space coordinates. When the request (3.26) turns out to be a condition on the Lagrange multipliers one obtains, under general assumptions, a split dependence

$$\lambda^k \approx \Lambda^k + l^r L_r^k, \tag{3.29}$$

where L_r^k are independent solutions of the homogeneous equation associated to (3.26), l^r are arbitrary coefficients, while Λ^k are the solutions of the non-homogeneous form (3.26), *i.e.*

$$L_r^k \left[\phi_k, \phi_i\right]_{q,p} \approx 0, \qquad \left[\phi_k, H\right]_{q,p} + \Lambda^i \left[\phi_k, \phi_i\right]_{q,p} \approx 0. \tag{3.30}$$

As for the expression (3.30) we want to stress that all the constraints are just the primary ones.

3.2.2 *First- and second-class constraints*

Let us now describe a different classification of the complete set of constraints which has not the dynamic character of the previous one. A generic function f of the phase-space coordinates is said to be *first-class* if and only if its Poisson brackets with the whole set of constraints weakly vanish

$$\left[f, \phi_i\right]_{q,p} \approx 0, \qquad i = 1, \ldots, M + M' \tag{3.31}$$

while a function g is said to be *second-class* if there is at least one constraint for which this does not happen. In particular this classification applies also to the constraints themselves. We now use the relation (3.29) in order to define a Hamiltonian which is first-class: the equations of motion for a regular function f are

$$\frac{d}{dt} f \approx \left[f, H\right]_{q,p} + \Lambda^k \left[f, \phi_k\right]_{q,p} + l^r L_r^k \left[f, \phi_k\right]_{q,p}, \tag{3.32}$$

and since all terms proportional to any constraint weakly vanish, we can write

$$\left[f, \Lambda^k \phi_k\right]_{q,p} \approx \Lambda^k \left[f, \phi_k\right]_{q,p}, \qquad \left[f, l^r L_r^k \phi_k\right]_{q,p} \approx l^r L_r^k \left[f, \phi_k\right]_{q,p}, \tag{3.33}$$

which allow us to define the so-called total Hamiltonian

$$H_T = H + \Lambda^k \phi_k + l^r L_r^k \phi_k = H' + l^r \phi_r, \tag{3.34}$$

where $H' = H + \Lambda^k \phi_k$ and $\phi_r = L_r^k \phi_k$, enabling us to write the equations of motion in the usual fashion

$$\frac{d}{dt} f \approx \left[f, H_T\right]. \tag{3.35}$$

The total Hamiltonian (3.34) is first-class since it is composed of first-class terms

$$\begin{aligned} \left[\phi_i, H'\right] &= \left[\phi_k, H + \Lambda^k \phi_k\right] \approx \left[\phi_i, H\right] + \Lambda^k \left[\phi_i, \phi_k\right] \approx 0, \\ \left[\phi_i, \phi_r\right] &= \left[\phi_k, L_r^k \phi_k\right] \approx L_r^k \left[\phi_i, \phi_k\right] \approx 0, \end{aligned} \tag{3.36}$$

which directly follow from (3.30). We now use the total Hamiltonian to show that first-class constraints are the generator of gauge transformations: since the coefficients l^a of (3.29) are completely arbitrary we may choose two different sets of them, $\{l^a\}$ and $\{\bar{l}^a\}$ which lead to two different equations of motion for a generic function f

$$\begin{aligned} \delta_t f &\approx [f, H_T]\,\delta t \approx \left[f, H'\right]\delta t + l^a\,[f, \phi_a]\,\delta t, \\ \overline{\delta_t f} &\approx \left[f, \bar{H}_T\right]\delta t \approx \left[f, H'\right]\delta t + \bar{l}^a\,[f, \phi_a]\,\delta t. \end{aligned} \tag{3.37}$$

We define the variation $\delta_\varepsilon f$, of parameters $\varepsilon^r = \delta t\left(l^r - \bar{l}^r\right)$, as

$$\delta_\varepsilon f = \delta_t f - \overline{\delta_t f} \approx \varepsilon^r\,[f, \phi_r]\,, \tag{3.38}$$

which is not physical since it is due to the arbitrary choice of the coefficients $\{l^r\}$ and $\{\bar{l}^r\}$. We then call the variation (3.38), gauge transformation of infinitesimal parameters ε^r, which is generated by the first-class constraints ϕ_r. Hence, *first-class constraints are the generators of gauge transformations* in a Hamiltonian constrained theory. This is the major importance of this classification.

From a geometrical point of view the transformations generated by first-class constraints are tangential to the constraints hypersurface while those generated by second-class constraints are not. Not all the transformations which are generated by first-class constraints represent the usual gauge transformations one is used to recognize for some field theories. However, some algorithms [87] exist in order to construct the generator of gauge transformations in the usual form (see section 3.4.3).

Finally we define the extended Hamiltonian as the total Hamiltonian to which we add the set of secondary first-class constraints which will be denoted as $\{\gamma_r\}$

$$H_E = H' + l^r\gamma_r, \tag{3.39}$$

which clearly generates the right equations of motion through its Poisson brackets action

$$\frac{d}{dt}f \approx [f, H_E]\,. \tag{3.40}$$

We conclude this section with the following consideration: once the complete set of constraints is known, then it is possible to count the physical degrees of freedom. Indeed, any physical observable will be described by some phase-space function lying on the constraints hypersurface, which means that each constraint reduces by one the number of the actual degrees of freedom. This is true for both first- and second-class constraints.

Moreover, for the first-class ones another condition must hold: the observable must have weakly vanishing Poisson brackets with each first-class constraint. This last requirement implies the gauge-invariance of the observable. Hence, as it is often said, first-class constraints always strike twice. Being N_F and N_S the number of first- and second-class constraints respectively, and N_T the total number of phase-space coordinates, the number of physical degrees of freedom N_{Phys} can be written as

$$N_{Phys} = \frac{1}{2}\left(N_T - N_S - 2N_F\right). \tag{3.41}$$

3.3 Canonical transformations

We already mentioned that in the Lagrangian formulation it is always possible to write the action functional in a new set of coordinates through an invertible transformation $Q^k = Q^k\left(q^i\right)$ without modifying the physical content. We now want to study what are the conditions to be imposed on some transformation of the phase-space variables

$$Q^k = Q^k\left(q^i, p_i\right), \qquad P_k = P_k\left(q^i, p_i\right), \tag{3.42}$$

for the stationarity condition to determine the same physical trajectories. We discuss the usual conditions for unconstrained systems and their generalization in the presence of constraints.

3.3.1 *Strongly canonical transformations*

Let us define two action functionals for the two sets of phase-space coordinates $\left\{Q^k, P_k\right\}$ and $\left\{q^k, p_k\right\}$ related to the Hamiltonians K and H respectively

$$\begin{aligned} S_1\left[Q^k, P_k\right] &= \int_{t_1}^{t_2} dt \left[P_k\dot{Q}^k - K\left(Q^i, P_i\right)\right], \\ S_2\left[q^k, p_k\right] &= \int_{t_1}^{t_2} dt \left[p_k\dot{q}^k - H\left(q^i, p_i\right)\right]. \end{aligned} \tag{3.43}$$

For the stationarity condition to determine the same physical trajectories we need the integrands to differ at most for some boundary term

$$P_k\dot{Q}^k - K\left(Q^i, P_i\right) = p_k\dot{q}^k - H\left(q^i, p_i\right) - \frac{d}{dt}F. \tag{3.44}$$

We can split the request (3.44) in a condition on the terms containing time derivatives and another one on the Hamiltonians

$$P_k \dot{Q}^k - p_k \dot{q}^k = \frac{d}{dt} F, \tag{3.45}$$

$$K\left(Q^i, P_i\right) = H\left(q^i, p_i\right). \tag{3.46}$$

Focusing on (3.46) we can derive some other equations to be satisfied by the derivatives of (3.42), in fact one has that for the total time derivatives $\dot{Q}$ and $\dot{P}$

$$\begin{aligned}\dot{Q}^k &= \frac{\partial K}{\partial q^i}\frac{\partial q^i}{\partial P_k} + \frac{\partial K}{\partial p_i}\frac{\partial p_i}{\partial P_k} = \frac{\partial Q^k}{\partial q^i}\frac{\partial H}{\partial p_i} - \frac{\partial Q^k}{\partial p_i}\frac{\partial H}{\partial q^i},\\ \dot{P}_k &= -\frac{\partial K}{\partial q^i}\frac{\partial q^i}{\partial Q^k} - \frac{\partial K}{\partial p_i}\frac{\partial p_i}{\partial Q^k} = \frac{\partial P_k}{\partial q^i}\frac{\partial H}{\partial p_i} - \frac{\partial P_k}{\partial p_i}\frac{\partial H}{\partial q^i},\end{aligned} \tag{3.47}$$

and since $H = K$ we obtain four conditions for the derivatives of the transformation

$$\frac{\partial q^i}{\partial P_k} = -\frac{\partial Q^k}{\partial p_i}, \quad \frac{\partial p_i}{\partial P_k} = \frac{\partial Q^k}{\partial q^i}, \quad \frac{\partial q^i}{\partial Q^k} = \frac{\partial P_k}{\partial p_i}, \quad \frac{\partial p_i}{\partial Q^k} = -\frac{\partial P_k}{\partial q^i}, \tag{3.48}$$

which are called *direct canonicity conditions* (see [153]). We denote those transformations satisfying (3.48) as *strongly canonical.* For an invertible transformation (3.42) these can be satisfied and they lead to a very important property of the Poisson brackets: let us write the fundamental Poisson brackets for the variables $\{Q^k, P_k\}$ with respect to the other set $\{q^k, p_k\}$, applying the derivatives chain rule we obtain

$$\begin{aligned}\left[Q^i, P_j\right]_{q,p} &= \sum_{k=1}^{N}\left\{\frac{\partial Q^i}{\partial q^k}\frac{\partial P_j}{\partial p_k} - \frac{\partial P_j}{\partial q^k}\frac{\partial Q^i}{\partial p_k}\right\}\\ &= \sum_{k=1}^{N}\left\{\frac{\partial Q^i}{\partial q^k}\frac{\partial q^k}{\partial Q_j} + \frac{\partial p_k}{\partial Q_j}\frac{\partial Q^i}{\partial p_k}\right\} = \delta^i_j,\\ \left[Q^i, Q^j\right]_{q,p} &= \sum_{k=1}^{N}\left\{\frac{\partial Q^i}{\partial q^k}\frac{\partial Q^j}{\partial p_k} - \frac{\partial Q^j}{\partial q^k}\frac{\partial Q^i}{\partial p_k}\right\}\\ &= \sum_{k=1}^{N}\left\{\frac{\partial p_k}{\partial P_i}\frac{\partial Q^j}{\partial p_k} + \frac{\partial Q^j}{\partial q^k}\frac{\partial q^k}{\partial P_i}\right\} = 0,\\ \left[P_i, P_j\right]_{q,p} &= \sum_{k=1}^{N}\left\{\frac{\partial P_i}{\partial q^k}\frac{\partial P_j}{\partial p_k} - \frac{\partial P_j}{\partial q^k}\frac{\partial P_i}{\partial p_k}\right\}\\ &= \sum_{k=1}^{N}\left\{\frac{\partial P_i}{\partial q^k}\frac{\partial q^k}{\partial Q^j} + \frac{\partial p_k}{\partial Q^j}\frac{\partial P_i}{\partial p_k}\right\} = 0,\end{aligned} \tag{3.49}$$

which in tensor language translates for the symplectic tensor as

$$\sigma^{\mu\nu} = \sigma^{\rho\sigma} \frac{\partial y^\mu}{\partial x^\rho} \frac{\partial y^\nu}{\partial x^\sigma}, \tag{3.50}$$

where with x^μ we indicate the set $\{q^k, p_k\}$ and with y^μ the set $\{Q^k, P_k\}$. Hence, the symplectic tensor is invariant under strongly canonical transformations, just like the Minkowski metric is under Lorentz transformations, meaning that Poisson brackets are also invariant.

Hence, considering only strongly canonical transformations we can drop the indices from the Poisson brackets

$$\left[A\left(q^i, p_i\right), B\left(q^i, p_i\right)\right]_{q,p} = \left[A\left(q^i, p_i\right), B\left(q^i, p_i\right)\right]_{Q,P} = \left[A\left(q^i, p_i\right), B\left(q^i, p_i\right)\right]. \tag{3.51}$$

Since these transformations involve, in general, both the coordinates and conjugate momenta they possess a much wider freedom of choice with respect to the transformations of the configuration coordinates of the Lagrangian formulation. Finally, we would like to stress that any boundary term introducing first order time derivatives in the Lagrangian density results in a strong canonical transformation acting only on the conjugate momenta.

3.3.2 *Weakly canonical transformations*

Let us now discuss the same arguments for constrained systems. We write the two action functionals using the extended Hamiltonian (3.39)

$$\begin{aligned} S_1\left[Q^k, P_k, \lambda^i\right] &= \int_{t_1}^{t_2} dt \left[P_k \dot{Q}^k - K'\left(Q^i, P_i\right) + \lambda^i \gamma_i\left(Q^i, P_i\right)\right], \\ S_2\left[q^k, p_k, \mu^j\right] &= \int_{t_1}^{t_2} dt \left[p_k \dot{q}^k - H'\left(q^i, p_i\right) + \mu^j \xi_j\left(Q^i, P_i\right)\right], \end{aligned} \tag{3.52}$$

where the two sets of constraints $\{\gamma_i\}$ and $\{\xi_i\}$ are all first-class. Now, considering an invertible and differentiable transformation (3.42) for the canonical variables it should be possible to transform one set of constraints in the other via a linear combination

$$\gamma_i = Z_i^k \xi_k, \tag{3.53}$$

and since the coefficients μ^i and λ^i are arbitrary it is possible to link them through the transformation

$$\mu^k = Z_i^k \lambda^i, \tag{3.54}$$

from which it follows that the two extended Hamiltonians should be equal

$$H_E = K_E. \tag{3.55}$$

The equations of motion generated by the extended Hamiltonians hold weakly which also means that the direct canonicity conditions should hold weakly

$$\frac{\partial q^i}{\partial P_k} \approx -\frac{\partial Q^k}{\partial p_i}, \quad \frac{\partial p_i}{\partial P_k} \approx \frac{\partial Q^k}{\partial q^i}, \quad \frac{\partial q^i}{\partial Q^k} \approx \frac{\partial P_k}{\partial p_i}, \quad \frac{\partial p_i}{\partial Q^k} \approx -\frac{\partial P_k}{\partial q^i}, \tag{3.56}$$

which define a *weakly canonical transformation* [181]. It then follows that in the general case an invertible transformation of the phase-space coordinates will be canonical *at least* weakly, *i.e.* Poisson brackets will be conserved modulo some linear combinations of the constraints

$$\left[A\left(q^i, p_i\right), B\left(q^i, p_i\right)\right]_{q,p} \approx \left[A\left(q^i, p_i\right), B\left(q^i, p_i\right)\right]_{Q,P}. \tag{3.57}$$

3.3.3 *Gauged canonical transformations*

As expected the presence of constraints loosens the conditions to be imposed to an invertible coordinate transformation (3.42) but still, we can renounce to invertibility in some special cases. Let us consider the case in which some of the conjugate momenta appear only linearly in some first-class constraints, $\{p_R\}$ and $\{P_R\}$, while some others appear quadratically or derived in others, $\{p_r\}$ and $\{P_r\}$. Let us define the invertible part of the transformation as

$$Q^k = Q^k\left(q^i\right), \qquad P_r = P_r\left(p_s\right), \tag{3.58}$$

while the two sets $\{p_R\}$ and $\{P_R\}$ are not expressible as a function of one another. This means that for the first-class constraints $\{\gamma_R\}$ and $\{\xi_R\}$, containing linear terms in $\{P_R\}$ and $\{p_R\}$ respectively, there will be no precise transformation linking them. It is then possible to write a completely arbitrary linear combination linking the two sets of first-class constraints

$$\gamma_R \approx 0 \approx W_R^S \xi_S, \tag{3.59}$$

where the coefficients W_R^S are arbitrary functions, while for the other constraints there will be some exact relation like (3.53). It follows for the extended Hamiltonians that

$$H_E \approx K_E. \tag{3.60}$$

Since the direct canonicity conditions (3.48) depend on Hamilton's equations of motion, because of (3.60) these conditions will be modified by some derivatives of first-class constraints, *i.e.* by some gauge transformation. Hence, fundamental Poisson brackets are changed as

$$\left[q^i, p_j\right]_{q,p} \approx \left[q^i, p_j\right]_{Q,P} + c^A \left[f, \gamma_A\right]_{Q,P} \neq \delta^i_j. \tag{3.61}$$

The invariance of the Poisson brackets is altered by a gauge transformation. These are the *gauged canonical transformations* since they differ from strongly canonical case because of a gauge transformation but can be regarded as "canonical" in the sense that they preserve the dynamic content of the theory. We will examine this property further with the example of the electromagnetic field. This kind of canonical transformations was firstly introduced in [90].

3.4 Electromagnetic field

In this section we will study most of the concepts we have just described, using the example of the electromagnetic field. We will first discuss the Lagrangian formulation and propose a modified Lagrangian density which helps us appreciating the freedom of definition of the action functional. We will then treat the Hamiltonian formulation.

3.4.1 *Modified Lagrangian formulation*

Electrodynamics represents one of the simplest cases of gauge theories. We will use here the covariant notation where the fields are represented by the four-vector $A^\mu = A^\mu(x)$ which is related to classical definitions of the electromagnetic fields as usual

$$E^i(x) = \partial^i A^0(x) - \partial_0 A^i(x), \qquad B^i(x) = \varepsilon^{ijk}\partial_j A_k(x). \tag{3.62}$$

Here we are dealing with the Minkowskian metric tensor, defined as $\eta_{\mu\nu} = \mathrm{diag}\{1, -1, -1, -1\}$, because of which the rules to apply to indices changes are, for the generic vector V^μ,

$$V^0 = V_0, \qquad V^i = -V_i. \tag{3.63}$$

There is a well-known Lagrangian density from which it is possible to derive the equations of motions

$$\mathcal{L}_{EM} = -\frac{1}{4}F^{\mu\nu}F_{\mu\nu}, \qquad F_{\mu\nu} = \partial_\nu A_\mu - \partial_\mu A_\nu = -F_{\nu\mu}, \tag{3.64}$$

which we derive requiring the stationarity of the action under arbitrary variations δA^{μ}

$$\begin{aligned}\delta S_{EM}\left[A^{\mu}\right] &= \frac{1}{4}\delta\int_{\mathcal{M}} d^4x F^{\mu\nu}F_{\nu\mu} = \int_{\mathcal{M}} d^4x F^{\mu\nu}\partial_{\mu}\delta A_{\nu} \\ &= \int_{\partial\mathcal{M}} d^3x \eta_{\mu}F^{\mu\nu}\delta A_{\nu} - \int_{\mathcal{M}} d^4x \partial_{\mu}F^{\mu\nu}\delta A_{\nu} = 0,\end{aligned} \tag{3.65}$$

where η^{μ} is the normal to the boundary hypersurface. We mostly used the antisymmetry of $F^{\mu\nu}$ and some index relabelling. At the last line we explicitly wrote the result of the integration of the boundary term $\partial_{\mu}\left(F^{\mu\nu}\delta A_{\nu}\right)$, which vanishes under the condition $\delta A_{\nu}|_{\partial\mathcal{M}} = 0$. Hence, the stationarity of the action functional leads to the well-known equations of motion for the electromagnetic field in the vacuum:

$$\partial_{\mu}F^{\mu\nu} = 0. \tag{3.66}$$

These equations can be easily mapped into some of Maxwell equations using (3.62). The four remnant Maxwell's equations directly follow from an algebraic identity which is due to the antisymmetry of $F^{\mu\nu}$

$$\partial_{\mu}F_{\nu\rho} + \partial_{\nu}F_{\rho\mu} + \partial_{\rho}F_{\mu\nu} = 0. \tag{3.67}$$

Since all the indices must differ, the equations (3.67) represent $4!/3! = 4$ independent equations to be satisfied by any given solution of (3.66). We finally note a certain similarity between (3.67) and Bianchi's identity (1.96) we described in the previous chapter. Let us observe that, by virtue of the antisymmetry of $F^{\mu\nu}$, these equations of motion are invariant under arbitrary transformations, the so-called *gauge transformations*, of the type

$$A^{\mu} \to A^{\mu} + \partial^{\mu}\phi, \tag{3.68}$$

where ϕ is a regular scalar function. Now, it also happens that the action functional we defined in (3.64) is invariant under such gauge transformations (3.68). However, for the equations of motion to be invariant under some transformations the invariance of the action functional is just a sufficient but not necessary condition. We can easily check this for the electromagnetic field. Since the Lagrangian density is defined up to some boundary terms we can always add one which is not gauge-invariant so that the invariance of the action functional is lost but the equations of motion would still be gauge invariant. Let us then define a parametrized Lagrangian density

$$\mathcal{L}_{EM}\left(\alpha\right) = -\frac{1}{4}F^{\mu\nu}F_{\mu\nu} + \alpha\partial_{\mu}\boldsymbol{\mathcal{B}}^{\mu}\left(A^{\nu}\right), \tag{3.69}$$

where the parameter α will allow us to track all the changes induced by the boundary term in the calculations. We choose to consider only boundary terms that do not depend on the fields derivatives so that the usual boundary conditions $\delta A_\nu|_{\partial\mathcal{M}} = 0$ are enough to guarantee that the equations of motion are not modified. Clearly, one recovers the usual formulation for $\alpha = 0$. We invite the reader to continuously observe how the results change in the $\alpha = 0$ case throughout the calculations. Finally we choose a special form for the boundary term, namely

$$\partial_\mu \boldsymbol{\mathcal{B}}^\mu \left(A^\nu\right) = \partial_0 \left(A^0 \partial_i A^i\right) - \partial_i \left(A^0 \partial_0 A^i\right), \tag{3.70}$$

which is not gauge-invariant nor covariant under Lorentz transformations. Clearly, also the covariance under Lorentz transformations is recovered in the equations of motion, *i.e.* when imposing the stationarity of the action functional. It is easy to check that there are no second order derivatives and that this boundary term can be written in two different ways

$$\begin{aligned}\partial_\mu \boldsymbol{\mathcal{B}}^\mu \left(A^\nu\right) =& \partial_0 \left(A^0 \partial_i A^i\right) - \partial_i \left(A^0 \partial_0 A^i\right) \\ =& \partial_0 A^0 \partial_i A^i - \partial_i A^0 \partial_0 A^i \\ =& \partial_i \left(A^i \partial_0 A^0\right) - \partial_0 \left(A^i \partial_i A^0\right)\end{aligned} \tag{3.71}$$

so that it is always possible to put the variations δA^0 and δA^k outside the derivative which appears in the boundary term, *i.e.*

$$\begin{aligned}\delta_{A^0} \partial_\mu \boldsymbol{\mathcal{B}}^\mu =& \partial_0 \left(\delta A^0 \partial_i A^i\right) - \partial_i \left(\delta A^0 \partial_0 A^i\right), \\ \delta_{A^i} \partial_\mu \boldsymbol{\mathcal{B}}^\mu =& \partial_i \left(\delta A^i \partial_0 A^0\right) - \partial_0 \left(\delta A^i \partial_i A^0\right).\end{aligned} \tag{3.72}$$

Even if (3.70) has a peculiar form it still fulfils all the requirements for it not to change the equations of motion.

3.4.2 *Hamiltonian formulation*

In this subsection we will follow all the steps needed to obtain a complete derivation of the Hamiltonian formulation of a constrained system. Each step is entitled and described. Let us now deal with the Hamiltonian formulation of the modified theory (3.69)

3.4.2.1 *Conjugate momenta and primary constraints*

The first step is to calculate the conjugate momenta to the velocities $\partial_0 A_\mu$. In order to do this we need to separate spatial and temporal derivatives in the Lagrangian density (3.69)

$$\mathcal{L}_{EM}\left(\alpha\right) = -\frac{1}{2} F^{0i} F_{0i} - \frac{1}{4} F^{ij} F_{ij} + \alpha \partial_0 A^0 \partial_i A^i - \alpha \partial_i A^0 \partial_0 A^i. \tag{3.73}$$

Hence, we calculate the conjugate momenta, starting from the conjugate to A^0

$$\Pi^0 = \frac{\partial \mathcal{L}_{EM}(\alpha)}{\partial \partial_0 A_0} = \alpha \partial_i A^i, \tag{3.74}$$

which clearly shows that Π^0 is not expressible as a function $\partial_0 A_0$ so that we define the primary constraint as

$$\varphi(\alpha) = \Pi^0 - \alpha \partial_i A^i \approx 0. \tag{3.75}$$

This result could be foreseen since in (3.69) there cannot be, generically, quadratic terms in $\partial_\mu A_\mu$, and specifically in $\partial_0 A_0$, because of the antisymmetry of $F^{\mu\nu}$. Considering the other conjugate momenta we obtain

$$\Pi^i = \frac{\partial \mathcal{L}_{EM}(\alpha)}{\partial \partial_0 A_i} = -F^{0i} \frac{\partial F_{0i}}{\partial \partial_0 A_i} - \alpha \partial^i A^0 = F^{0i} - \alpha \partial^i A^0, \tag{3.76}$$

which in this case are invertible functions of the velocities leading to

$$F^{0i} = \Pi^i + \alpha \partial^i A^0, \qquad \partial^0 A^i = (1-\alpha)\partial^i A^0 - \Pi^i. \tag{3.77}$$

3.4.2.2 *Hamiltonian density*

Now we can rewrite the Lagrangian density as a function of the fields A_μ and their conjugate momenta Π^μ being careful to substitute every $\partial_0 A_0$ with a Lagrange multiplier λ

$$\begin{aligned}
\mathcal{L}_{EM}(\alpha) &= -\frac{1}{2} F^{0i} F_{0i} - \frac{1}{4} F^{ij} F_{ij} + \alpha \partial_0 A^0 \partial_i A^i - \alpha \partial_i A_0 \partial^0 A^i \\
&= -\frac{1}{2}\left(\Pi^i + \alpha \partial^i A^0\right)\left(\Pi_i + \alpha \partial_i A_0\right) - \alpha \partial_i A_0 \left[(1-\alpha)\partial^i A^0 - \Pi^i\right] \\
&\quad - \frac{1}{4} F^{ij} F_{ij} + \alpha \lambda \partial_i A^i \\
&= -\frac{1}{2}\Pi^i \Pi_i - \alpha\left(1 - \frac{1}{2}\alpha\right)\partial_i A_0 \partial^i A^0 + \alpha \lambda \partial_i A^i - \frac{1}{4} F^{ij} F_{ij},
\end{aligned} \tag{3.78}$$

and finally we can calculate the Hamiltonian density

$$\begin{aligned}
\mathcal{H}_{EM}(\alpha) &= \Pi^k \partial_0 A_k + \lambda \Pi^0 - \mathcal{L} \\
&= -\frac{1}{2}\Pi^k \Pi_k + (1-\alpha)\Pi^k \partial_k A_0 + \alpha\left(1 - \frac{1}{2}\alpha\right)\partial_i A_0 \partial^i A^0 \\
&\quad + \lambda \varphi(\alpha) + \frac{1}{4} F^{ij} F_{ij}.
\end{aligned} \tag{3.79}$$

Now that we have calculated the Hamiltonian containing all primary constraints we can look for the secondary ones. In order to use the same notation we used at the beginning of this chapter we define from (3.79) a *canonical* part which contains the invertible conjugate momenta

$$\mathcal{H}^{C}_{EM}(\alpha) = -\frac{1}{2}\Pi^k\Pi_k + (1-\alpha)\,\Pi^k\partial_k A_0 + \alpha\left(1-\frac{1}{2}\alpha\right)\partial_i A_0 \partial^i A^0 + \frac{1}{4}F^{ij}F_{ij}, \tag{3.80}$$

so that (3.79) can be written as

$$\mathcal{H}_{EM}(\alpha) = \mathcal{H}^{C}_{EM}(\alpha) + \lambda\varphi(\alpha). \tag{3.81}$$

3.4.2.3 *Secondary constraints*

The next step is to look for secondary constraints, *i.e.* to impose the conservation along the dynamics of the primary constraint we defined in (3.75). In order to do this we need to give the definition of Poisson brackets in the case of field theories. Given two arbitrary functions of the phase-space coordinates $M(x) = M(A(x), \Pi(x), \partial_i A(x), \partial_i \Pi(x))$ and $N = N(A(x), \Pi(x), \partial_i A(x), \partial_i \Pi(x))$, their Poisson bracket is defined as

$$\begin{aligned}
&[M, N]_{A,\Pi} \\
&= \int d^3x d^3y \left[m(x)\, M(x), n(x)\, N(x)\right]_{A,\Pi} \\
&= \int d^3x d^3y d^3z \left\{ \frac{\delta m(x)\, M(x)}{\delta A_\mu(z)} \frac{\delta n(x)\, N(x)}{\delta \Pi^\mu(z)} - \frac{\delta m(x)\, M(x)}{\delta \Pi^\mu(z)} \frac{\delta n(x)\, N(x)}{\delta A_\mu(z)} \right\},
\end{aligned} \tag{3.82}$$

where the regular functions of spatial coordinates $m = m(x)$ and $n = n(x)$ are the so-called test functions or smearing functions. Their main property is to go to zero fast enough on the boundary so that any integral on the boundary containing a test function would automatically vanish. This property will often be used in order to discard boundary terms. The test functions are needed because the functional derivative of a function is a distribution which has to act, by definition, on the space of test functions. In the case of Poisson brackets between integrated functionals the test functions do not play any role, hence, they can be avoided. However, their presence is crucial for all the other calculations. The evolution of any dynamic quantity, say $M(x) = M(A(x), \Pi(x), \partial_i A(x), \partial_i \Pi(x))$, is then defined by its Poisson bracket with the Hamiltonian functional which does not need any test function

$$\int d^3xm(x)\,\partial_0 M(x) = \int d^3x\,[m(x)\,M(x)\,,H_{EM}(\alpha)]$$
$$= \int d^3xd^3y\,[m(x)\,M(x)\,,\mathcal{H}_{EM}(y;\alpha)]$$
$$= \int d^3xd^3y\,\{[m(x)\,M(x)\,,\mathcal{H}^C_{EM}(y;\alpha)] + [m(x)\,M(x)\,,\lambda(y)\,\varphi(y;\alpha)]\}\,. \tag{3.83}$$

Now we can calculate the secondary constraint which arises from the request of conservation of the primary constraint (3.75) during the dynamic evolution

$$\int d^3xf(x)\,\partial_0\varphi(x;\alpha)$$
$$= \int d^3xd^3y\,[f(x)\,\varphi(x;\alpha)\,,\mathcal{H}_{EM}(y;\alpha)]$$
$$= \int d^3xd^3y\,\{[f(x)\,\varphi(x;\alpha)\,,\mathcal{H}^C_{EM}(y;\alpha)] + [f(x)\,\varphi(x;\alpha)\,,\lambda(y)\,\varphi(y;\alpha)]\}\,. \tag{3.84}$$

Now, the second Poisson bracket in the last line of (3.84) vanishes because Poisson brackets are antisymmetric. Hence, we are only left with the Poisson bracket with the canonical part of the Hamiltonian density[2]

$$\int d^3xf(x)\,\partial_0\varphi(x;\alpha)$$
$$= \int d^3xd^3y\,\left[f(x)\,\Pi^0(x) - \alpha f(x)\,\partial^x_i A^i(x)\,,\mathcal{H}^C_{EM}(y;\alpha)\right]$$
$$= \int d^3xd^3y\,\left[f(x)\,\Pi^0(x)\,,(1-\alpha)\,\Pi^k(y)\,\partial^y_k A_0(y)\right]$$
$$+ \int d^3xd^3y\left[f(x)\,\Pi^0(x)\,,\alpha\left(1-\frac{1}{2}\alpha\right)\partial^y_i A_0(y)\,\partial^i_y A^0(y)\right]$$
$$- \int d^3xd^3y\left[\alpha f(x)\,\partial^x_i A^i(x)\,,-\frac{1}{2}\Pi^k(y)\,\Pi_k(y) + (1-\alpha)\,\Pi^k(y)\,\partial^y_k A_0(y)\right]$$

[2]Only in this case we perform all the calculations explicitly so that the reader can get acquainted with the formalism.

$$
\begin{aligned}
&= -\int d^3xd^3yd^3zf(x)\,\delta^{(3)}(x-z)\,(1-\alpha)\,\Pi^k(y)\,\partial_k^y\delta^{(3)}(y-z) \\
&- \int d^3xd^3yd^3zf(x)\,\delta^{(3)}(x-z)\,\alpha\,(2-\alpha)\,\partial_i^y A_0(y)\,\partial_y^i\delta^{(3)}(y-z) \\
&- \int d^3xd^3yd^3z\alpha f(x)\,\partial_k^x\delta_i^k\delta^{(3)}(x-z)\,\delta^{(3)}(y-z) \\
&\times\left[-\Pi^k(y)\,\delta_k^i + (1-\alpha)\,\delta_k^i\partial_y^k A_0(y)\right] \\
&= \int d^3zf(z)\,(1-\alpha)\,\partial_k\Pi^k(z) \\
&- \int d^3zf(z)\,\alpha\,(2-\alpha)\,\partial_i\partial^i A_0(z) \\
&+ \int d^3zf(z)\left[\alpha\partial_i\Pi^i(y) - \alpha\,(1-\alpha)\,\partial_i\partial^i A_0(z)\right] \\
&= \int d^3zf(z)\left[\partial_k\Pi^k(z) + \alpha\partial^i\partial_i A_0(z)\right] \approx 0.
\end{aligned}
\tag{3.85}
$$

Hence we have derived the secondary constraint which we write as

$$\chi(\alpha) = \partial_i\Pi^i + \alpha\partial^i\partial_i A_0 \approx 0. \tag{3.86}$$

3.4.2.4 *Constraints algebra*

Now that we have a secondary constraint it is useful to analyze its Poisson bracket with the primary one[3]

$$
\begin{aligned}
&\int d^3xd^3y\,[f(x)\,\varphi(x;\alpha)\,,g(y)\,\chi(y;\alpha)] \\
&= \int d^3xd^3y\left[f(x)\left(\Pi^0(x) - \alpha\partial_i^x A^i(x)\right),g(y)\left(\partial_i^y\Pi^i(y) + \alpha\partial_y^i\partial_i^y A_0(y)\right)\right] \\
&= \int d^3z\alpha\partial^i f(z)\,\partial_i g(z) - \int d^3z\alpha\partial^k f(z)\,\partial_k g(y) = 0,
\end{aligned}
\tag{3.87}
$$

which we shortly write as

$$[\varphi(\alpha)\,,\chi(\alpha)] = 0. \tag{3.88}$$

In this case the result is quite simple. Generally speaking the Poisson brackets for first-class constraints are expected either to be linear combinations

[3]Generally speaking we expect the boundary term not to change the constraints algebra since it induces a strong canonical transformation.

of constraints, *i.e.* to weakly vanish, or to identically be equal to zero as it happens in this case. Is this sufficient to affirm that the constraints (3.75) and (3.86) are first-class? Not yet. We still need to impose the conservation along the dynamics of the secondary constraint in order to see if it is automatically guaranteed or we need to impose a tertiary constraint.

3.4.2.5 *Tertiary constraint?*

Let us look for a tertiary constraint

$$\begin{aligned}
&\int d^3x f(x)\, \partial_0 \chi(x;\alpha) \\
&= \int d^3x d^3y \left[f(x)\, \chi(x;\alpha), \mathcal{H}_{EM}(y;\alpha)\right] \\
&= \int d^3x d^3y \left\{\left[f(x)\, \chi(x;\alpha), \mathcal{H}^C_{EM}(y;\alpha)\right] + \left[f(x)\, \chi(x;\alpha), \lambda(y)\, \varphi(y;\alpha)\right]\right\},
\end{aligned} \tag{3.89}$$

because of (3.88) the second term in the last line of (3.89) vanishes[4]. Hence, we are left with the Poisson bracket involving only the canonical Hamiltonian density $\mathcal{H}^C_{EM}(\alpha)$ and the term $\partial_i \Pi^i$ of the secondary constraint (3.86) since the term $\partial_i \partial^i A^0$ concerns only the Poisson bracket with the primary constraints which vanishes. Hence, we write

$$\begin{aligned}
&\int d^3x f(x)\, \partial_0 \chi(x;\alpha) \\
&= \int d^3x d^3y \left[f(x)\, \partial^x_i \Pi^i(x), \frac{1}{4} F^{ij}(y)\, F_{ij}(y)\right] \\
&= \int d^3z \left[f(z)\, \partial_j \partial_k F^{kj}(z)\right] = 0.
\end{aligned} \tag{3.90}$$

Since $\partial_i \partial_j F^{ij} = 0$ identically, because F^{ij} is antisymmetric, we have that no tertiary constraint arises which means that the primary and secondary constraints are both first-class.

3.4.2.6 *Equations of motion*

Now that we have derived all the basic properties of the Hamiltonian formulation of the theory we are left with the computation of the equa-

[4]Boundary terms arising in the integrations implying the Lagrange multiplier $\lambda = \lambda(x)$ vanish because of the localization of Dirac delta and not because of some specified behavior of λ at the boundary as it happens in the case of test functions. Thus, the result (3.88) still holds.

tions of motion. For convenience we now write once again the canonical Hamiltonian density and the two constraints

$$\mathcal{H}^{C}_{EM}(\alpha) = -\frac{1}{2}\Pi^k\Pi_k + (1-\alpha)\Pi^k\partial_k A_0 + \alpha\left(1-\frac{1}{2}\alpha\right)\partial_i A_0 \partial^i A^0 + \frac{1}{4}F^{ij}F_{ij}, \tag{3.91}$$

$$\varphi(\alpha) = \Pi^0 - \alpha\partial_i A^i \approx 0, \tag{3.92}$$

$$\chi(\alpha) = \partial_i \Pi^i + \alpha\partial^i\partial_i A_0 \approx 0. \tag{3.93}$$

Now, the equations of motion for the fields A^μ are

$$\int d^3x f(x)\partial_0 A^0(x) = \int d^3x d^3y \left[f(x) A^0(x), \lambda(x)\varphi(y;\alpha)\right] = \int d^3z f(z)\lambda(z) \tag{3.94}$$

$$\begin{aligned}&\int d^3x f^i(x)\partial_0 A_i(x)\\ &= \int d^3x d^3y \left[f_i(x) A^i(x), (1-\alpha)\Pi^i(y)\partial^y_i A_0(y) - \frac{1}{2}\Pi^i(y)\Pi_i(y)\right]\\ &= \int d^3z f^i(z)\left[(1-\alpha)\partial_i A_0(z) - \Pi_i(z)\right],\end{aligned} \tag{3.95}$$

which are shortly written as

$$\partial_0 A^0 = \lambda, \tag{3.96}$$

$$\partial_0 A_i = (1-\alpha)\partial_i A_0 - \Pi_i. \tag{3.97}$$

Indeed, from the equation of motion of A^0 (3.96) we recover the convention we adopted when writing the Hamiltonian density, while from the equation of motion of A^i we recover, as expected, the definition of the conjugate momenta (3.77). Let us now derive the equations of motion for the invertible conjugate momenta

$$\begin{aligned}&\int d^3x f_i(x)\partial_0 \Pi^i(x)\\ &= \int d^3x d^3y \left[f_i(x)\Pi^i(x), \frac{1}{4}F^{ij}(y)F_{ij}(y) - \alpha\lambda(y)\partial^i_y A_i(y)\right]\\ &= \int d^3z f_k(z)\left[\partial_j F^{kj}(y) - \alpha\partial^k\lambda(z)\right]\end{aligned} \tag{3.98}$$

which we rewrite for the sake of clearness

$$\partial_0 \Pi^i = \partial_j F^{ij} - \alpha \partial^i \lambda. \tag{3.99}$$

The calculation of $\partial_0 \Pi^0$ has already been performed in order to obtain the secondary constraint. We now recover the Lagrangian equations of motion (3.66) in order to verify the consistency of our calculations. Putting together (3.96), (3.97) and (3.99) we obtain

$$\begin{aligned}
\partial_0 \Pi^i =& \partial_0 \left[(1-\alpha) \partial^i A^0 - \partial^0 A^i \right] = \partial_0 \left[F^{0i} - \alpha \partial^i A^0 \right] \\
\partial_0 \Pi^i =& \partial_j F^{ij} - \alpha \partial^i \lambda = \partial_j F^{ij} - \alpha \partial^i \partial_0 A^0 \\
& \partial_0 F^{0i} - \alpha \partial^i \partial_0 A^0 = \partial_j F^{ij} - \alpha \partial^i \partial_0 A^0
\end{aligned} \tag{3.100}$$

which, after one exchange of indices in F^{0i} and taking into account the minus sign, turns out to be

$$\partial_\nu F^{i\nu} = 0. \tag{3.101}$$

With respect to (3.66) there is still one equation missing which comes from the secondary constraint (3.86). Plugging (3.97) into the secondary constraint (3.93) we get

$$\chi(\alpha) = \partial_i \Pi^i + \alpha \partial^i \partial_i A_0 = \partial_i F^{0i} = 0, \tag{3.102}$$

which turns out to be the Gauss' law for the electric field

$$\partial_i F^{0i} = \partial_i E^i = 0, \tag{3.103}$$

which is the reason why the secondary constraint (3.86) (usually in the form with $\alpha = 0$) is commonly called *Gauss' constraint*. Now, considering (3.101) together with (3.103) we obtain the Lagrangian equations of motion, *i.e.* the second order ones

$$\partial_\nu F^{\mu\nu} = 0. \tag{3.104}$$

This final step helps us noticing how the parameter α completely disappears passing to the Lagrangian equations of motion. Indeed this is the expected result since α is the prefactor of the boundary term whose presence do not change the stationarity condition of the action functional for the Lagrangian density (3.69). We chose to specify a particular form of the boundary term (3.70) for the sake of calculations and because we will encounter a similar boundary term in Dirac Lagrangian formulation of General Relativity.

On the other hand all the most interesting Hamiltonian quantities do depend on the parameter α, which is understandable since the boundary term (3.70) modifies the definitions of the conjugate momenta. Interestingly, the number of constraints and their properties do not change.

3.4.3 *Gauge transformations*

We now want to show, in this simple case, how to construct the generator of the gauge transformations for the variables of the phase-space. We briefly report here the basic notions of the work by Castellani [87].

Now, as we stressed before, a gauge theory has an invariance of the equations of motion under the action of some transformations. This means that, for a generic system described by the canonical coordinates $q^i = q^i(t)$ and $p_i = p_i(t)$, if we change them by some functions

$$q^i(t) \to q^i(t) + \eta^i(t), \qquad p_i(t) \to p_i(t) + \xi_i(t), \tag{3.105}$$

the transformed Hamiltonian equations of motion and the transformed constraints should hold for the transformed variables (3.105) as well if we want η^i and ξ_i to be gauge functions. This requirement, together with some other hypotheses on the constraints, lead to define a phase-space generator of gauge transformations which is defined by the following algorithm: one starts with a primary first-class constraint as the first term of a chain of unspecified length k, which we will call G_k, then one looks for the *previous* term of the chain, G_{k-1}, requiring it to satisfy the relation

$$G_{k-1} + [G_k, H]_{q,p} = \text{primary}, \tag{3.106}$$

where 'primary' means any arbitrary linear combination of primary constraints. One continues backwards in the construction of this chain up to the last term G_0 which is the only one in the chain satisfying the requirement

$$[G_0, H]_{q,p} = \text{primary}. \tag{3.107}$$

All terms of the chain have to be first-class. With the set $\{G_i | i = 1, \dots, k\}$ it is possible to define the first-class function where $\varepsilon^{(i)}$ is the i-th time derivative of the transformation parameter $\varepsilon = \varepsilon(t)$

$$G = \sum_{i=0}^{k} \varepsilon^{(i)} G_i, \tag{3.108}$$

which is the generator of the gauge transformations in the phase-space. In the general case this procedure should be repeated for each primary first-class constraint of the theory and it is not granted that the choices made in order to fulfil the requirements (3.106) and (3.107) lead to chains of minimal length[5]. In [87] the chains of minimal length are conjectured to be the one generating the usual gauge transformations.

[5] In order to satisfy (3.106) and (3.107) one usually chooses arbitrarily some functions. This choice might lead to a more G_k with respect to other choices of the arbitrary functions. The set $\{G_k\}$ with the minimal number of elements is called *chain of minimal length*. In [87] Castellani discussed how to split non-minimal chains.

Let us now try to apply this formalism to the Hamiltonian formulation of the electromagnetism in order to retrieve the transformations (3.68) and the yet unknown transformations on the conjugate momenta. We start by defining the last term of the chain as the only primary first-class constraint of the theory (3.75)

$$G_k(\alpha) = \varphi(\alpha) = \Pi^0 - \alpha\partial_i A^i, \tag{3.109}$$

and we continue defining the previous term of the chain as prescribed by (3.106) in the notation of distribution we used so far

$$\int d^3x f(x)\left\{G_{k-1}(x;\alpha) + \left[\varphi(x;\alpha), H_{EM}\right]\right\} = \int d^3x f(x)\,\Phi(x)\,\varphi(x;\alpha), \tag{3.110}$$

where $\Phi(x) = \Phi(A^\mu(x), \Pi_\mu(x), \partial_i A^\mu(x), \partial_i \Pi_\mu(x))$ is an arbitrary function to be determined which stands for the only possibility of linear combinations of the primary constraint which is required in (3.106). Continuing the calculations we get

$$\int d^3x f(x)\left\{G_{k-1}(x;\alpha) + \chi(x;\alpha)\right\} = \int d^3x f(x)\,\Phi(x)\,\varphi(x;\alpha),$$

$$\int d^3x f(x)\, G_{k-1}(x;\alpha) = \int d^3x f(x)\left\{\Phi(x)\,\varphi(x;\alpha) - \chi(x;\alpha)\right\}. \tag{3.111}$$

Let us see if we can end the chain here (of course we can since the constraints are two and are both first-class) by imposing the second condition (3.107) which in this case reads

$$\int d^3x f(x)\left[G_{k-1}(x;\alpha), H_{EM}\right] = \int d^3x f(x)\,\Xi(x)\,\varphi(x;\alpha), \tag{3.112}$$

where $\Xi(x) = \Xi(A^\mu(x), \Pi_\mu(x), \partial_i A^\mu(x), \partial_i \Pi_\mu(x))$. Writing more carefully we get

$$\begin{aligned}
&\int d^3x f(x)\left[G_{k-1}(x;\alpha), H_{EM}\right] \\
&= \int d^3x f(x)\{\varphi(x;\alpha)\left[\Phi(x), H_{EM}\right] + \Phi(x)\left[\varphi(x;\alpha), H_{EM}\right] - \left[\chi(x;\alpha), H_{EM}\right]\} \\
&= \int d^3x f(x)\{\varphi(x;\alpha)\left[\Phi(x), H_{EM}\right] + \Phi(x)\,\chi(x;\alpha)\} = \int d^3x f(x)\,\Xi(x)\,\varphi(x;\alpha),
\end{aligned} \tag{3.113}$$

which can be satisfied only for $\Phi = 0$. Going back to (3.111) we obtain the second, and last, term of the chain

$$G_0(\alpha) = -\chi(\alpha). \tag{3.114}$$

We can now write the generator of gauge transformations ($k = 1$) as

$$G(\alpha) = \varepsilon G_0(\alpha) + \partial_0 \varepsilon G_1(\alpha) = -\varepsilon \chi(\alpha) + \partial_0 \varepsilon \varphi(\alpha), \tag{3.115}$$

and control that it generates the right gauge transformations on A^μ

$$\begin{aligned} \int d^3x f_\mu(x)\, \delta A^\mu(x) &= \int d^3x d^3y f_\mu(x) \left[A^\mu(x), G(y;\alpha)\right] \\ &= \int d^3z f_\mu(z)\, \partial^\mu \varepsilon(z), \end{aligned} \tag{3.116}$$

$$\delta A^\mu = \partial^\mu \varepsilon, \tag{3.117}$$

which exactly reproduces the gauge transformations (3.68) since $\varepsilon = \varepsilon(x)$ is an arbitrary function. We notice that the action on A^μ is independent of the value of the parameter α. Let us check the transformations of the conjugate momenta

$$\begin{aligned} &\int d^3x f(x)\, \delta\Pi^0(x) \\ &= \int d^3x d^3y f(x) \left[\Pi^0(x), G(y;\alpha)\right] \\ &= \int d^3z f(z)\, \alpha \partial_i \partial^i \varepsilon(z), \end{aligned} \tag{3.118}$$

$$\begin{aligned} \int d^3x f_i(x)\, \delta\Pi^i(x) &= \int d^3x d^3y f_i(x) \left[\Pi^i(x), G(y;\alpha)\right] \\ &= -\int d^3z f_i(z)\, \alpha \partial^i \partial_0 \varepsilon(z). \end{aligned} \tag{3.119}$$

Thus, we obtain

$$\delta\Pi^0 = \alpha \partial_i \partial^i \varepsilon, \qquad \delta\Pi^i = -\alpha \partial^i \partial^0 \varepsilon, \tag{3.120}$$

which are both depending on the value of the boundary parameter α, so that if there was no boundary term, $\alpha = 0$, the conjugate momenta would be gauge invariant, as one should expect since in this case $\Pi^i = F^{0i} = E^i$, being the electric field gauge invariant. It is easy to check that the transformation satisfies the expressions defining the conjugate momenta

$$\Pi^0 = \alpha \partial_i A^i \to \left(\Pi^0\right)' = \alpha \partial_i A^i + \alpha \partial_i \partial^i \varepsilon, \Rightarrow \delta\Pi^0 = \alpha \partial_i \partial^i \varepsilon, \tag{3.121}$$

$$\Pi^i = F^{0i} - \alpha \partial^i A^0, \to \left(\Pi^i\right)' = F^{0i} - \alpha \partial^i A^0 - \alpha \partial^i \partial^0 \varepsilon \Rightarrow \delta\Pi^i = -\alpha \partial^i \partial^0 \varepsilon. \tag{3.122}$$

With these results we are now able to gauge transform any regular function of the phase-space coordinates. Hence, even though we lost both covariance and gauge-invariance of the action functional we are still able to obtain (trivially) the right covariant and gauge-invariant set of second order equations of motion and (much less trivially) the right generator of the gauge transformations in the phase-space. This shows the robustness of the Hamiltonian formulation of gauge theories.

3.4.4 *Gauged canonicity*

We now want to study how a transformation of variables for the electromagnetic field might turn out to be gauged canonical. Indeed, this happens in a rather simple way. Let us consider the transformation of coordinates

$$A^0 = N^2 + N^i N_i, \qquad A^i = N^i. \tag{3.123}$$

Note that the quantities $\{N, N^i\}$ do not compose a covariant vector. There are two ways of obtaining a Hamiltonian formulation of electromagnetism in this new set of variables: either we directly transform the Lagrangian density (3.69) or we impose the canonicity of the transformation starting directly from the Hamiltonian formulation in the $\{A^\mu, \Pi_\mu\}$. These two procedures will not lead to the same result since the first one will turn out to be a non-invertible transformation of the phase-space variables while the second will be by definition invertible. Let us transform the Lagrangian density (3.69) and calculate the new conjugate momenta π^μ

$$\begin{aligned}\tilde{\mathcal{L}}_{EM}(\alpha) = &-\frac{1}{2}F^{0i}F_{0i} - \frac{1}{4}F^{ij}F_{ij} + 2\alpha N\partial_0 N \partial_i N^i \\ &- 2\alpha\left(N\partial^i N + N_k \partial^i N^k - N^i \partial_k N^k\right)\partial_0 N_i.\end{aligned} \tag{3.124}$$

New conjugate momenta will be denoted by π_L and π_L^i in order to indicate that their definition follows from the Lagrangian density (3.69) after the transformation (3.123) has been imposed

$$\pi_L = \frac{\partial \tilde{\mathcal{L}}_{EM}(\alpha)}{\partial \partial_0 N} = 2\alpha N \partial_i N^i, \qquad \phi(\alpha) = \pi_L - 2\alpha N \partial_i N^i \approx 0, \tag{3.125}$$

$$\pi_L^i = \frac{\partial \tilde{\mathcal{L}}_{EM}(\alpha)}{\partial \partial_0 N_i} = F^{0i} - 2\alpha\left(N\partial^i N + N_k \partial^i N^k - N^i \partial_k N^k\right). \tag{3.126}$$

In order to compare these conjugate momenta with those of the usual formulation, (3.75) and (3.77), the only thing we can do is to compare the definitions of the velocities. This fact should be clear since we are starting from the same configuration space described by the two sets of coordinates $\{A^\mu, \partial_0 A^\mu\}$ and $\{N, N^i, \partial_0 N, \partial_0 N^i\}$, which can be mapped onto each other, but we are ending in the two *a priori* different phase-spaces described by the conjugate variables $\{A^\mu, \Pi_\mu\}$ and $\{N, N^i, \pi_L, \pi_L^i\}$ respectively. These two phase-spaces are a priori different because the transformation leading to them from the configuration spaces are not invertible: they can only be rendered invertible thanks to arbitrary functions, *i.e.* the Lagrange multipliers. It is from this arbitrariness that the definition of gauged canonicity arises. Let us write the invertible velocities as functions of the canonical variables

$$\begin{aligned}\partial^0 N^i =& 2(1-\alpha)\left(N\partial^i N + N_k\partial^i N^k\right) + 2\alpha N^i\partial_k N^k - \pi_{L}^i,\\ \partial^0 A^i =& (1-\alpha)\partial^i A^0 - \Pi^i\\ =& 2(1-\alpha)\left(N\partial^i N + N_k\partial^i N^k\right) - \Pi^i,\end{aligned} \tag{3.127}$$

and since $\partial_0 N^i = \partial_0 A^i$, because of (3.123), we can write

$$\pi_{L}^i = \Pi^i + 2\alpha N^i\partial_k N^k = \Pi^i + 2\alpha A^i\partial_k A^k. \tag{3.128}$$

However, if we apply the same procedure to the arbitrary velocities $\partial_0 A^0 = \lambda$ and $\partial_0 N = \Lambda$ we obtain a redefinition of the Lagrange multipliers which are already arbitrary functions

$$\begin{aligned}\partial_0 A^0 =& 2N\partial_0 N + 2N^k\partial_0 N_k\\ \lambda =& 2N\Lambda + 4(1-\alpha)\left(NN^k\partial_k N + N^i N^k\partial_k N_i\right) + 4\alpha N^k N_k\partial_i N^i - 2N^k\pi_k^{L},\end{aligned} \tag{3.129}$$

so that some transformation linking π_{L} to, at least, Π^0 is missing: the phase-space transformation defined by (3.123) and (3.128) is not invertible. With the invertible part of the transformation for the conjugate momenta we can check that the transformation is not strongly canonical

$$\left[N, \pi_{L}^i\right]_{A,\Pi} = \left[\sqrt{A^0 - A^k A_k}, \Pi^i + 2\alpha A^i\partial_k A^k\right]_{A,\Pi} - \frac{A^i}{\sqrt{A^0 - A^k A_k}} \neq 0. \tag{3.130}$$

In order to better understand the reason why this happens we can manipulate the definitions of the primary constraints (3.75) and (3.125) by writing

$$\begin{aligned}\pi_{L} =& 2\alpha\sqrt{A^0 - A^k A_k}\partial_i A^i + \Phi\varphi(\alpha) - \Phi\varphi(\alpha)\\ =& \Phi\Pi^0 + \alpha\partial_i A^i\left(2\sqrt{A^0 - A^k A_k} - \Phi\right) - \Phi\varphi(\alpha),\end{aligned} \tag{3.131}$$

where $\Phi(x) = \Phi\left(A^\mu(x), \Pi_\mu(x), \partial_i A^\mu(x), \partial_i\Pi_\mu(x)\right)$ is an arbitrary function of the phase-space coordinates. Of course the new expression still does not depend on Π^0. We now make a very peculiar choice, among the infinite possible ones, for the arbitrary function Φ

$$\Phi = 2\sqrt{A^0 - A^k A_k}, \tag{3.132}$$

which identically leads us to

$$\pi_{L} = 2\sqrt{A^0 - A^k A_k}\Pi^0 - 2\sqrt{A^0 - A^k A_k}\varphi(\alpha), \tag{3.133}$$

that again does not depend on Π^0. However, we can decide to evaluate this expression on the hypersurface defined by the primary constraint $\varphi(\alpha)$ giving

$$\pi_L = 2\alpha\sqrt{A^0 - A^k A_k}\partial_i A^i \approx 2\sqrt{A^0 - A^k A_k}\Pi^0, \tag{3.134}$$

so that we have introduced the dependence on Π^0 modulo some combination of the primary constraint. We evaluate now a fundamental Poisson bracket with respect to the old set of canonical variables $\{A_\mu, \Pi^\mu\}$

$$[N, \pi_L]_{A,\Pi} = 1 - [A^0, \varphi(\alpha)]_{A,\Pi} = 0. \tag{3.135}$$

Hence, transformation (3.123) turns out not to be strongly canonical because of the action of a first-class constraint, *i.e.* it is gauged canonical.

Let us now follow the other road and directly impose the canonicity of the transformation (3.123). The strong canonicity condition can be satisfied imposing the condition $\Pi^\mu \delta A_\mu = \pi_C \delta N + \pi_C^i \delta N_i$

$$\begin{aligned} &\pi_C \delta N + \pi_C^i \delta N_i \\ &= \pi_C \frac{1}{2N}\delta A^0 + \left(\pi_C^i - \frac{1}{N}A^i \pi_C\right)\delta A_i \\ &= \Pi^\mu \delta A_\mu, \end{aligned} \tag{3.136}$$

where we mixed old and new variables in order to lighten the notation, which gives for the new conjugate momenta

$$\Pi^0 = \frac{1}{2N}\pi_C, \qquad \Pi^i = \pi_C^i - \frac{1}{N}A^i \pi_C, \tag{3.137}$$

$$\pi_C = 2N\Pi^0 = 2\sqrt{A^0 - A^k A_k}\Pi^0, \qquad \pi_C^i = \Pi^i + 2A^i \Pi^0, \tag{3.138}$$

which we can easily check gives the right result for (3.135)

$$[N, \pi_C]_{A,\Pi} = \frac{1}{2N}\left[A^0, 2\sqrt{A^0 - A^k A_k}\Pi^0\right]_{A,\Pi} = [A^0, \Pi^0]_{A,\Pi} = 1. \tag{3.139}$$

Indeed the two definitions of the transformation for the conjugate momenta (3.128), (3.133) and (3.137) differ for a combination of the primary first-class constraint, hence they are weakly equal

$$\pi_L = 2N\Pi^0 - 2N\varphi(\alpha) \approx 2N\Pi^0 = \pi_C, \tag{3.140}$$

$$\begin{aligned} \pi_L^i &= \Pi^i + 2\alpha A^i \partial_k A^k \\ &= \Pi^i + 2A^i \Pi^0 - 2A^i \varphi(\alpha) \\ &\approx \Pi^i + 2A^i \Pi^0 = \pi_C^i, \end{aligned} \tag{3.141}$$

which is however a change that does not affect the physical content of the theory since it vanishes on the constraints hypersurface, *i.e.* on the dynamics.

Chapter 4

Lagrangian Formulations

With respect to other Lagrangian formulations of gauge theories, General Relativity is characterized by a peculiar feature: it is not possible to write a gauge-invariant action functional without making some assumptions on the boundary of the four-dimensional manifold taken as integration domain. As we shall see, the only Lagrangian density containing only first order derivatives of the metric tensor, hence free from any boundary assumption, is non-covariant.

Because of this property, which should play some important role on the quantum level, it is possible to choose among several different Lagrangian formulations for General Relativity, even in the same coordinate representation, according to the boundary conditions one chooses to deal with.

Along this chapter we will review in some detail several Lagrangian formulations pointing out their symmetry properties and the needed hypothesis on the boundaries. We will also pay attention to the boundary terms either proper of the given formulation or arising from the stationarity condition. They will be useful in connecting Lagrangian densities defined by different procedures and in different representations.

This chapter is structured as follows: we will discuss the most commonly used Lagrangian formulation and how the coupling to matter can be treated. We will then analyze the $\Gamma\Gamma$ Lagrangian density (proposed by Einstein in [118]), and Dirac's. We will examine a class of generalized Lagrangian formulations, called $f(R)$, and the first-order Palatini formulation. We will review the ADM formulation using the embedding technique we introduced in Chapter 1. Finally, we will analyze all the collected boundary terms and define a transformation linking ADM formulation to Dirac's and $\Gamma\Gamma$.

4.1 Metric representation

The Lagrangian densities we will present in this section are all written in terms of the metric tensor and differ for boundary terms, with the exceptions of the $f(R)$ and Palatini formulations. As we have seen in the previous chapter any Lagrangian density is defined up to a boundary term. Some of those do involve second order derivatives, hence they do not vanish under variation when imposing the condition $\delta g|_{\partial\mathcal{M}} = 0$. However, a suitable choice of the manifold renders all these formulations perfectly equivalent, and may turn out to make the computations much easier. This discussion might seem a little bit too subtle in a classical framework. Though, if one would try to compute a path-integral formulation the choice of the boundary term turns out to be non-trivial. Of course, all these formulations are equivalent when considering compact manifolds without boundaries (*e.g.* a closed universe).

4.1.1 *Einstein-Hilbert formulation*

We start from the most common and most used Lagrangian density which is the so-called *Einstein-Hilbert* one which reads

$$\mathcal{L}_{EH} = \sqrt{-g}R, \tag{4.1}$$

which leads to the action functional

$$S_{EH}\left[g_{\mu\nu}\right] = \frac{c^3}{16\pi G}\int_{\mathcal{M}} d^4x\sqrt{-g}R. \tag{4.2}$$

Let us now calculate the equations of motion. The variation of the action functional (4.2) with respect to the inverse metric tensor $\delta g^{\mu\nu}$ is

$$\delta S_{EH}\left[g_{\mu\nu}\right] = \frac{c^3}{16\pi G}\int_{\mathcal{M}} d^4x\left[\left(\delta\sqrt{-g}\right)R + \sqrt{-g}R_{\mu\nu}\delta g^{\mu\nu} + \sqrt{-g}g^{\mu\nu}\delta R_{\mu\nu}\right]. \tag{4.3}$$

Starting from the first term and recalling equation (1.73) we write

$$\left(\delta\sqrt{-g}\right)R = -\frac{\delta g}{2\sqrt{-g}}R = -\frac{g g^{\mu\nu}\delta g_{\mu\nu}}{2\sqrt{-g}}R = \frac{g g_{\mu\nu}\delta g^{\mu\nu}}{2\sqrt{-g}}R = -\frac{1}{2}\sqrt{-g}g_{\mu\nu}R\delta g^{\mu\nu}, \tag{4.4}$$

while the second term is already proportional to $\delta g^{\mu\nu}$. The third term is the one that requires a bit more attention. In order to evaluate the variation of the Ricci tensor we start by calculating the variation for the Riemann tensor as defined in (1.89):

$$\begin{aligned}\delta R^{\rho}{}_{\sigma\mu\nu} =& \partial_\mu \delta\Gamma^{\rho}_{\nu\sigma} - \partial_\nu \delta\Gamma^{\rho}_{\mu\sigma} + \delta\Gamma^{\tau}_{\nu\sigma}\Gamma^{\rho}_{\mu\tau} + \Gamma^{\tau}_{\nu\sigma}\delta\Gamma^{\rho}_{\mu\tau} - \delta\Gamma^{\tau}_{\mu\sigma}\Gamma^{\rho}_{\nu\tau} - \Gamma^{\tau}_{\mu\sigma}\delta\Gamma^{\rho}_{\nu\tau}\\ =& \partial_\mu \delta\Gamma^{\rho}_{\nu\sigma} + \Gamma^{\rho}_{\mu\tau}\delta\Gamma^{\tau}_{\nu\sigma} - \Gamma^{\tau}_{\mu\sigma}\delta\Gamma^{\rho}_{\nu\tau} - \Gamma^{\tau}_{\mu\nu}\delta\Gamma^{\rho}_{\tau\sigma}\\ & -\partial_\nu \delta\Gamma^{\rho}_{\mu\sigma} - \Gamma^{\rho}_{\nu\tau}\delta\Gamma^{\tau}_{\mu\sigma} + \Gamma^{\tau}_{\nu\sigma}\delta\Gamma^{\rho}_{\mu\tau} + \Gamma^{\tau}_{\mu\nu}\delta\Gamma^{\rho}_{\tau\sigma}\end{aligned} \tag{4.5}$$

where we added and subtracted the last terms in the last two lines. Now it is easy to recognize the covariant derivatives of the variations of the Christoffel symbols[1] allowing us to write in a compact form the variation of the Riemann tensor

$$\delta R^{\rho}{}_{\sigma\mu\nu} = \nabla_\mu \delta\Gamma^{\rho}_{\nu\sigma} - \nabla_\nu \delta\Gamma^{\rho}_{\mu\sigma}, \tag{4.6}$$

from which we can directly state the variation for the Ricci tensor

$$\delta R_{\sigma\nu} = \delta R^{\rho}{}_{\sigma\rho\nu} = \nabla_\rho \delta\Gamma^{\rho}_{\nu\sigma} - \nabla_\nu \delta\Gamma^{\rho}_{\sigma\rho}. \tag{4.7}$$

Now we rewrite the last term of (4.3) as a covariant boundary term (see (1.76))

$$\begin{aligned}\sqrt{-g}g^{\mu\nu}\delta R_{\mu\nu} =& \sqrt{-g}g^{\mu\nu}\left(\nabla_\tau \delta\Gamma^{\tau}_{\mu\nu} - \nabla_\nu \delta\Gamma^{\tau}_{\mu\tau}\right)\\ =& \sqrt{-g}\nabla_\mu\left(g^{\rho\sigma}\delta\Gamma^{\mu}_{\rho\sigma} - g^{\mu\nu}\delta\Gamma^{\tau}_{\nu\tau}\right)\\ =& \partial_\mu\left[\sqrt{-g}\left(g^{\rho\sigma}\delta\Gamma^{\mu}_{\rho\sigma} - g^{\mu\nu}\delta\Gamma^{\tau}_{\nu\tau}\right)\right],\end{aligned} \tag{4.8}$$

which we explicitly define as

$$\partial_\mu \boldsymbol{\delta\mathcal{EH}}^\mu = \partial_\mu\left[\sqrt{-g}\left(g^{\rho\sigma}\delta\Gamma^{\mu}_{\rho\sigma} - g^{\mu\nu}\delta\Gamma^{\tau}_{\nu\tau}\right)\right] = \sqrt{-g}g^{\mu\nu}\delta R_{\mu\nu}. \tag{4.9}$$

It clearly appears that this boundary term arising from the variation contains first derivatives of the variation of the metric tensor, through the variation of the Christoffel symbols. Hence, the requirement $\delta g^{\mu\nu}|_{\partial\mathcal{M}} = 0$ is not enough to assure the vanishing of (4.9) on the boundary. However if we choose to treat only those manifolds that are asymptotically Minkowskian we can safely affirm that (4.9) evaluates to zero. We now rewrite the stationarity condition (4.3) inserting the results (4.4) and (4.9) as

$$\begin{aligned}\delta S_{EH}\left[g_{\mu\nu}\right] =& \frac{c^3}{16\pi G}\int_{\mathcal{M}} d^4x\sqrt{-g}\left(R_{\mu\nu} - \frac{1}{2}g_{\mu\nu}R\right)\delta g^{\mu\nu}\\ & + \frac{c^3}{16\pi G}\int_{\mathcal{M}} d^4x\partial_\mu \boldsymbol{\delta\mathcal{EH}}^\mu = 0,\end{aligned} \tag{4.10}$$

and considering only a class of possible manifolds (for example asymptotically Minkowskian) we obtain the vanishing of the boundary term and Einstein's field equations

$$R_{\mu\nu} - \frac{1}{2}g_{\mu\nu}R = 0. \tag{4.11}$$

[1] The covariant derivative of the variation of the Christoffel is well-defined since $\delta\Gamma^{\rho}_{\mu\nu}$ is a tensor as we will see later.

4.1.2 *Stress-Energy tensor*

Before moving on let us analyze how to insert the coupling to other fields which here we indicate with ϕ. Generally speaking one should write a Lagrangian density $\mathcal{L}_\phi = \mathcal{L}_\phi\left(\phi, \partial_\mu\phi, g_{\rho\sigma}, \partial_\mu g_{\rho\sigma}\right)$, which behaves as a scalar density of weight $1/2$ under diffeomorphisms. We then write the action for the field ϕ as

$$S_\phi\left[\phi, g_{\rho\sigma}\right] = \int_{\mathcal{M}} d^4x \mathcal{L}_\phi\left(\phi, \partial_\mu\phi, g_{\rho\sigma}, \partial_\mu g_{\rho\sigma}\right), \tag{4.12}$$

and the stationarity condition under arbitrary variations of the inverse metric tensor

$$\delta_g S_\phi\left[\phi, g_{\rho\sigma}\right] = \int_{\mathcal{M}} d^4x \left(\frac{\partial\mathcal{L}_\phi}{\partial g^{\rho\sigma}} - \partial_\mu \frac{\partial\mathcal{L}_\phi}{\partial\partial_\mu g^{\rho\sigma}}\right)\delta g^{\rho\sigma} = 0, \tag{4.13}$$

since the dependence of $\mathcal{L}_\phi$ on $g^{\rho\sigma}$ comprises first derivatives of the metric tensor the vanishing of the boundary term arising in the variation procedure is assured. We then define the symmetric stress-energy tenor as

$$\frac{\partial\mathcal{L}_\phi}{\partial g^{\rho\sigma}} - \partial_\mu \frac{\partial\mathcal{L}_\phi}{\partial\partial_\mu g^{\rho\sigma}} = \sqrt{-g}T_{\rho\sigma}. \tag{4.14}$$

Clearly, the stationarity condition on (4.12) for arbitrary variations $\delta\phi$ leads to the dynamic equations for the field ϕ on curved background. In relativistic field theory one usually defines the stress-energy tensor through the invariance of the action functional under spacetime translations. Indeed, because of Noether theorem, this symmetry leads to determine a rank two tensor with vanishing ordinary divergence which is the stress-energy tensor. However this tensor is not, in general, symmetric while the one we have just defined in (4.14) is automatically symmetric since the derivatives are calculated with respect to the symmetric metric tensor. Hence each result of the derivations has to be symmetrized with respect to the couple of indices μ and ν. In the general relativistic case it is possible to show that the invariance of the action functional (4.12) under diffeomorphisms leads to the vanishing of the covariant divergence of the stress-energy tensor (4.14). Let us define an infinitesimal coordinate transformation of parameters ξ^μ

$$y^\mu = x^\mu + \xi^\mu, \tag{4.15}$$

for which the inverse metric tensor transforms, at the first order in ξ^μ, as

$$\begin{aligned} g'^{\rho\sigma}\left(y^\tau\right) =& g^{\mu\nu}\left(x^\tau\right)\frac{\partial y^\rho}{\partial x^\mu}\frac{\partial y^\sigma}{\partial x^\nu} = g^{\mu\nu}\left(x^\tau\right)\left(\delta^\rho_\mu + \frac{\partial\xi^\rho}{\partial x^\mu}\right)\left(\delta^\sigma_\nu + \frac{\partial\xi^\sigma}{\partial x^\nu}\right) \\ \simeq& g^{\rho\sigma}\left(x^\tau\right) + g^{\rho\nu}\left(x^\tau\right)\frac{\partial\xi^\sigma}{\partial x^\nu} + g^{\sigma\mu}\left(x^\tau\right)\frac{\partial\xi^\rho}{\partial x^\mu}. \end{aligned} \tag{4.16}$$

We also expand at the first order in ξ^μ the transformed metric tensor

$$g'^{\rho\sigma}(y^\tau) \simeq g'^{\rho\sigma}(x^\tau) + \frac{\partial g^{\rho\sigma}(x^\tau)}{\partial x^\mu}\xi^\mu, \tag{4.17}$$

so that we write the functional variation of the inverse metric tensor at a given point under an infinitesimal diffeomorphism as

$$\begin{aligned}\delta g^{\rho\sigma}(x^\tau) =& g'^{\rho\sigma}(x^\tau) - g^{\rho\sigma}(x^\tau)\\ =& g^{\rho\nu}(x^\tau)\frac{\partial \xi^\sigma}{\partial x^\nu} + g^{\sigma\mu}(x^\tau)\frac{\partial \xi^\rho}{\partial x^\mu} - \frac{\partial g^{\rho\sigma}(x^\tau)}{\partial x^\mu}\xi^\mu,\end{aligned} \tag{4.18}$$

or more compactly as

$$\delta g^{\rho\sigma} = g^{\rho\nu}\partial_\nu \xi^\sigma + g^{\sigma\mu}\partial_\mu \xi^\rho - \xi^\mu \partial_\mu g^{\rho\sigma}. \tag{4.19}$$

The variation of a tensor still behaves as a tensor. It is possible to write (4.19) in a manifestly covariant form as

$$\begin{aligned}\delta g^{\rho\sigma} =& g^{\rho\nu}\partial_\nu \xi^\sigma + g^{\sigma\mu}\partial_\mu \xi^\rho - \xi^\mu \partial_\mu g^{\rho\sigma}\\ =& g^{\rho\nu}\partial_\nu \xi^\sigma + g^{\rho\nu}\Gamma^\sigma_{\tau\nu}\xi^\tau + g^{\mu\sigma}\partial_\mu \xi^\rho + g^{\mu\sigma}\Gamma^\rho_{\mu\tau}\xi^\tau,\end{aligned} \tag{4.20}$$

where we applied the compatibility of the covariant derivative with the metric tensor (1.32) in order to write $\partial_\mu g^{\rho\sigma}$ as a combination of the Christoffel symbols. We easily recognize in the last line of (4.20) two covariant derivatives applied to the infinitesimal parameters ξ^μ

$$\delta g^{\rho\sigma} = g^{\rho\mu}\nabla_\mu \xi^\sigma + g^{\mu\sigma}\nabla_\mu \xi^\rho = 2g^{\mu(\rho}\nabla_\mu \xi^{\sigma)}. \tag{4.21}$$

Here, a special remark is needed: if it was possible to find a vector field ξ^μ for which the functional variation $\delta g^{\mu\nu}$ vanishes this would represent, at each point of the spacetime, a stationary direction (for example such a direction for time independent metrics is the time direction, or for spherically symmetric metrics the angular direction). In order to determine this special vector field one should solve the so-called Killing equation

$$\delta g^{\rho\sigma} = 2g^{\mu(\rho}\nabla_\mu \xi^{\sigma)} = 0. \tag{4.22}$$

Resuming our analysis of the stress-energy tensor we now plug in (4.13) the result (4.21) obtaining

$$\begin{aligned}\delta_g S_\phi[\phi, g_{\mu\nu}] =& \int_{\mathcal{M}} d^4x\sqrt{-g}T_{\mu\nu}\delta g^{\mu\nu} = 2\int_{\mathcal{M}} d^4x\sqrt{-g}T_{\mu\nu}\nabla_\rho(g^{\rho\mu}\xi^\nu)\\ =& 2\int_{\mathcal{M}} d^4x\sqrt{-g}\nabla_\rho(T_{\mu\nu}g^{\rho\mu}\xi^\nu) - 2\int_{\mathcal{M}} d^4x\sqrt{-g}\nabla_\rho(g^{\rho\mu}T_{\mu\nu})\xi^\nu,\end{aligned} \tag{4.23}$$

where we removed the symmetrization since the stress-energy tensor is symmetric by construction. The boundary term vanishes since the condition $\delta g^{\mu\nu}|_{\partial\mathcal{M}} = 0$ implies $\xi^{\mu}|_{\partial\mathcal{M}} = 0$. Hence, the stationarity of the action under diffeomorphisms implies that the covariant divergence of the stress-energy tensor vanishes

$$\nabla_{\rho} T^{\rho}{}_{\nu} = \nabla_{\rho} \left(g^{\rho\mu} T_{\mu\nu} \right) = 0, \tag{4.24}$$

just as the stationarity of the action of a relativistic field theory under spacetime translations leads to the conservation of the subsequent non-symmetric stress-energy tensor.

We finally consider an action functional composed by the Einstein-Hilbert (4.2) and the matter (4.12) ones which we write as

$$S\left[g_{\mu\nu}, \phi\right] = \frac{c^3}{16\pi G} \int d^4x \left[\sqrt{-g} R - \frac{8\pi G}{c^4} \mathcal{L}_{\phi} \right], \tag{4.25}$$

for which the stationarity condition under arbitrary variations of the inverse metric tensor leads to Einstein's field equations coupled to matter

$$R_{\mu\nu} - \frac{1}{2} g_{\mu\nu} R = \frac{8\pi G}{c^4} T_{\mu\nu}. \tag{4.26}$$

Keeping in mind this way of coupling external fields to the metric tensor we now continue discussing, for the sake of clearness, just the vacuum case.

Indeed the result (4.24) could have been foreseen since also vacuum Einstein's equations have a vanishing covariant divergence because of Bianchi identity (1.96).

4.1.3 $\Gamma\Gamma$ *formulation*

As we studied for the electromagnetic field in the previous chapter the covariance of the action functional is a sufficient but not a necessary condition for equations of motion to be covariant. It is possible to determine a non-covariant Lagrangian density starting from the Einstein-Hilbert one for which no troubling boundary terms emerge when imposing the stationarity condition. More technically, it is possible to write a Lagrangian density which does not contain any second order derivative of the metric tensor, thus a non-covariant one, which leads to a stationarity condition that holds without any specific hypothesis on the manifold. However, being non-covariant, one has to deal with a greater difficulty in the calculations. Let us write down more carefully the Einstein-Hilbert Lagrangian density

$$\begin{aligned}
\mathcal{L}_{EH} =&\sqrt{-g}R = \sqrt{-g}g^{\mu\nu}R_{\mu\nu} \\
=&\sqrt{-g}g^{\mu\nu}\left(\partial_\rho\Gamma^\rho_{\mu\nu} - \partial_\nu\Gamma^\rho_{\mu\rho} + \Gamma^\rho_{\mu\nu}\Gamma^\sigma_{\rho\sigma} - \Gamma^\sigma_{\mu\rho}\Gamma^\rho_{\nu\sigma}\right) \\
=&\sqrt{-g}g^{\mu\nu}\partial_\rho\Gamma^\rho_{\mu\nu} - \sqrt{-g}g^{\mu\nu}\partial_\nu\Gamma^\rho_{\mu\rho} + \sqrt{-g}g^{\mu\nu}\left(\Gamma^\rho_{\mu\nu}\Gamma^\sigma_{\rho\sigma} - \Gamma^\sigma_{\mu\rho}\Gamma^\rho_{\nu\sigma}\right).
\end{aligned} \tag{4.27}$$

From the first two terms of the last line of (4.27) we can isolate a boundary term and two terms proportional to the Christoffel symbol

$$\begin{aligned}
\sqrt{-g}g^{\mu\nu}\partial_\rho\Gamma^\rho_{\mu\nu} &= \partial_\rho\left(\sqrt{-g}g^{\mu\nu}\Gamma^\rho_{\mu\nu}\right) - \Gamma^\rho_{\mu\nu}\partial_\rho\left(\sqrt{-g}g^{\mu\nu}\right), \\
\sqrt{-g}g^{\mu\nu}\partial_\mu\Gamma^\rho_{\nu\rho} &= \partial_\nu\left(\sqrt{-g}g^{\mu\nu}\Gamma^\rho_{\mu\rho}\right) - \Gamma^\rho_{\mu\rho}\partial_\nu\left(\sqrt{-g}g^{\mu\nu}\right).
\end{aligned} \tag{4.28}$$

Hence, subtracting the two ordinary derivatives in (4.28) we obtain the non-variational boundary term of the Einstein-Hilbert Lagrangian density

$$\partial_\mu\boldsymbol{\mathcal{EH}}^\mu = \partial_\mu\left[\sqrt{-g}\left(g^{\sigma\rho}\Gamma^\mu_{\sigma\rho} - g^{\mu\nu}\Gamma^\rho_{\nu\rho}\right)\right], \tag{4.29}$$

which does not reduce under variation to (4.9):

$$\delta\left(\partial_\mu\boldsymbol{\mathcal{EH}}^\mu\right) \neq \partial_\mu\delta\boldsymbol{\mathcal{EH}}^\mu. \tag{4.30}$$

We notice that the term $\boldsymbol{\mathcal{EH}}^\mu$ is not a vector but it behaves as such under variation. Now, let us focus on the two terms proportional to the Christoffel symbols in (4.28): for the first one we have

$$\begin{aligned}
\Gamma^\rho_{\mu\nu}\partial_\rho\left(\sqrt{-g}g^{\mu\nu}\right) =&\sqrt{-g}\Gamma^\rho_{\mu\nu}\left(\partial_\rho g^{\mu\nu} + \frac{1}{2}g^{\sigma\tau}\partial_\rho g_{\sigma\tau}g^{\mu\nu}\right) \\
=&\sqrt{-g}\Gamma^\rho_{\mu\nu}\left(-g^{\mu\sigma}g^{\nu\tau}\partial_\rho g_{\sigma\tau} + \Gamma^\sigma_{\rho\sigma}g^{\mu\nu}\right) \\
=&\sqrt{-g}\left(\Gamma^\rho_{\mu\nu}\Gamma^\sigma_{\rho\sigma}g^{\mu\nu} - 2\Gamma^\rho_{\mu\nu}\Gamma^\mu_{\rho\sigma}g^{\sigma\nu}\right),
\end{aligned} \tag{4.31}$$

where in the first line we used $\partial_\rho g^{\mu\nu} = -g^{\mu\sigma}g^{\nu\tau}\partial_\rho g_{\sigma\tau}$ which directly follows from $\partial_\rho\left(g^{\mu\sigma}g_{\sigma\nu}\right) = 0$, and in the second line we used Ricci theorem (1.32) in order to rewrite the ordinary derivative of the metric tensor via the Christoffel symbols. Following the same approach we obtain for the second term proportional to the Christoffel symbol in (4.28)

$$\Gamma^\rho_{\mu\rho}\partial_\nu\left(\sqrt{-g}g^{\mu\nu}\right) = -\sqrt{-g}\Gamma^\rho_{\mu\rho}\Gamma^\mu_{\nu\sigma}g^{\nu\sigma}. \tag{4.32}$$

Now we can subtract the results of (4.32) and (4.31) and after relabeling dummy indices we get

$$\Gamma^\rho_{\mu\rho}\partial_\nu\left(\sqrt{-g}g^{\mu\nu}\right) - \Gamma^\rho_{\mu\nu}\partial_\rho\left(\sqrt{-g}g^{\mu\nu}\right) = 2\sqrt{-g}g^{\mu\nu}\left(\Gamma^\sigma_{\mu\rho}\Gamma^\rho_{\nu\sigma} - \Gamma^\rho_{\mu\nu}\Gamma^\sigma_{\rho\sigma}\right). \tag{4.33}$$

We finally write down the decomposition we started to make in (4.27):

$$\begin{aligned}\mathcal{L}_{EH} =&\sqrt{-g}g^{\mu\nu}\left(\Gamma^{\rho}_{\mu\nu}\Gamma^{\sigma}_{\rho\sigma}-\Gamma^{\sigma}_{\mu\rho}\Gamma^{\rho}_{\nu\sigma}\right)+\sqrt{-g}g^{\mu\nu}\partial_{\rho}\Gamma^{\rho}_{\mu\nu}-\sqrt{-g}g^{\mu\nu}\partial_{\nu}\Gamma^{\rho}_{\mu\rho}\\ =&\sqrt{-g}g^{\mu\nu}\left(\Gamma^{\rho}_{\mu\nu}\Gamma^{\sigma}_{\rho\sigma}-\Gamma^{\sigma}_{\mu\rho}\Gamma^{\rho}_{\nu\sigma}\right)+2\sqrt{-g}g^{\mu\nu}\left(\Gamma^{\sigma}_{\mu\rho}\Gamma^{\rho}_{\nu\sigma}-\Gamma^{\rho}_{\mu\nu}\Gamma^{\sigma}_{\rho\sigma}\right)+\partial_{\mu}\boldsymbol{\mathcal{E}\mathcal{H}}^{\mu}\\ =&\sqrt{-g}g^{\mu\nu}\left(\Gamma^{\sigma}_{\mu\rho}\Gamma^{\rho}_{\nu\sigma}-\Gamma^{\rho}_{\mu\nu}\Gamma^{\sigma}_{\rho\sigma}\right)+\partial_{\mu}\boldsymbol{\mathcal{E}\mathcal{H}}^{\mu}.\end{aligned} \tag{4.34}$$

If we only consider the term bilinear in the Christoffel symbols we can define the so-called $\Gamma\Gamma$ *Lagrangian density*:

$$\mathcal{L}_{\Gamma\Gamma}=\sqrt{-g}g^{\mu\nu}\left(\Gamma^{\sigma}_{\mu\rho}\Gamma^{\rho}_{\nu\sigma}-\Gamma^{\rho}_{\mu\nu}\Gamma^{\sigma}_{\rho\sigma}\right), \tag{4.35}$$

which is the first one ever proposed by Einstein in general coordinate systems [118]. Now, the derivation of the equations of motion is a bit more convoluted than how it was for the previous action functional (4.2), since this Lagrangian density is not covariant.

We now derive the equations of motion giving some milestones along the way. Let us write the variation of the action functional defined by the Lagrangian density (4.35)

$$\begin{aligned}\delta S_{\Gamma\Gamma}\left[g_{\mu\nu}\right]=&\int_{\mathcal{M}}d^4x\sqrt{-g}\left(\Gamma^{\sigma}_{\mu\rho}\Gamma^{\rho}_{\nu\sigma}-\Gamma^{\rho}_{\mu\nu}\Gamma^{\sigma}_{\rho\sigma}\right)\delta g^{\mu\nu}\\ &-\int_{\mathcal{M}}d^4x\sqrt{-g}\left(\Gamma^{\sigma}_{\tau\rho}\Gamma^{\rho}_{\pi\sigma}-\Gamma^{\rho}_{\tau\pi}\Gamma^{\sigma}_{\rho\sigma}\right)g^{\tau\pi}\frac{1}{2}g_{\mu\nu}\delta g^{\mu\nu}\\ &+\int_{\mathcal{M}}d^4x\sqrt{-g}\left(2\Gamma^{\sigma}_{\mu\rho}\delta\Gamma^{\rho}_{\nu\sigma}-\delta\Gamma^{\rho}_{\mu\nu}\Gamma^{\sigma}_{\rho\sigma}-\Gamma^{\rho}_{\mu\nu}\delta\Gamma^{\sigma}_{\rho\sigma}\right)g^{\mu\nu}.\end{aligned} \tag{4.36}$$

Now we need to compute the Christoffel symbols variations. Let us begin by writing explicitly the variation

$$\begin{aligned}\delta\Gamma^{\rho}_{\mu\nu}=&\frac{1}{2}\delta g^{\rho\sigma}\left(\partial_{\mu}g_{\nu\sigma}+\partial_{\nu}g_{\sigma\mu}-\partial_{\sigma}g_{\mu\nu}\right)\\ &+\frac{1}{2}g^{\rho\sigma}\left(\partial_{\mu}\delta g_{\nu\sigma}+\partial_{\nu}\delta g_{\sigma\mu}-\partial_{\sigma}\delta g_{\mu\nu}\right).\end{aligned} \tag{4.37}$$

One needs to write the variations with respect to the same type of metric tensor which we choose to be the direct metric $g_{\mu\nu}$. Moreover we expect the variation of the Christoffel symbols to be tensorial quantities. One can easily figure this out thinking about the transformation law for the Christoffel symbols: an infinitesimal functional variation would transform as a tensor since the terms in the second derivatives of the coordinate transformation would cancel out. In order to write this variation in a manifestly covariant form we just need to rewrite the variation $\delta g^{\delta\gamma}$ as $\delta g^{\delta\gamma}=-g^{\mu\delta}g^{\nu\gamma}\delta g_{\mu\nu}$, to transform the derivatives of the first line of (4.37) via the compatibility condition (1.32) and to substitute the ordinary derivatives of the second line via covariant derivatives and the Christoffel symbols. Preserving the order of the lines of (4.37) one obtains

$$\begin{aligned}\delta\Gamma^\rho_{\mu\nu} =& -\Gamma^\tau_{\mu\nu} g^{\rho\sigma}\delta g_{\tau\sigma} \\ &+ \frac{1}{2} g^{\rho\sigma}\left(\nabla_\mu \delta g_{\nu\sigma} + \nabla_\nu \delta g_{\sigma\mu} - \nabla_\sigma \delta g_{\mu\nu} + 2\Gamma^\tau_{\mu\nu}\delta g_{\tau\sigma}\right),\end{aligned} \tag{4.38}$$

which reduces to

$$\delta\Gamma^\rho_{\mu\nu} = \frac{1}{2} g^{\rho\sigma}\left(\nabla_\mu \delta g_{\nu\sigma} + \nabla_\nu \delta g_{\sigma\mu} - \nabla_\sigma \delta g_{\mu\nu}\right). \tag{4.39}$$

Expression (4.39) is clearly covariant. Let us now recover the calculations for the stationarity condition of the $\Gamma\Gamma$ action functional. Taking the first term of the last line of (4.36) we write using (4.39)

$$\begin{aligned}&2\sqrt{-g} g^{\mu\nu}\Gamma^\sigma_{\mu\rho}\delta\Gamma^\rho_{\nu\sigma} \\ =&\sqrt{-g} g^{\mu\nu}\Gamma^\sigma_{\mu\rho} g^{\rho\tau}\left(\nabla_\nu \delta g_{\sigma\tau} + \nabla_\sigma \delta g_{\tau\nu} - \nabla_\tau \delta g_{\nu\sigma}\right) \\ =&\sqrt{-g}\Gamma^\sigma_{\mu\rho}\left(-g^{\mu\nu} g_{\sigma\tau}\nabla_\nu \delta g^{\rho\tau} - g^{\mu\nu} g_{\tau\nu}\nabla_\sigma \delta g^{\rho\tau} + g^{\rho\tau} g_{\nu\sigma}\nabla_\tau \delta g^{\mu\nu}\right)\end{aligned} \tag{4.40}$$

where we used the property $\delta\left(g_{\sigma\tau} g^{\rho\tau}\right) = 0$. At this point we cannot just execute derivations by part since, being $\Gamma^\rho_{\nu\sigma}$ a pseudo-tensor, we do not know how to apply the covariant derivative on it. Thus we need to express the covariant derivatives in terms of ordinary derivatives and the Christoffel symbols. As an example let us take the first term of (4.40)

$$\begin{aligned}&-\sqrt{-g}\Gamma^\sigma_{\mu\rho} g^{\mu\nu} g_{\sigma\tau}\nabla_\nu \delta g^{\rho\tau} \\ =& -\sqrt{-g}\Gamma^\sigma_{\mu\rho} g^{\mu\nu} g_{\sigma\tau}\left(\partial_\nu \delta g^{\rho\tau} + \Gamma^\rho_{\pi\lambda}\delta g^{\pi\tau} + \Gamma^\tau_{\nu\pi}\delta g^{\rho\pi}\right).\end{aligned} \tag{4.41}$$

Now, we can perform the derivations by part since we have an expression with ordinary derivatives and we are no longer linked to the geometrical meaning of the expression. The best one can do is to repeat the same procedure for the other terms in (4.40) and find similar terms before executing derivations by part which would rapidly increase the number of terms. Once all similar terms have been recognized and erased one obtains

$$2\sqrt{-g} g^{\mu\nu}\Gamma^\sigma_{\mu\rho}\delta\Gamma^\rho_{\nu\sigma} = -\sqrt{-g}\left(\Gamma^\rho_{\mu\nu}\partial_\rho \delta g^{\mu\nu} + 2\Gamma^\sigma_{\mu\rho}\Gamma^\rho_{\sigma\nu}\delta g^{\mu\nu}\right), \tag{4.42}$$

which is a fair reduction of the previous expression. At this stage we perform the derivations by parts obtaining at last

$$\begin{aligned}2\sqrt{-g} g^{\mu\nu}\Gamma^\sigma_{\mu\rho}\delta\Gamma^\rho_{\nu\sigma} =&\sqrt{-g}\left(\partial_\sigma \Gamma^\sigma_{\mu\nu} + \Gamma^\sigma_{\rho\sigma}\Gamma^\rho_{\mu\nu} - 2\Gamma^\sigma_{\mu\rho}\Gamma^\rho_{\nu\sigma}\right)\delta g^{\mu\nu} \\ &-\partial_\mu\left(\sqrt{-g}\Gamma^\mu_{\rho\sigma}\delta g^{\rho\sigma}\right),\end{aligned} \tag{4.43}$$

where in the first line we recognize terms which will appear in $\sqrt{-g}R_{\mu\nu}\delta g^{\mu\nu}$ (with the right sign). We report now the result of the rather long, but completely similar, calculations for the last two terms of the last line of (4.36)

$$\begin{aligned}
-\sqrt{-g}g^{\mu\nu}\delta\Gamma^{\rho}_{\mu\nu}\Gamma^{\sigma}_{\rho\sigma} =& \sqrt{-g}\left(\frac{1}{2}g_{\mu\nu}g^{\tau\rho}\partial_{\tau}\Gamma^{\sigma}_{\rho\sigma} - \partial_{\mu}\Gamma^{\sigma}_{\nu\sigma}\right)\delta g^{\mu\nu} \\
&+\sqrt{-g}\left(\Gamma^{\sigma}_{\rho\sigma}\Gamma^{\rho}_{\mu\nu} - \frac{1}{2}g_{\mu\nu}g^{\pi\tau}\Gamma^{\sigma}_{\rho\sigma}\Gamma^{\rho}_{\pi\tau}\right)\delta g^{\mu\nu} \\
&+\partial_{\mu}\left[\sqrt{-g}\left(\Gamma^{\sigma}_{\rho\sigma}\delta g^{\rho\mu} - \frac{1}{2}\Gamma^{\sigma}_{\rho\sigma}g^{\rho\mu}g_{\nu\tau}\delta g^{\nu\tau}\right)\right],
\end{aligned} \tag{4.44}$$

$$\begin{aligned}
-\sqrt{-g}g^{\mu\nu}\Gamma^{\rho}_{\mu\nu}\delta\Gamma^{\sigma}_{\rho\sigma} =& \sqrt{-g}\left(g^{\pi\tau}\Gamma^{\rho}_{\pi\sigma}\Gamma^{\sigma}_{\rho\tau} - \frac{1}{2}g^{\pi\tau}\Gamma^{\sigma}_{\rho\sigma}\Gamma^{\rho}_{\pi\tau} - \frac{1}{2}g^{\pi\tau}\partial_{\rho}\Gamma^{\rho}_{\pi\tau}\right)g_{\mu\nu}\delta g^{\mu\nu} \\
&+\frac{1}{2}\partial_{\mu}\left(\sqrt{-g}\Gamma^{\mu}_{\rho\sigma}g^{\rho\sigma}g_{\nu\tau}\delta g^{\nu\tau}\right).
\end{aligned} \tag{4.45}$$

We can finally compose these three results obtaining

$$\begin{aligned}
&\sqrt{-g}\left(2\Gamma^{\sigma}_{\mu\rho}\delta\Gamma^{\rho}_{\nu\sigma} - \delta\Gamma^{\rho}_{\mu\nu}\Gamma^{\sigma}_{\rho\sigma} - \Gamma^{\rho}_{\mu\nu}\delta\Gamma^{\sigma}_{\rho\sigma}\right)g^{\mu\nu} \\
=&\sqrt{-g}\left(R_{\mu\nu} - \frac{1}{2}g_{\mu\nu}R\right)\delta g^{\mu\nu} \\
&+\sqrt{-g}\left(\Gamma^{\sigma}_{\rho\sigma}\Gamma^{\rho}_{\mu\nu} - \Gamma^{\sigma}_{\mu\rho}\Gamma^{\rho}_{\nu\sigma}\right)\delta g^{\mu\nu} \\
&+\sqrt{-g}\left(\Gamma^{\rho}_{\pi\sigma}\Gamma^{\sigma}_{\rho\tau} - \Gamma^{\rho}_{\pi\tau}\Gamma^{\sigma}_{\rho\sigma}\right)g^{\pi\tau}\frac{1}{2}g_{\mu\nu}\delta g^{\mu\nu} \\
&+\partial_{\mu}\left[\sqrt{-g}\left(\Gamma^{\sigma}_{\rho\sigma}\delta g^{\rho\mu} - \frac{1}{2}\Gamma^{\sigma}_{\rho\sigma}g^{\rho\mu}g_{\nu\tau}\delta g^{\nu\tau} - \Gamma^{\mu}_{\rho\sigma}\delta g^{\rho\sigma} + \frac{1}{2}\Gamma^{\mu}_{\rho\sigma}g^{\rho\sigma}g_{\nu\tau}\delta g^{\nu\tau}\right)\right].
\end{aligned} \tag{4.46}$$

We define the boundary term which arises variating the action functional as

$$\begin{aligned}
&\partial_{\mu}\boldsymbol{\delta\Gamma\Gamma}^{\mu} \\
=&\partial_{\mu}\left[\sqrt{-g}\left(\Gamma^{\sigma}_{\rho\sigma}\delta g^{\rho\mu} - \frac{1}{2}\Gamma^{\sigma}_{\rho\sigma}g^{\rho\mu}g_{\nu\tau}\delta g^{\nu\tau} - \Gamma^{\mu}_{\rho\sigma}\delta g^{\rho\sigma} + \frac{1}{2}\Gamma^{\mu}_{\rho\sigma}g^{\rho\sigma}g_{\nu\tau}\delta g^{\nu\tau}\right)\right]
\end{aligned} \tag{4.47}$$

so that we obtain for the stationarity of the action functional (4.36)

$$\delta S_{\Gamma\Gamma}\left[g_{\mu\nu}\right] = \int_{\mathcal{M}} d^4x\sqrt{-g}\left(R_{\mu\nu} - \frac{1}{2}g_{\mu\nu}R\right)\delta g^{\mu\nu} + \int_{\mathcal{M}} d^4x\partial_{\mu}\boldsymbol{\delta\Gamma\Gamma}^{\mu} = 0. \tag{4.48}$$

We immediately see that unlike (4.9) the variational boundary term (4.47) poses no problems since it does not contain any variation of the first derivatives of the metric tensor and it automatically vanishes on the boundary just as a consequence of $\delta g^{\mu\nu}|_{\partial\mathcal{M}} = 0$: there is no need to specify any peculiar property of the manifold or any asymptotic behavior of the metric tensor. Hence, we obtain once again the Einstein's field equations.

It is interesting to notice that the variation of the Einstein-Hilbert boundary term, $\partial_\mu \boldsymbol{\mathcal{EH}}^\mu$ (4.29), corresponds to the sum of the variational boundary terms of the Einstein-Hilbert Lagrangian density, $\partial_\mu \boldsymbol{\delta\mathcal{EH}}^\mu$ (4.9), and of the $\Gamma\Gamma$ one, $\partial_\mu \boldsymbol{\delta\Gamma\Gamma}^\mu$ (4.47)

$$\delta\left(\partial_\mu \boldsymbol{\mathcal{EH}}^\mu\right) = \partial_\mu \boldsymbol{\delta\mathcal{EH}}^\mu + \partial_\mu \boldsymbol{\delta\Gamma\Gamma}^\mu. \tag{4.49}$$

4.1.4 *Dirac formulation*

Here we follow the steps taken by Dirac [113] in order to obtain one of the first Hamiltonian formulations of General Relativity, although we review just the Lagrangian formulation part. Let us take as a starting point the $\Gamma\Gamma$ Lagrangian density (4.35) and write it in the usual form of other field theories: the kinetic term, quadratic in the first derivatives, is explicitly written contracted with some products of three inverse metric tensors. Indeed, after relabeling all the indices one obtains

$$\mathcal{L}_{\Gamma\Gamma} = \frac{1}{4}\sqrt{-g}\partial_\rho g_{\mu\nu}\partial_\sigma g_{\pi\tau}\left[g^{\rho\sigma}\left(g^{\pi\tau}g^{\mu\nu} - g^{\mu\pi}g^{\nu\tau}\right) + 2g^{\sigma\nu}\left(g^{\rho\tau}g^{\mu\pi} - g^{\rho\mu}g^{\pi\tau}\right)\right]. \tag{4.50}$$

Now, we separate all the terms quadratic, $\mathcal{L}_{\Gamma\Gamma}(2)$, and linear, $\mathcal{L}_{\Gamma\Gamma}(1)$, in the velocities from the homogeneous ones $\mathcal{L}_{\Gamma\Gamma}(0)$. The velocities can be calculated with respect to any coordinate x^μ so we just name the time variable as x^0 and the other three spatial coordinates labeled with Latin indices. For the quadratic part we obtain

$$\begin{aligned}
\mathcal{L}_{\Gamma\Gamma}(2) &= \frac{1}{4}\sqrt{-g}\partial_0 g_{\mu\nu}\partial_0 g_{\pi\tau}\left[g^{00}g^{\pi\tau}g^{\mu\nu} - g^{00}g^{\mu\pi}g^{\nu\tau} + 2g^{0\nu}g^{0\tau}g^{\mu\pi} - 2g^{0\nu}g^{0\mu}g^{\pi\tau}\right] \\
&= \frac{1}{4}\sqrt{-g}\partial_0 g_{\mu\nu}\partial_0 g_{\pi\tau}g^{00}\left[g^{\pi\tau}g^{\mu\nu} - g^{\mu\pi}g^{\nu\tau} + 2\frac{g^{0\nu}g^{0\tau}}{g^{00}}g^{\mu\pi} - 2\frac{g^{0\pi}g^{0\tau}}{g^{00}}g^{\mu\nu}\right] \\
&= \frac{1}{4}\sqrt{-g}\partial_0 g_{\mu\nu}\partial_0 g_{\pi\tau}g^{00}\left(\frac{g^{0\mu}g^{0\pi}}{g^{00}} - g^{\mu\pi}\right)\left(g^{\nu\tau} - \frac{g^{0\nu}g^{0\tau}}{g^{00}}\right) \\
&\quad + \frac{1}{4}\sqrt{-g}\partial_0 g_{\mu\nu}\partial_0 g_{\pi\tau}g^{00}\left(g^{\pi\tau} - \frac{g^{0\pi}g^{0\tau}}{g^{00}}\right)\left(g^{\mu\nu} - \frac{g^{0\nu}g^{0\mu}}{g^{00}}\right),
\end{aligned} \tag{4.51}$$

where from the second to the third line we relabeled some indices and we summed and subtracted the quantity $g^{0\mu}g^{0\pi}g^{0\nu}g^{0\tau}/\left(g^{00}\right)^2$. Dirac defined a symmetric rank two tensor which vanishes whenever one of the indices equals 0

$$e^{\mu\nu} = g^{\mu\nu} - \frac{g^{0\nu}g^{0\mu}}{g^{00}} \quad \Rightarrow \quad e^{0\nu} = 0, \tag{4.52}$$

through which we can write the quadratic term (4.51) in a more compact way

$$\mathcal{L}_{\Gamma\Gamma}(2) = \frac{1}{4}\sqrt{-g}\partial_0 g_{\mu\nu}\partial_0 g_{\pi\tau} g^{00}\left(e^{\mu\nu}e^{\pi\tau} - e^{\mu\pi}e^{\nu\tau}\right). \tag{4.53}$$

It is clearly visible that there are no quadratic terms in the velocities $\partial_0 g_{0\nu}$ from which the constrained nature of the Hamiltonian formulation follows. We write down the linear term

$$\begin{aligned}\mathcal{L}_{\Gamma\Gamma}(1) =& \frac{1}{2}\sqrt{-g}\partial_0 g_{\mu\nu}\partial_i g_{\pi\tau}\left(g^{0i}g^{\pi\tau}g^{\mu\nu} - g^{0i}g^{\mu\pi}g^{\nu\tau}\right) \\ &+ \frac{1}{2}\sqrt{-g}\partial_0 g_{\mu\nu}\partial_i g_{\pi\tau}\left(2g^{i\nu}g^{0\tau}g^{\mu\pi} - g^{i\nu}g^{0\mu}g^{\pi\tau} - g^{0\tau}g^{i\pi}g^{\mu\nu}\right),\end{aligned} \tag{4.54}$$

and the homogeneous one

$$\mathcal{L}_{\Gamma\Gamma}(0) = \frac{1}{4}\sqrt{-g}\partial_i g_{\mu\nu}\partial_j g_{\pi\tau}\left[g^{ij}\left(g^{\pi\tau}g^{\mu\nu} - g^{\mu\pi}g^{\nu\tau}\right) + 2g^{j\nu}\left(g^{i\tau}g^{\mu\pi} - g^{i\mu}g^{\pi\tau}\right)\right]. \tag{4.55}$$

Dirac managed to notice that all terms proportional to $\partial_0 g_{0\nu}$ in (4.54) can be deleted by a non-covariant boundary term that he found to be

$$\begin{aligned}\partial_\mu \mathcal{D}^\mu =& \partial_i\left[\frac{g^{0i}}{g^{00}}\partial_0\left(\sqrt{-g}g^{00}\right)\right] - \partial_0\left[\frac{g^{0i}}{g^{00}}\partial_i\left(\sqrt{-g}g^{00}\right)\right] \\ =& \partial_i\left(\frac{g^{0i}}{g^{00}}\right)\partial_0\left(\sqrt{-g}g^{00}\right) - \partial_0\left(\frac{g^{0i}}{g^{00}}\right)\partial_i\left(\sqrt{-g}g^{00}\right).\end{aligned} \tag{4.56}$$

Indeed, this boundary term does not contain second order derivatives of the metric tensor and it does not modify the equations of motion. We already encountered (on purpose) this kind of boundary term in the previous chapter when we modified the Lagrangian density of the electromagnetic field (see (3.69) and (3.70)). Thus, we define *Dirac Lagrangian density* as

$$\mathcal{L}_{\mathcal{D}} = \mathcal{L}_{\Gamma\Gamma} + \partial_\mu \mathcal{D}^\mu. \tag{4.57}$$

In order to compute the modification we explicitly write Dirac boundary term as

$$\begin{aligned}\partial_\mu \mathcal{D}^\mu =& \frac{1}{2}\sqrt{-g}\partial_0 g_{\mu\nu}\partial_i g_{\pi\tau}\left(g^{i\tau}g^{0\pi}g^{\mu\nu} - g^{i\nu}g^{0\mu}g^{\pi\tau}\right) \\ &+ \frac{1}{2}\sqrt{-g}\partial_0 g_{\mu\nu}\partial_i g_{\pi\tau}\left(g^{\mu\nu}g^{0\tau}g^{0\pi} - g^{\pi\tau}g^{0\nu}g^{0\mu}\right)\frac{g^{0i}}{g^{00}} \\ &+ \sqrt{-g}\partial_0 g_{\mu\nu}\partial_i g_{\pi\tau}\left(g^{i\pi}g^{0\nu} - g^{i\nu}g^{0\pi}\right)\frac{g^{0\tau}g^{0\mu}}{g^{00}},\end{aligned} \tag{4.58}$$

and we subtract it to (4.54) obtaining

$$\begin{aligned}\mathcal{L}_{\mathcal{D}}(1) =& \mathcal{L}_{\Gamma\Gamma}(1) - \partial_\mu \mathcal{D}^\mu \\ =& \frac{1}{2}\sqrt{-g}\partial_0 g_{\mu\nu}\partial_i g_{\pi\tau}\left[\left(e^{\pi\tau}e^{\mu\nu} - e^{\mu\pi}e^{\nu\tau}\right)g^{0i} + 2\left(e^{i\nu}e^{\mu\tau} - e^{i\tau}e^{\mu\nu}\right)g^{0\pi}\right].\end{aligned} \tag{4.59}$$

It appears in the previous expression (4.59) that whenever $\mu, \nu = 0$ the whole term vanishes. We can now sum (4.59) to the quadratic part in the velocities (4.53) which is unchanged and, since for a generic tensor $A_{\mu\nu}e^{\mu\nu} = A_{rs}e^{rs}$, we write

$$\begin{aligned}&\mathcal{L}_{\Gamma\Gamma}(2) + \mathcal{L}_{\mathcal{D}}^{*}(1) \\ =& \frac{1}{4}\sqrt{-g}\left(e^{rs}e^{mn} - e^{rm}e^{sn}\right) \\ \times& \left[\partial_0 g_{rs}\partial_0 g_{mn}g^{00} + 2\partial_0 g_{rs}\partial_i g_{mn}g^{0i} - 4\partial_0 g_{rs}\partial_n g_{m\sigma}g^{0\sigma}\right].\end{aligned} \tag{4.60}$$

The final steps towards Dirac Lagrangian density are the following: we write the Christoffel symbol Γ^0_{rs}

$$-2\Gamma^0_{rs} = \partial_0 g_{rs}g^{00} + \partial_i g_{rs}g^{0i} - \left(\partial_r g_{s\sigma} + \partial_s g_{\sigma r}\right)g^{0\sigma}, \tag{4.61}$$

since it is linear in the velocities $\partial_0 g_{rs}$ we wish to use it in order to write in a more compact way (4.60). Thus, we try to write the expression

$$\begin{aligned}&\frac{4}{g^{00}}\left(e^{rs}e^{mn} - e^{rm}e^{sn}\right)\Gamma^0_{rs}\Gamma^0_{mn} \\ =& \left(e^{rs}e^{mn} - e^{rm}e^{sn}\right)\left[\partial_0 g_{rs}\partial_0 g_{mn}g^{00} + 2\partial_0 g_{mn}\partial_i g_{rs}g^{0i} - 4\partial_0 g_{mn}\partial_r g_{s\sigma}g^{0\sigma}\right] \\ &+ \frac{1}{g^{00}}\left(e^{rs}e^{mn} - e^{rm}e^{sn}\right)\left[\partial_i g_{rs}g^{0i} - \left(\partial_r g_{s\sigma} + \partial_s g_{\sigma r}\right)g^{0\sigma}\right] \\ &\times \left[\partial_k g_{mn}g^{0k} - \left(\partial_m g_{n\rho} + \partial_n g_{\rho m}\right)g^{0\rho}\right],\end{aligned} \tag{4.62}$$

recognizing in the first line of (4.62) something proportional to (4.60). We write

$$\mathcal{L}_{\Gamma\Gamma}(2) + \mathcal{L}_{\mathcal{D}}^{*}(1) = \frac{\sqrt{-g}}{g^{00}}\left(e^{rs}e^{mn} - e^{rm}e^{sn}\right)\Gamma^0_{rs}\Gamma^0_{mn} + \mathcal{L}_X, \tag{4.63}$$

where $\mathcal{L}_X$ is a new homogeneous term in the velocities containing the product of the square brackets in (4.62)

$$\begin{aligned}\mathcal{L}_X =& -\frac{\sqrt{-g}}{4g^{00}}\left(e^{rs}e^{mn} - e^{rm}e^{sn}\right)\left[\partial_i g_{rs}g^{0i} - \left(\partial_r g_{s\sigma} + \partial_s g_{\sigma r}\right)g^{0\sigma}\right] \\ &\times \left[\partial_k g_{mn}g^{0k} - \left(\partial_m g_{n\rho} + \partial_n g_{\rho m}\right)g^{0\rho}\right].\end{aligned} \tag{4.64}$$

We finally write Dirac Lagrangian density as

$$\mathcal{L}_{\mathcal{D}} = \frac{\sqrt{-g}}{g^{00}} \left(e^{rs}e^{mn} - e^{rm}e^{sn}\right) \Gamma^0_{rs}\Gamma^0_{mn} + \mathcal{L}_X + \mathcal{L}_{\Gamma\Gamma}(0) . \tag{4.65}$$

We conclude this review of Dirac formulation with the very simple calculation of the momenta conjugated to $\partial_0 g_{ij}$, since in (4.65) there are no longer terms proportional to $\partial_0 g_{0\mu}$

$$p^{mn} = \frac{\partial \mathcal{L}_{\mathcal{D}}}{\partial \partial_0 g_{mn}} = \frac{c^3}{16\pi G}\sqrt{-g}\left(e^{rs}e^{mn} - e^{rm}e^{sn}\right)\Gamma^0_{rs}, \tag{4.66}$$

recovering the dimensional constant we neglected so far. We can then state that the Hamiltonian formulation of General Relativity is a constrained one and the four primary constraints of Dirac formulation are

$$\varphi^\mu = p^{0\mu} \approx 0. \tag{4.67}$$

4.1.5 *General $f(R)$ Lagrangian densities*

Let us now discuss a class of Lagrangian densities which prvide different equations of motion with respect to General Relativity. This class is expressed by an action functional of the form (see [269] for a review)

$$S\left[g_{\mu\nu}\right] = \frac{c^3}{16\pi G}\int_{\mathcal{M}} d^4x\sqrt{-g}f(R), \tag{4.68}$$

where f is a smooth function of the scalar curvature R. In order to obtain the equations of motion we impose the stationarity condition on this functional and following the same steps as for the Einstein-Hilbert Lagrangian density we obtain

$$\delta S\left[g_{\mu\nu}\right] = \frac{c^3}{16\pi G}\int_{\mathcal{M}} d^4x\sqrt{-g}\left\{f'(R)\left(R_{\mu\nu}\delta g^{\mu\nu} + g^{\mu\nu}\delta R_{\mu\nu}\right) - \frac{1}{2}g_{\mu\nu}f(R)\,\delta g^{\mu\nu}\right\}, \tag{4.69}$$

where $f'(R) = \partial f/\partial R$. As for the Einstein-Hilbert Lagrangian density the most troubling term is the one proportional to $\delta R_{\mu\nu}$. In order to factorize the metric variation we need to rewrite some derivatives. Exploiting the variation of the Christoffel symbols (4.39)

$$\begin{aligned}\sqrt{-g}f'(R)\,g^{\mu\nu}\delta R_{\mu\nu} =& \sqrt{-g}g^{\mu\nu}g^{\rho\sigma}\left[f'(R)\nabla_\rho\nabla_\mu\delta g_{\sigma\nu} - f'(R)\nabla_\mu\nabla_\nu\delta g_{\rho\sigma}\right]\\ =& \sqrt{-g}g^{\mu\nu}g^{\rho\sigma}\left[\nabla_\mu\left(f'(R)\nabla_\rho\delta g_{\sigma\nu} - f'(R)\nabla_\nu\delta g_{\rho\sigma}\right)\right]\\ &-\sqrt{-g}g^{\mu\nu}g^{\rho\sigma}\left[\nabla_\mu\left(\nabla_\rho f'(R)\,\delta g_{\sigma\nu} - \nabla_\nu f'(R)\,\delta g_{\rho\sigma}\right)\right]\\ &+\sqrt{-g}g^{\mu\nu}g^{\rho\sigma}\left[\nabla_\mu\nabla_\rho f'(R)\,\delta g_{\sigma\nu} - \nabla_\mu\nabla_\nu f'(R)\,\delta g_{\rho\sigma}\right],\end{aligned} \tag{4.70}$$

where we used the symmetry of the metric tensor and some relabeling of indices. Since we want to factorize the variations of the inverse metric tensor we exploit the relation $\delta\left(g_{\rho\mu}g^{\mu\nu}\right)=\delta\left(\delta^{\nu}_{\rho}\right)=0$ and write

$$\begin{aligned}\sqrt{-g}f'(R)\,g^{\mu\nu}\delta R_{\mu\nu} =&\sqrt{-g}\nabla_{\mu}\left(g^{\mu\nu}g_{\rho\sigma}f'(R)\,\nabla_{\nu}\delta g^{\rho\sigma}-f'(R)\,\nabla_{\sigma}\delta g^{\mu\sigma}\right)\\ &-\sqrt{-g}\nabla_{\mu}\left(g^{\mu\nu}g_{\rho\sigma}\nabla_{\nu}f'(R)\,\delta g^{\rho\sigma}-\nabla_{\sigma}f'(R)\,\delta g^{\mu\sigma}\right)\\ &+\sqrt{-g}\left(g_{\rho\sigma}g^{\mu\nu}\nabla_{\mu}\nabla_{\nu}f'(R)-\nabla_{\rho}\nabla_{\sigma}f'(R)\right)\delta g^{\rho\sigma}.\end{aligned} \tag{4.71}$$

Now we recognize in the first two lines a covariant divergence since all quantities are covariant. Recalling the property of the covariant divergence (1.76) we define the variational boundary term

$$\begin{aligned}\partial_{\mu}\boldsymbol{\delta\mathcal{FR}}^{\mu} =&\partial_{\mu}\left[\sqrt{-g}\left(g^{\mu\nu}g_{\rho\sigma}f'(R)\,\nabla_{\nu}\delta g^{\rho\sigma}-f'(R)\,\nabla_{\sigma}\delta g^{\mu\sigma}\right)\right]\\ &-\partial_{\mu}\left[\sqrt{-g}\left(g^{\mu\nu}g_{\rho\sigma}\nabla_{\nu}f'(R)\,\delta g^{\rho\sigma}-\nabla_{\sigma}f'(R)\,\delta g^{\mu\sigma}\right)\right].\end{aligned} \tag{4.72}$$

After relabeling the indices we write (4.70) as

$$\begin{aligned}\sqrt{-g}f'(R)\,g^{\mu\nu}\delta R_{\mu\nu} =&\sqrt{-g}\left[g_{\mu\nu}g^{\rho\sigma}\nabla_{\rho}\nabla_{\sigma}f'(R)-\nabla_{\mu}\nabla_{\nu}f'(R)\right]\delta g^{\mu\nu}\\ &+\partial_{\mu}\boldsymbol{\delta\mathcal{FR}}^{\mu}.\end{aligned} \tag{4.73}$$

Finally we obtain the complete variation of the action (4.69) using (4.73)

$$\begin{aligned}\delta S\left[g_{\mu\nu}\right] =&\frac{c^{3}}{16\pi G}\int_{\mathcal{M}}d^{4}x\sqrt{-g}\left[f'(R)\,R_{\mu\nu}-\frac{1}{2}g_{\mu\nu}f(R)\right]\delta g^{\mu\nu}\\ &+\frac{c^{3}}{16\pi G}\int_{\mathcal{M}}d^{4}x\sqrt{-g}\left[g_{\mu\nu}g^{\rho\sigma}\nabla_{\rho}\nabla_{\sigma}f'(R)-\nabla_{\mu}\nabla_{\nu}f'(R)\right]\delta g^{\mu\nu}\\ &+\frac{c^{3}}{16\pi G}\int_{\mathcal{M}}d^{4}x\partial_{\mu}\boldsymbol{\delta\mathcal{FR}}^{\mu}.\end{aligned} \tag{4.74}$$

Once again, the condition $\delta g^{\mu\nu}|_{\partial\mathcal{M}}=0$ is not sufficient for the boundary term (4.72) to vanish since it contains derivatives of the variated metric tensor $\delta g^{\mu\nu}$. Hence, some other hypotheses on $\partial\mathcal{M}$ must be considered. Once the boundary term has been handled, it is easy to read that the stationarity condition $\delta S\left[g_{\mu\nu}\right]=0$ implies the generalized vacuum field equations

$$f'(R)\,R_{\mu\nu}-\frac{1}{2}g_{\mu\nu}f(R)+g_{\mu\nu}g^{\rho\sigma}\nabla_{\rho}\nabla_{\sigma}f'(R)-\nabla_{\mu}\nabla_{\nu}f'(R)=0. \tag{4.75}$$

Why should we be interested in this kind of field equations? Indeed, there are some behaviors both at a galactic and at a cosmological scale which are usually interpreted as evidence of dark matter or dark energy (see section 2.4). It is possible to write the generic function $f(R)$ as a power series in R, ranging from both negative, for large scale corrections, and positive powers, which become relevant in a quantum setting.

These models are equivalent to General Relativity plus additional fields [206]; in fact by the following conformal rescaling of the metric tensor

$$g_{\mu\nu} \to f'(R)g_{\mu\nu} = e^{\sqrt{2/3}\varphi}g_{\mu\nu}, \tag{4.76}$$

the Lagrangian density becomes

$$\mathcal{L} = \frac{c^3}{16\pi G}(R - g^{\mu\nu}\partial_\mu\varphi\partial_\nu\varphi - V(\varphi)), \tag{4.77}$$

where the potential V is the only relic of the function f and it reads

$$V = e^{-\sqrt{2/3}\varphi}R(e^{-\sqrt{2/3}\varphi}g_{\mu\nu}) - e^{-2\sqrt{2/3}\varphi}f(R(e^{-\sqrt{2/3}\varphi}g_{\mu\nu})). \tag{4.78}$$

As for the expression of f there are plenty of different proposals in literature, from $\frac{1}{R}$ terms, which are able to explain dark energy [292](but these models suffer of instabilities, while the Newtonian limit [116] and the evolution of scalar cosmological perturbations [47] are not properly reproduced), to Lagrangian $R + \alpha R^2$ with α relevant in the early universe dynamics. Also non-analytical and non-local f have been considered. However, no $f(R)$ theory exists, up to now, which is able to pass all experimental tests [12] and to predict a sensible deviation from General Relativity.

4.1.6 *Palatini formulation*

We have discussed in the previous sections how we must add some boundary terms to the Einstein-Hilbert action in order to achieve a formulation with first order derivatives only. However, it is possible to see the Einstein-Hilbert action as a first order action by itself (*i.e.* as containing only first order derivatives) by promoting the Levi-Civita connections $\Gamma^\rho_{\mu\nu}$ to be independent variables. This is the so-called *Palatini formulation.* Hence, let us perform the variation of the action with respect to independent variation of $\Gamma^\rho_{\mu\nu}$ and $g_{\mu\nu}$, so getting two contributions:

$$\delta S = \frac{c^3}{16\pi G}\int\left\{\frac{\delta[\sqrt{-g}R(g,\Gamma)]}{\delta g_{\mu\nu}}\delta g_{\mu\nu} + \frac{\delta[\sqrt{-g}R(g,\Gamma)]}{\delta\Gamma^\rho_{\mu\nu}}\delta\Gamma^\rho_{\mu\nu}\right\}d^4x, \tag{4.79}$$

and thus two sets of equations of motion

$$\frac{\delta[\sqrt{-g}R(g,\Gamma)]}{\delta\Gamma^\rho_{\mu\nu}} = \sqrt{-g}g^{\sigma\tau}\frac{\delta R_{\sigma\tau}}{\delta\Gamma^\rho_{\mu\nu}} = 0, \tag{4.80}$$

$$\frac{\delta[\sqrt{-g}R(g,\Gamma)]}{\delta g_{\mu\nu}} = -\sqrt{-g}\left[R_{\mu\nu}(\Gamma) - \frac{1}{2}g_{\mu\nu}g^{\rho\sigma}R_{\rho\sigma}(\Gamma)\right] = 0. \tag{4.81}$$

In the absence of matter this formulation is completely equivalent to GR. In fact, it can be shown that equation (4.80) imply that connections are equal to Christoffel symbols, $\Gamma^{\rho}_{\mu\nu} = \Gamma^{\rho}_{\mu\nu}(g)$ and by substituting their expression into (4.81) Einstein's equations come out.

In the presence of matter, the Palatini formulation gives different results with respect to the standard metric formulation only if the matter Lagrangian contains the connections.

The fact that connections are not fundamental fields in General Relativity suggests that a second order formulation is the right one. However, we can enrich the affine structure of spacetime, such that the connections become independent from the metric. A possibility is to introduce *torsion*. Torsion is defined as the antisymmetric part of the Levi-Civita connections

$$T^{\rho}_{\mu\nu} = \frac{1}{2}(\Gamma^{\rho}_{\mu\nu} - \Gamma^{\rho}_{\nu\mu}), \tag{4.82}$$

and it is a tensor.

If torsion is present, the connections differ from Christoffel symbol, since we get (by assuming (1.32))

$$\Gamma^{\rho}_{\mu\nu} = \{^{\rho}_{\mu\nu}\} - K^{\rho}_{\mu\nu}, \qquad K^{\rho}_{\mu\nu} = -\frac{1}{2}(T^{\rho}_{\mu\nu} - T_{\mu\nu}{}^{\rho} - T_{\nu\mu}{}^{\rho}), \tag{4.83}$$

$K^{\rho}_{\mu\nu}$ being the contortion tensor. The previous relation clarifies that the introduction of torsion provides new degrees of freedom, such that connections are no longer determined only by the metric tensor. Moreover, we want to stress that, in general, a modification is produced in the symmetric part of the Levi-Civita connection, too.

From a geometrical point of view, the presence of torsion implies that infinitesimal parallelograms do not close.

An objection against the Palatini formulation is the use of non-tensorial variables, thus of quantities with an unclear geometrical meaning, which can be made to vanish by a diffeomorphism in any given spacetime point.
Therefore, it looks more appropriate to work with other variables containing first-order derivatives of the metric and transforming as tensors. This is the case for a formulation based on Lorentz connections $\omega^{\alpha\beta}_{\mu}$ and on vierbein vectors e^{α}_{μ}.
The action (4.2) can be rewritten via (1.134) as

$$S_G = -\frac{c^3}{16\pi G}\int d^4x\, e\, e^{\mu}_{\alpha}\, e^{\nu}_{\beta}\, R^{\alpha\beta}_{\mu\nu}. \tag{4.84}$$

A Palatini-like framework in terms of $(e^{\mu}_{\alpha}, \omega^{\alpha\beta}_{\mu})$ is equivalent to the metric formulation in vacuum. In fact, the vanishing of the variation of the action (4.84) with respect to $\omega_{\mu}^{\alpha\beta}$ gives

$$\partial_{[\mu}e^{\alpha}_{\nu]} - \omega^{\alpha\gamma}_{[\mu}e_{\nu]\gamma} = 0, \tag{4.85}$$

whose solution is the condition (1.124). In fact, $\omega_{\mu}^{\alpha\beta}$ must be some vectors containing first derivatives of vierbein, while, from the properties of the Riemann tensor, it follows that they have to be antisymmetric in indices α and β. The only expression satisfying such requests is given by (1.124), hence by spin connections.

This is equivalent to the torsionless condition. In fact, by substituting the expression for $\omega_{\mu}^{\alpha\beta}$ in (4.85), the torsion vanishes, *i.e.*

$$\partial_{[\mu}e^{\alpha}_{\nu]} - e^{\gamma\rho}(\nabla_{[\mu}e^{\alpha}_{\rho})e_{\nu]\gamma} = \Gamma^{\rho}_{[\mu\nu]}e^{\alpha}_{\rho} = 0. \tag{4.86}$$

The remaining equations of motion (obtained by varying (4.84) with respect to e^{α}_{μ}) reduce to Einstein's ones.

Generically, in a Palatini formulation in terms of vierbein and spin connections, torsion is present if the right-hand side of (4.85) does not vanish. This is the case for spinors (see section 7.2.1).

4.2 ADM formalism

In order to infer a dynamic description for the gravitational field a proper time parameter has to be fixed with respect to which the evolution occurs. This point seems to conflict with the request of General Covariance, but this is not the case since by a formal splitting of the spacetime a diffeomorphisms-invariant dynamic treatment can be achieved (*ADM splitting*, named after Arnowitt, Deser and Misner [14–16], see [17] for a review).

Such a splitting works in a global hyperbolic spacetime, *i.e.* a manifold whose metric tensor can be inferred by the evolution backwards and forward of initial conditions on a particular hypersurface (Cauchy hypersurface). Global hyperbolicity is a fundamental requirement, because it ensures the possibility of having a well-posed initial-value formulation for the gravitational field. Geroch demonstrated that a global hyperbolic spacetime is diffeomorphic to a manifold $\Sigma_{x^0} \otimes \mathbb{R}$, where Σ_{x^0} denotes the hypersurfaces of equal time [140].

The identification of spatial hypersurfaces Σ_{x^0}, giving a slicing of the full spacetime, simply requires the introduction of a time-like vector field η; in fact, the Frobenius theorem ensures that, under very general assumptions, vectors orthogonal to η globally define a sub-manifold. Our goal is to introduce the new ADM variables and define all the needed quantities in order to write the Einstein-Hilbert Lagrangian density.

4.2.1 *Spacetime foliation and extrinsic curvature*

Let us define the family of hypersurfaces each of which is identified by the parameter x^0 through the parametric equations

$$u^\mu = u^\mu\left(x^i; x^0\right), \qquad i = 1, 2, 3. \tag{4.87}$$

Given the set of vectors $\left\{\vec{f}_\mu\right\}$ as a basis defined over the entire manifold we can define a new basis which is tangent to one hypersurface at fixed x^0 the same way we did in the first chapter for an embedded manifold

$$\vec{b}_i = \frac{\partial u^\mu}{\partial x^i}\vec{f}_\mu. \tag{4.88}$$

This basis can be completed using $\vec{\eta} = \eta^\mu \vec{f}_\mu$ as a fourth vector orthogonal to $\left\{\vec{b}_i\right\}$ by definition, *i.e.* $\vec{\eta}\cdot\vec{b}_i = 0$.

We define now the so-called *deformation vector* $\vec{b}_0$ which can be projected on the complete basis $\left\{\vec{b}_i, \vec{\eta}\right\}$

$$\vec{b}_0 = \frac{\partial u^\mu}{\partial x^0}\vec{f}_\mu = N\vec{\eta} + N^i\vec{b}_i, \tag{4.89}$$

where N and N^i are respectively called *lapse function* and *shift vector* (see figure 4.1). We now impose that the normal vector be normalized to one and time-like, *i.e.* $\vec{\eta}\cdot\vec{\eta} = -1$, so that the hypersurfaces are space-like. Thus, we have defined a new basis, $\left\{\vec{b}_i, \vec{b}_0\right\}$, which we can use in order to write down the metric tensor as

$$\begin{aligned}
g_{ij} =& \vec{b}_i\cdot\vec{b}_j = h_{ij},\\
g_{0i} =& \vec{b}_0\cdot\vec{b}_i = \left(N\vec{\eta} + N^k\vec{b}_k\right)\cdot\vec{b}_i = N^k h_{ki},\\
g_{00} =& \vec{b}_0\cdot\vec{b}_0 = \left(N\vec{\eta} + N^i\vec{b}_i\right)\cdot\left(N\vec{\eta} + N^j\vec{b}_j\right)\\
=& -N^2 + N^iN^jh_{ij},
\end{aligned} \tag{4.90}$$

we then write the direct and inverse metric tensors as

$$\begin{aligned}
g_{00} &= -N^2 + N^iN^jh_{ij}, \quad g_{0i} = N^kh_{ki}, \quad g_{ij} = h_{ij},\\
g^{00} &= -\frac{1}{N^2}, \quad g^{0i} = \frac{N^i}{N^2}, \quad g^{ij} = h^{ij} - \frac{N^iN^j}{N^2},
\end{aligned} \tag{4.91}$$

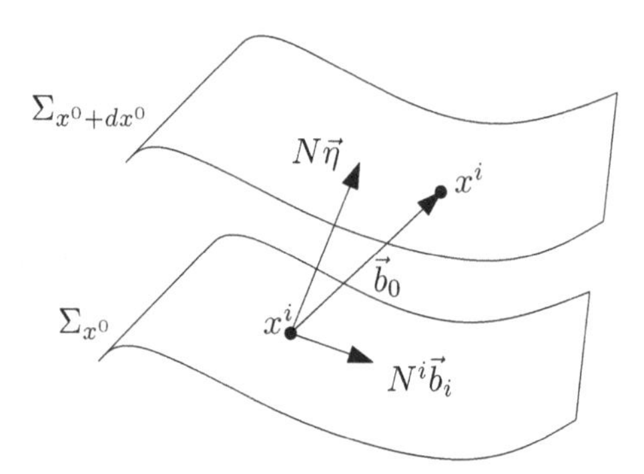

Figure 4.1 ADM foliation of the spacetime manifold by means of space-like hypersurfaces Σ_{x^0} where x^0 is the chosen time variable. The deformation vector $\vec{b}_0$, connecting the points with the same coordinates x^i of two slices infinitesimally closed, has non-vanishing projections on the normal vector $\vec{\eta}$, *i.e.* the lapse function N, and on the tangent basis vectors $\vec{b}_i$, *i.e.* the shift vector N^i. This happens because of the intrinsic and extrinsic curvature properties of the different hypersurfaces.

and we recognize h_{ij} as the induce metric on the foliation hypersurface, and h^{ij} as its inverse. We also write, for the sake of completeness, the inverse transformation for ADM variables as functions of the metric tensor

$$N = \frac{1}{\sqrt{-g^{00}}}, \qquad N^k = -\frac{g^{0k}}{g^{00}}, \qquad h_{ij} = g_{ij}. \tag{4.92}$$

We notice here that we must fix the sign of N and we choose it to be positive and that of g^{00} to be negative for the ADM transformation to be invertible. This restriction reflects the choice of a foliation that evolves towards the future. At this point we can calculate the determinant of the metric tensor which will be needed for the Lagrangian density. As a first step we diagonalize the metric tensor writing the four-dimensional line element ds^2 as

$$\begin{aligned} ds^2 &= \left(-N^2 + N^r N^s h_{rs}\right) dx^0 dx^0 + 2N^k h_{ki} dx^0 dx^i + h_{ij} dx^i dx^j \\ &= -\,N^2 dy^0 dy^0 + h_{ij} dy^i dy^j , \end{aligned} \tag{4.93}$$

so that the transformed metric tensor reads

$$g'_{\mu\nu} = \begin{pmatrix} -N^2 & 0 \\ 0 & h_{ij} \end{pmatrix}, \tag{4.94}$$

whose determinant clearly is $g' = -N^2 h$. In order to write down the determinant of the original metric tensor $g_{\mu\nu}$ we need to calculate the Jacobian of the coordinate transformation we used in order to get the diagonalized metric (4.94), $dy^0 = dx^0, \quad dy^i = dx^i + N^i dx^0$. The Jacobian matrix is

$$\mathcal{J}^\alpha_\beta = \frac{\partial x^\alpha}{\partial y^\beta} \begin{pmatrix} 1 & 0\,0\,0 \\ -N^1 & 1\,0\,0 \\ -N^2 & 0\,1\,0 \\ -N^3 & 0\,0\,1 \end{pmatrix}, \tag{4.95}$$

and calculating the determinant from the first row one immediately sees that $J = \det \mathcal{J} = 1$. Hence the determinants coincide, $g = g^{'}$, and we write

$$\sqrt{-g} = N\sqrt{h}. \tag{4.96}$$

Now that we have defined some metric properties of the hypersurfaces we continue by linking their extrinsic and intrinsic properties to those of the four-dimensional manifold $\mathcal{M}$. Let us begin by defining the projector on the hypersurfaces. Being the set of vectors $\left\{\vec{\eta}, \vec{b}_i\right\}$ a complete basis we can write the completeness relation

$$-\eta^\mu \eta^\nu + h^{ij} b^\mu_i b^\nu_j = g^{\mu\nu}. \tag{4.97}$$

One easily recognizes that the projector is given by $q^{\mu\nu} = h^{ij} b^\mu_i b^\nu_j$ since any component parallel to η^μ vanishes when contracted with it.

Let us now define the covariant derivative on the hypersurfaces (the steps we follow are completely similar to those of the first chapter). For a four-dimensional vector belonging to a specific hypersurface $\vec{A} = A^k \vec{b}_k \in \Sigma_{x^0}$ we write

$$\partial_i \vec{A} = \partial_i \left(A^k \vec{b}_k\right) = \left(\partial_i A^k\right) + A^r \partial_i \vec{b}_r. \tag{4.98}$$

Once again, we can write the derivative of the tangent basis vector as a linear combination of the same basis, whose coefficients are $\bar{\Gamma}$, plus a component along the normal to Σ_{x^0}, whose coefficient is $\Pi_{ir} = \Pi_{ri}$

$$\partial_i \vec{b}_r = \bar{\Gamma}^k_{ir} \vec{b}_k + \Pi_{ir} \vec{\eta}. \tag{4.99}$$

Hence, we write

$$\partial_i \vec{A} = \left(\partial_i A^k + \bar{\Gamma}^k_{ir} A^r\right) \vec{b}_k + A^r \Pi_{ir} \vec{\eta}, \tag{4.100}$$

from which the projection of the derivative on the hypersurface, *i.e.* the covariant derivative on the hypersurface, reads

$$D_i A^k = \partial_i A^k + \bar{\Gamma}^k_{ir} A^r. \tag{4.101}$$

As for the covariant derivative of covariant vectors one can start from the derivative of the scalar quantity[2] $A_k A^k$ and obtain

$$\begin{aligned} D_i \left(A^k A_k\right) =& A_k \left(D_i A^k\right) + A^k \left(D_i A_k\right) \\ =& A_k \left(\partial_i A^k + \bar{\Gamma}^k_{ir} A^r\right) + A^k \left(D_i A_k\right) \\ =& \partial_i \left(A^k A_k\right) \\ =& A_k \partial_i A^k + A^k \partial_i A_k, \end{aligned} \tag{4.102}$$

from which one reads the definition of the covariant derivative

$$D_i A_k = \partial_i A_k - \bar{\Gamma}^r_{ik} A_r. \tag{4.103}$$

The three-dimensional Christoffel symbols $\bar{\Gamma}^k_{ij}$ are defined by the usual relation

$$\bar{\Gamma}^k_{ij} = \frac{1}{2} h^{kr} \left(\partial_i h_{jr} + \partial_j h_{ri} - \partial_r h_{ij}\right). \tag{4.104}$$

Let us link the covariant derivative on the hypersurface with the covariant derivatives in the four-dimensional manifold. Let us begin by writing the three-dimensional covariant derivative in terms of four-dimensional indices

$$D_\mu A_\nu = b^i_\mu b^j_\nu D_i A_j, \qquad b^i_\mu = \frac{\partial x^i}{\partial u^\mu}. \tag{4.105}$$

Hence, we look for the relation with the four-dimensional covariant derivative

$$\begin{aligned} D_\mu A_\nu =& b^i_\mu b^j_\nu D_i A_j = b^i_\mu b^j_\nu \vec{b}_j \cdot \left(\partial_i \vec{A}\right) = b^i_\mu b^j_\nu \vec{b}_j \cdot \left(b^\rho_i \partial_\rho \vec{A}\right) \\ =& b^i_\mu b^j_\nu b^\rho_i \vec{b}_j \cdot \left(\nabla_\rho A^\gamma\right) \vec{f}_\gamma = b^i_\mu b^\rho_i b^j_\nu b^\sigma_i \nabla_\rho A^\gamma \vec{f}_\sigma \cdot \vec{f}_\gamma \\ =& q^\rho_\mu q^\sigma_\nu \nabla_\rho A^\gamma g_{\gamma\sigma} = q^\rho_\mu q^\sigma_\nu \nabla_\rho A_\sigma, \end{aligned} \tag{4.106}$$

from which it follows that the covariant derivative on Σ_{x^0} is given by the projection on the hypersurfaces of the four-dimensional covariant derivative. Let us now define the so-called *extrinsic curvature*

$$K_{\mu\nu} = q^\rho_\mu q^\sigma_\nu \nabla_\rho \eta_\sigma = K_{\nu\mu}, \tag{4.107}$$

which accounts for the variation of the normal to the hypersurface as projected on them. Indeed this is a quantity which accounts for a curvature of the hypersurface as seen from the four-dimensional manifold, hence it is extrinsic. Let us demonstrate the symmetry property of this tensor on its three-dimensional definition

$$\begin{aligned} K_{ij} =& \vec{b}_i \cdot \left(\partial_j \vec{\eta}\right) = \partial_j \left(\vec{b}_i \cdot \vec{\eta}\right) - \vec{\eta} \partial_j \vec{b}_i \\ =& - \vec{\eta} \cdot \left(\bar{\Gamma}^r_{ji} \vec{b}_r + \Pi_{ji} \vec{\eta}\right) = - \vec{\eta} \cdot \left(\bar{\Gamma}^r_{ij} \vec{b}_r + \Pi_{ij} \vec{\eta}\right) \\ =& \vec{b}_j \cdot \left(\partial_i \vec{\eta}\right) = K_{ji}, \end{aligned} \tag{4.108}$$

[2] $A_k A^k$ is a scalar quantity under three-dimensional diffeomorphisms on the hypersurface.

since $K_{\mu\nu} = b^i_\mu b^j_\nu K_{ij}$ the result automatically holds for $K_{\mu\nu}$. From (4.89) it is possible to write the normal vector as

$$\vec{\eta} = \frac{1}{N}\left(\vec{b}_0 - N^k \vec{b}_k\right), \tag{4.109}$$

from which one can easily check that[3]

$$K_{ij} = -\partial_j \log(N)\, \vec{\eta}\cdot\vec{b}_i + \frac{1}{N}\left[\Gamma^\mu_{0(i} g_{j)\mu} - h_{k(i} D_{j)} N^k\right], \tag{4.110}$$

where we took the symmetric part of the expression with respect to indices i and j. Since by definition $\vec{\eta}\cdot\vec{b}_i = 0$ we obtain

$$K_{ij} = \frac{1}{2N}\left(\partial_0 h_{ij} - D_i N_j - D_j N_i\right). \tag{4.111}$$

It is interesting to notice that for a generic vector $\vec{V} = V\vec{\eta} + V^k \vec{b}_k$ one finds

$$\vec{b}_j \cdot \left(\partial_i \vec{V}\right) = V K_{ij} + D_i V_j, \tag{4.112}$$

which means that the variation of the normal component with respect to the tangential direction to the hypersurface, once projected, is always proportional to the extrinsic curvature.

4.2.2 *Gauss-Codazzi equation*

We can now aim directly at writing the four-dimensional scalar curvature R in terms of the three-dimensional one $\bar{R}$ and some function of the extrinsic curvature $K_{\mu\nu}$ so that in the end it will be possible to write the Einstein-Hilbert Lagrangian density in terms of the ADM variables. We now write the completeness relation (4.97) in a slightly modified way

$$g^{\mu\nu} = q^{\mu\nu} + \varepsilon \eta^\mu \eta^\nu, \tag{4.113}$$

where $\varepsilon = \eta^\mu \eta_\mu = \pm 1$ according to the choice of the foliation. Let us now define the Riemann tensor on Σ_{x^0} as we did in the first chapter for the four-dimensional manifold. Given a vector $A_\sigma \in \Sigma_{x^0}$, *i.e.* $A_\sigma \eta^\sigma = 0$, the three-dimensional Riemann tensor is defined by the commutator of the three-dimensional covariant derivatives

$$\left[D_\mu, D_\nu\right] A_\rho = -\,\bar{R}^\sigma{}_{\rho\mu\nu} A_\sigma. \tag{4.114}$$

We begin with the first term of the commutator (4.114)

$$D_\mu\left(D_\nu A_\rho\right) = q^\pi_\mu q^\tau_\nu q^\sigma_\rho \left[\nabla_\pi\left(q^\phi_\tau\right) q^\omega_\sigma \nabla_\phi A_\omega + q^\phi_\tau \nabla_\pi\left(q^\omega_\sigma\right)\nabla_\phi A_\omega + q^\phi_\tau q^\omega_\sigma \nabla_\pi \nabla_\phi A_\omega\right], \tag{4.115}$$

[3]We use the four-dimensional Christoffel symbols since $\vec{b}_0 \notin \Sigma_{x^0}$.

and from the definition of the projecting tensor one has $q^{\nu}_{\mu} = \delta^{\nu}_{\mu} - \varepsilon\eta_{\mu}\eta^{\nu}$, hence

$$\begin{aligned} D_{\mu}\left(D_{\nu}A_{\rho}\right) =& q^{\pi}_{\mu}q^{\tau}_{\nu}q^{\sigma}_{\rho}\left[-\varepsilon\left(\nabla_{\pi}\eta_{\tau}\right)\eta^{\phi}q^{\omega}_{\sigma}\nabla_{\phi}A_{\omega} - \varepsilon\eta_{\tau}\left(\nabla_{\pi}\eta^{\phi}\right)q^{\omega}_{\sigma}\nabla_{\phi}A_{\omega}\right] \\ &+q^{\pi}_{\mu}q^{\tau}_{\nu}q^{\sigma}_{\rho}\left[-\varepsilon q^{\phi}_{\tau}\left(\nabla_{\pi}\eta_{\sigma}\right)\eta^{\omega}\nabla_{\phi}A_{\omega} - \varepsilon q^{\phi}_{\tau}\eta_{\sigma}\left(\nabla_{\pi}\eta^{\omega}\right)\nabla_{\phi}A_{\omega}\right] \\ &+q^{\pi}_{\mu}q^{\tau}_{\nu}q^{\sigma}_{\rho}q^{\phi}_{\tau}q^{\omega}_{\sigma}\nabla_{\pi}\nabla_{\phi}A_{\omega}. \end{aligned} \tag{4.116}$$

Since by definition $q^{\mu\nu}\eta_{\nu} = 0$ two terms in the previous expression cancel out. Moreover, for the projecting tensor one has that $q^{\pi}_{\mu}q^{\nu}_{\pi} = q^{\nu}_{\mu}$, and recalling the definition of the extrinsic curvature (4.107) we finally obtain

$$D_{\mu}\left(D_{\nu}A_{\rho}\right) = -\varepsilon K_{\mu\nu}\eta^{\pi}q^{\sigma}_{\rho}\nabla_{\pi}A_{\sigma} + \varepsilon K_{\mu\rho}K_{\nu\sigma}A^{\sigma} + q^{\tau}_{\mu}q^{\pi}_{\nu}q^{\sigma}_{\rho}\nabla_{\tau}\nabla_{\pi}A_{\sigma}. \tag{4.117}$$

For the second term of the commutator (4.114) we obtain

$$D_{\nu}\left(D_{\mu}A_{\rho}\right) = -\varepsilon K_{\nu\mu}\eta^{\pi}q^{\sigma}_{\rho}\nabla_{\pi}A_{\sigma} + \varepsilon K_{\nu\rho}K_{\mu\sigma}A^{\sigma} + q^{\tau}_{\mu}q^{\pi}_{\nu}q^{\sigma}_{\rho}\nabla_{\pi}\nabla_{\tau}A_{\sigma}. \tag{4.118}$$

Hence, for the whole three-dimensional Riemann tensor (4.114) we get

$$\left[D_{\mu}, D_{\nu}\right]A_{\rho} = -q^{\pi}_{\mu}q^{\tau}_{\nu}q^{\phi}_{\rho}R_{\sigma\phi\pi\tau}A^{\sigma} - \varepsilon\left(K_{\nu\rho}K_{\mu\sigma} - K_{\mu\rho}K_{\nu\sigma}\right)A^{\sigma} = -\bar{R}_{\sigma\rho\mu\nu}A^{\sigma}. \tag{4.119}$$

Since $A^{\sigma} \in \Sigma_{x^0}$ it follows that $A^{\rho} = A^{\sigma}\delta^{\rho}_{\sigma} = A^{\sigma}q^{\rho}_{\sigma}$ we can finally write the three-dimensional Riemann tensor expressed through the four-dimensional one and the extrinsic curvature

$$\bar{R}_{\sigma\rho\mu\nu} = q^{\omega}_{\sigma}q^{\phi}_{\rho}q^{\pi}_{\mu}q^{\tau}_{\nu}R_{\omega\phi\pi\tau} + \varepsilon\left(K_{\nu\rho}K_{\mu\sigma} - K_{\mu\rho}K_{\nu\sigma}\right), \tag{4.120}$$

from which we derive the three-dimensional Ricci tensor

$$\bar{R}_{\rho\nu} = q^{\omega\pi}q^{\phi}_{\rho}q^{\tau}_{\nu}R_{\omega\phi\pi\tau} + \varepsilon\left(K_{\nu\rho}K - K_{\mu\rho}K^{\mu}_{\nu}\right), \tag{4.121}$$

and the three-dimensional scalar curvature

$$\bar{R} = g^{\nu\rho}\bar{R}_{\nu\rho} = q^{\omega\pi}q^{\phi\tau}R_{\omega\phi\pi\tau} + \varepsilon\left(K^2 - K_{\mu\nu}K^{\mu\nu}\right). \tag{4.122}$$

Rewriting the four-dimensional scalar curvature using the completeness relation (4.113) one obtains[4]

$$R = g^{\mu\rho}g^{\nu\sigma}R_{\mu\nu\rho\sigma} = q^{\mu\rho}q^{\nu\sigma}R_{\mu\nu\rho\sigma} + 2\varepsilon\eta^{\mu}\eta^{\rho}q^{\nu\sigma}R_{\mu\nu\rho\sigma}. \tag{4.123}$$

We continue by explicitly calculating the quantities K^2 and $K_{\mu\nu}K^{\mu\nu}$

$$K^2 = q^{\sigma\rho}q^{\mu\nu}\nabla_{\sigma}\eta_{\rho}\nabla_{\mu}\eta_{\nu}, \qquad K_{\mu\nu}K^{\mu\nu} = q^{\mu\rho}q^{\nu\sigma}\nabla_{\rho}\eta_{\sigma}\nabla_{\mu}\eta_{\nu}, \tag{4.124}$$

[4]Using the symmetry properties of the Riemann tensor for which $\eta^{\mu}\eta^{\nu}\eta^{\rho}\eta^{\sigma}R_{\mu\nu\rho\sigma} = 0$.

we also write the last term of (4.123), performing some integrations by part, as

$$\begin{aligned}2\varepsilon q^{\nu\sigma}\eta^{\rho}R_{\mu\nu\rho\sigma}\eta^{\mu} =&2\varepsilon q^{\nu\sigma}\eta^{\rho}\left[\nabla_{\sigma},\nabla_{\rho}\right]\eta_{\nu}\\ =&2\varepsilon\left[\nabla_{\sigma}\left(\eta^{\rho}\nabla_{\rho}\eta^{\sigma}-\eta^{\sigma}\nabla_{\rho}\eta^{\rho}\right)-\nabla_{\sigma}\eta^{\rho}\nabla_{\rho}\eta^{\sigma}+\nabla_{\rho}\eta^{\rho}\nabla_{\sigma}\eta^{\sigma}\right].\end{aligned} \tag{4.125}$$

It is useful to write down the following properties

$$\nabla_{\mu}\varepsilon=\nabla_{\mu}\left(\eta_{\sigma}\eta^{\sigma}\right)=2\eta^{\sigma}\nabla_{\mu}\eta_{\sigma}=0, \tag{4.126}$$

$$\nabla_{\rho}\eta^{\rho}=g^{\rho\sigma}\nabla_{\rho}\eta_{\sigma}=\left(q^{\rho\sigma}+\varepsilon\eta^{\rho}\eta^{\sigma}\right)\nabla_{\rho}\eta_{\sigma}=K. \tag{4.127}$$

Hence, for the second term of the last line of (4.125) we obtain

$$\nabla_{\sigma}\eta^{\rho}\nabla_{\rho}\eta^{\sigma}=g^{\rho\nu}g^{\sigma\mu}\nabla_{\sigma}\eta_{\nu}\nabla_{\rho}\eta_{\mu}=q^{\rho\nu}q^{\sigma\mu}\nabla_{\sigma}\eta_{\nu}\nabla_{\rho}\eta_{\mu}=K_{\mu\nu}K^{\mu\nu}, \tag{4.128}$$

which allows us to write the whole expression (4.125)

$$2\varepsilon q^{\nu\sigma}\eta^{\rho}R_{\mu\nu\rho\sigma}\eta^{\mu}=-2\varepsilon\left(K_{\mu\nu}K^{\mu\nu}-K^{2}\right)+2\varepsilon\nabla_{\mu}\left(\eta^{\nu}\nabla_{\nu}\eta^{\mu}-\eta^{\mu}K\right). \tag{4.129}$$

Plugging this result into (4.123) and using (4.120) we can finally write the four-dimensional scalar curvature as a function of the three-dimensional one and of the extrinsic curvature

$$R=\bar{R}-\varepsilon\left(K_{\mu\nu}K^{\mu\nu}-K^{2}\right)+2\varepsilon\nabla_{\mu}\left(\eta^{\nu}\nabla_{\nu}\eta^{\mu}-\eta^{\mu}K\right), \tag{4.130}$$

which is the so-called *Gauss-Codazzi* equation. Now, the last step is to take the right signature, *i.e.* $\varepsilon=-1$, and to see that since by construction $b_i^{\mu}b_{\mu}^{j}=\delta_i^j$ one has that $K_{\mu\nu}K^{\mu\nu}=K_{ij}K^{ij}$ and also $K=K_{\mu}^{\mu}=K_i^i$, thus leading to

$$R=\bar{R}+K_{ij}K^{ij}-K^{2}-2\nabla_{\mu}\left(\eta^{\nu}\nabla_{\nu}\eta^{\mu}-\eta^{\mu}K\right). \tag{4.131}$$

It is worth noticing that each term of the Gauss-Codazzi equation is covariant.

4.2.3 *ADM Lagrangian density*

We can immediately write down the Einstein-Hilbert Lagrangian density in terms of ADM variables

$$\mathcal{L}_{EH}=\sqrt{-g}R=N\sqrt{h}\left(\bar{R}+K_{ij}K^{ij}-K^{2}\right)+2N\sqrt{h}\nabla_{\mu}\left(\eta^{\nu}\nabla_{\nu}\eta^{\mu}-\eta^{\mu}K\right), \tag{4.132}$$

and we define, applying the results for the covariant divergence, the ADM boundary term as

$$\partial_{\mu}\boldsymbol{\mathcal{ADM}}^{\mu}=\partial_{\mu}\left[2N\sqrt{h}\left(\eta^{\mu}K-\eta^{\nu}\nabla_{\nu}\eta^{\mu}\right)\right]. \tag{4.133}$$

Once again, the treatment of this boundary term is not completely trivial since the derivatives of the variation of the variables arise on the boundary when imposing the stationarity condition. Moreover other terms of this kind emerge from the variation of $\bar{R}$ analogously to what happens for the variation of R. However, for some given boundary conditions it is possible to discard these terms and to have a well-posed variation principle for the ADM Lagrangian density which is defined as

$$\mathcal{L}_{ADM} = N\sqrt{h}\left(\bar{R} + K_{ij}K^{ij} - K^2\right) = N\sqrt{h}\left[\bar{R} + \left(h^{ir}h^{js} - h^{ij}h^{rs}\right)K_{ij}K_{rs}\right]. \tag{4.134}$$

We now briefly discuss some features of this Lagrangian density. First of all it is covariant under four-dimensional diffeomorphisms while nor the $\Gamma\Gamma$ nor Dirac Lagrangian densities are. Secondly it clearly appears that there are no velocities for N and N^i. Indeed there are two primary constraints

$$\Pi = \frac{\partial \mathcal{L}_{ADM}}{\partial \partial_0 N} \approx 0, \qquad \Pi_k = \frac{\partial \mathcal{L}_{ADM}}{\partial \partial_0 N^k} \approx 0. \tag{4.135}$$

Because of their link to g_{00} and g_{0i} this feature is closely related to the missing of terms proportional to $\partial_0 g_{0\mu}$ in Dirac Lagrangian density (4.65). The connection to Dirac formulation can be appreciated calculating the momenta conjugated to h_{ij} (taking into account also the dimensional constant)

$$\Pi^{ij} = \frac{\partial \mathcal{L}_{ADM}}{\partial \partial_0 h_{ij}} = \frac{c^3}{16\pi G}\sqrt{h}\left(h^{ri}h^{sj} - h^{ij}h^{rs}\right)K_{rs}, \tag{4.136}$$

and comparing them to those calculated from Dirac formulation (4.66). After some tedious, but not difficult, calculations one finds that

$$\frac{K_{ij}}{N} = \Gamma^0_{ij}, \tag{4.137}$$

and that according to the definition of the ADM transformation one has

$$e^{rs} = g^{rs} - \frac{g^{0r}g^{0s}}{g^{00}} = g^{rs} + \frac{N^r N^s}{N^2} = h^{rs}. \tag{4.138}$$

These results allow us to explicitly write the transformation between Dirac and ADM conjugate momenta to g_{ij}

$$\sqrt{-g}\left(e^{ri}e^{sj} - e^{rs}e^{ij}\right)\Gamma^0_{rs} = N\sqrt{h}\left(h^{ri}h^{sj} - h^{rs}h^{ij}\right)\frac{K_{rs}}{N}, \tag{4.139}$$

hence

$$\Pi^{ij} = p^{ij}. \tag{4.140}$$

This result was reported in [185] and leads to a strong non-canonicity of the transformation linking Dirac and ADM formulations. However, this turns out to be a gauged-canonical transformation as we discussed in the previous chapter. Let us analyze the fundamental Poisson bracket of ADM variables with respect to Dirac's

$$\left[N, \Pi^{ij}\right]_{g,p} = \frac{\partial}{\partial g_{ij}}\left(\frac{1}{\sqrt{-g^{00}}}\right) = -\frac{1}{2\left(-g^{00}\right)^{\frac{3}{2}}} g^{0i} g^{0j} \neq 0, \tag{4.141}$$

stating the non-strong canonicity of the transformation. In [90] it was proposed that this kind of non-canonicity does not affect the classical physical equivalence of the theories since (4.140) is not canonical because of some combination of primary first-class constraints[5]. Of course, a completely similar result still holds for the transformation linking the $\Gamma\Gamma$ Hamiltonian formulation to the ADM which is always reported in [90]. The canonical version of (4.140) is given by[6]

$$\Pi^{ij}_{C} = N^i N^j p^{00} + 2N^{(i}p^{j)0} + p^{ij} \approx p^{ij}. \tag{4.142}$$

The transformation linking Dirac variables $\{p^{\mu\nu}, g_{\mu\nu}\}$ to the ADM ones is not invertible since we do not know how to link the conjugate momenta of both formulations defining the primary constraints (4.67) and (4.135). Hence, this is exactly the same situation we produced in the electromagnetic field example in Chapter 3.

What could be the consequences on the quantization scheme is yet unknown.

4.3 Boundary terms

Here we briefly review all the boundary terms we encountered so far. We can distinguish two different types of Lagrangian densities:

- the covariant ones, such as the Einstein-Hilbert (4.2) and the ADM (4.134) Lagrangian densities, which allow for a geometric interpretation of the variables but necessarily contain second order derivatives of the metric tensor requiring some hypothesis on the boundary for the variational principle to be well-posed;

[5] As we will see in Chapter 6 all constraints of GR in the ADM formulation are first-class.

[6] We remember here that the strong canonicity condition is given by $p_{\mu\nu}\delta g^{\mu\nu} = \Pi_C \delta N + \Pi^C_i \delta N^i + \Pi^{ij}_C \delta h_{ij}$.

- the non-covariant ones, as the $\Gamma\Gamma$ and Dirac Lagrangian densities, which are treated in a purely algebraic fashion and do not contain any second order derivative thus being free from additional boundary conditions.

Hence, it seems that requiring an explicit covariance of the action functional leads to clearer calculations but to additional boundary conditions, while the non-covariant Lagrangian densities, which have a well-defined stationarity condition, are tougher to treat.

It is then interesting to show how to solve the problem of the variation of the boundary terms for the former and how the different boundary terms are linked to one another.

4.3.1 *Gibbons-York-Hawking boundary term*

Let us examine in more detail (along the steps of [294]) what is the trouble with the boundary term arising from the variation of the Einstein-Hilbert action functional:

$$\int_{\mathcal{M}} d^4x \partial_\mu \boldsymbol{\delta\mathcal{E}\mathcal{H}}^\mu = \int_{\partial\mathcal{M}} d^3x \sqrt{-g} \left(g^{\rho\sigma} \eta_\mu \delta\Gamma^\mu_{\rho\sigma} - \eta^\mu \delta\Gamma^\rho_{\mu\rho} \right), \tag{4.143}$$

where we used Gauss' theorem. Accounting for the vanishing of the variation on the boundaries, $\delta g_{\mu\nu}|_{\partial\mathcal{M}} = 0$, we write the variations of the Christoffel symbols as

$$\begin{aligned} \delta\Gamma^\mu_{\rho\sigma}\Big|_{\partial\mathcal{M}} &= \frac{1}{2} g^{\mu\nu} \left(\partial_\rho \delta g_{\sigma\nu} + \partial_\sigma \delta g_{\nu\rho} - \partial_\nu \delta g_{\rho\sigma} \right), \\ \delta\Gamma^\rho_{\mu\rho}\Big|_{\partial\mathcal{M}} &= \frac{1}{2} g^{\rho\sigma} \partial_\mu \delta g_{\rho\sigma}. \end{aligned} \tag{4.144}$$

We then split the metric tensor along the boundary as $g^{\mu\nu} = q^{\mu\nu} + \varepsilon\eta^\mu\eta^\nu$, where $q^{\mu\nu}$ is a projection operator on the boundary, η^μ is the normal vector at each point and ε accounts for the normal to be space-like, $\varepsilon = +1$, or time-like, $\varepsilon = -1$. The same procedure was performed for the ADM transformation and $q^{\mu\nu}$ is the induced metric on the boundary (4.97). Hence, for the determinant of the metric along the boundary we obtain $\sqrt{-g}|_{\partial\mathcal{M}} = -\varepsilon\sqrt{|q|}$. We now substitute these results in (4.143) and use the completeness relation

$$\begin{aligned} &- \int_{\partial\mathcal{M}} d^3x \varepsilon \sqrt{|q|} \left(g^{\rho\sigma} \eta_\mu \delta\Gamma^\mu_{\rho\sigma} - \eta^\mu \delta\Gamma^\rho_{\mu\rho} \right) \\ &= - \int_{\partial\mathcal{M}} d^3x \varepsilon \sqrt{|q|} \left(q^{\rho\sigma} \eta^\mu \partial_\rho \delta g_{\sigma\mu} - q^{\rho\sigma} \eta^\mu \partial_\mu \delta g_{\rho\sigma} \right). \end{aligned} \tag{4.145}$$

We now face the root of the problem: the troubling derivatives are those projected along the normal to the boundary. In fact as the variation of the metric tensor on the boundary is identically zero it follows that its tangential components, *i.e.* $q^{\rho\sigma}\partial_\rho\delta g_{\sigma\nu}$, vanish for continuity reasons, while those projected along the normal, *i.e.* $q^{\rho\sigma}\eta^\mu\partial_\mu\delta g_{\rho\sigma}$, do not. Hence, the only remaining term is

$$\int_{\mathcal{M}} d^4x\partial_\mu\boldsymbol{\delta\mathcal{E}\mathcal{H}}^\mu = \int_{\partial\mathcal{M}} d^3x\varepsilon\sqrt{|q|}q^{\rho\sigma}\eta^\mu\partial_\mu\delta g_{\rho\sigma}. \tag{4.146}$$

There is a way to modify the Lagrangian density which provides the counter-term needed to compensate the boundary contribution (4.146): let us write the extrinsic curvature (4.107) related to the boundary of the manifold

$$K = \nabla_\mu\eta^\mu = \partial_\mu\eta^\mu + \frac{1}{2}g^{\rho\sigma}\eta^\mu\partial_\mu g_{\rho\sigma}, \tag{4.147}$$

its variation on the boundary reads

$$\delta K\Big|_{\partial\mathcal{M}} = \frac{1}{2}q^{\rho\sigma}\eta^\mu\partial_\mu\delta g_{\rho\sigma}, \tag{4.148}$$

which is just half of the contribution (4.146) that we wanted to cancel. We can write a modified action functional

$$S_{EH-GYH}[g_{\mu\nu}] = \int_{\mathcal{M}} d^4x\sqrt{-g}R - 2\int_{\partial\mathcal{M}} d^3x\varepsilon\sqrt{|q|}K, \tag{4.149}$$

which properly accounts for the derivatives of the variations of the metric but in turn is not covariant because the second integral is performed on the boundary. The latter is covariant only for three-dimensional diffeomorphism performed on the boundary.

This new term is known as *Gibbons-York-Hawking boundary term* [143, 299] who first proposed it mainly to deal with a path-integral formulation in ADM variables.

4.3.2 *Comparison among different formulations*

So far we collected some boundary terms arising in the various derivations. Since it is not always obvious which is the relation among these boundary terms we write here some key results [90] which are fully in agreement with the general symmetry observation we already made about the covariance properties of the various Lagrangian densities. We write here the different boundary terms we encountered

$$\begin{aligned}\partial_\mu \boldsymbol{\mathcal{EH}}^\mu =&\partial_\mu\left[\sqrt{-g}\left(g^{\sigma\rho}\Gamma^\mu_{\sigma\rho}-g^{\mu\nu}\Gamma^\rho_{\nu\rho}\right)\right]\\ =&\partial_\mu\left[\sqrt{-g}g^{\rho\sigma}g^{\mu\nu}\left(\partial_\rho g_{\sigma\nu}-\partial_\nu g_{\rho\sigma}\right)\right],\end{aligned} \tag{4.150}$$

$$\partial_\mu \boldsymbol{\mathcal{D}}^\mu = \partial_i\left[\frac{g^{0i}}{g^{00}}\partial_0\left(\sqrt{-g}g^{00}\right)\right]-\partial_0\left[\frac{g^{0i}}{g^{00}}\partial_i\left(\sqrt{-g}g^{00}\right)\right], \tag{4.151}$$

$$\partial_\mu \boldsymbol{\mathcal{ADM}}^\mu = 2\partial_\mu\left[N\sqrt{h}\left(\eta^\mu K-\eta^\nu\nabla_\nu\eta^\mu\right)\right]. \tag{4.152}$$

As a choice, we propose the transformation of $\partial_\mu \boldsymbol{\mathcal{EH}}^\mu$ and $\partial_\mu \boldsymbol{\mathcal{D}}^\mu$ in ADM variables in order to make the comparison

$$\begin{aligned}\partial_\mu \boldsymbol{\mathcal{EH}}^\mu =& \partial_0\left(\frac{\sqrt{h}h^{ij}}{N}\partial_0 h_{ij}-\frac{\sqrt{h}}{N}\partial_i N^i-\frac{\sqrt{h}N^k h^{ij}}{N}\partial_k h_{ij}\right)\\ &+\partial_r\left(\frac{\sqrt{h}N^r N^k h^{ij}}{N}\partial_k h_{ij}+\frac{\sqrt{h}N^r}{N}\partial_i N^i-\frac{\sqrt{h}N^r h^{ji}}{N}\partial_0 h_{ij}\right)\\ &+\partial_r\left(-\frac{\sqrt{h}}{N}\partial_0 N^r-2\sqrt{h}h^{ri}\partial_i N+\frac{\sqrt{h}N^i}{N}\partial_i N^r\right)\\ &+\partial_r\left[N\sqrt{h}h^{ij}h^{rk}\left(\partial_i h_{jk}-\partial_k h_{ij}\right)\right],\end{aligned} \tag{4.153}$$

$$\partial_\mu \boldsymbol{\mathcal{D}}^\mu = \partial_0\left(\frac{\sqrt{h}}{N}\partial_i N^i\right)-\partial_i\left(\frac{\sqrt{h}}{N}\partial_0 N^i\right), \tag{4.154}$$

and we also explicitly write the expression for the ADM boundary term

$$\begin{aligned}\partial_\mu \boldsymbol{\mathcal{ADM}}^\mu =&\partial_0\left(\frac{\sqrt{h}}{N}h^{ij}\partial_0 h_{ij}-2\frac{\sqrt{h}}{N}\partial_i N^i-\frac{\sqrt{h}N^j h^{ik}}{N}\partial_j h_{ik}\right)\\ &+\partial_r\left(-\frac{\sqrt{h}N^r h^{ij}}{N}\partial_0 h_{ij}+2\frac{\sqrt{h}N^r}{N}\partial_i N^i\right)\\ &+\partial_r\left(\frac{\sqrt{h}N^r N^j h^{ik}}{N}\partial_j h_{ik}-2\sqrt{h}h^{ri}\partial_i N\right).\end{aligned} \tag{4.155}$$

We now subtract the Einstein-Hilbert boundary term to the ADM one and we obtain

$$\begin{aligned}&\partial_\mu \boldsymbol{\mathcal{ADM}}^\mu-\partial_\mu \boldsymbol{\mathcal{EH}}^\mu\\ =&-\partial_0\left(\frac{\sqrt{h}}{N}\partial_i N^i\right)+\partial_r\left(\frac{\sqrt{h}}{N}\partial_0 N^r\right)+\partial_r\left(\frac{\sqrt{h}N^r}{N}\partial_i N^i-\frac{\sqrt{h}N^i}{N}\partial_i N^r\right)\\ &-\partial_r\left[N\sqrt{h}h^{ij}h^{rk}\left(\partial_i h_{jk}-\partial_k h_{ij}\right)\right].\end{aligned} \tag{4.156}$$

We immediately recognize in the first two terms the opposite of Dirac boundary term. This fact fully explains why in the ADM Lagrangian density there are no time derivatives for N and N^k: Dirac boundary term is automatically generated by the ADM transformation performed on the Einstein-Hilbert Lagrangian density and it plays the same role as for the $\Gamma\Gamma$ Lagrangian density, it deletes all linear terms in $\partial_0 N$ and $\partial_0 N^k$.

The term in the last line of (4.156) is exactly the spatial analogous of the Einstein-Hilbert boundary term (4.150). We could expect this result since in the ADM Lagrangian density the three-dimensional scalar curvature appears bringing in second order derivatives which are completely missing in the $\Gamma\Gamma$ Lagrangian density. It is possible to repeat the procedure we applied in order to obtain the $\Gamma\Gamma$ Lagrangian density from the Einstein-Hilbert one for the term $N\sqrt{h}\bar{R}$,

$$N\sqrt{h}\bar{R} = N\sqrt{h}h^{ij}\left(\bar{\Gamma}^r_{sj}\bar{\Gamma}^s_{ri} - \bar{\Gamma}^r_{ij}\bar{\Gamma}^s_{rs}\right) + \sqrt{h}h^{ij}\bar{\Gamma}^r_{ir}\partial_j N - \sqrt{h}h^{ij}\bar{\Gamma}^r_{ij}\partial_r N + \partial_r \boldsymbol{\mathcal{R}}^r, \tag{4.157}$$

where we have defined a boundary term as the last term of (4.156)

$$\partial_r \boldsymbol{\mathcal{R}}^r = \partial_r \left[N\sqrt{h}h^{ij}h^{rk}\left(\partial_i h_{jk} - \partial_k h_{ij}\right)\right]. \tag{4.158}$$

Finally, we define the boundary term which is needed in order to obtain the quadratic term in the extrinsic curvature in the ADM Lagrangian density

$$\partial_r \boldsymbol{\mathcal{S}}^r = \partial_r \left(\frac{\sqrt{h}N^i}{N}\partial_i N^r - \frac{\sqrt{h}N^r}{N}\partial_i N^i\right). \tag{4.159}$$

All in all we can finally write the exact transformation linking the boundary terms we encountered so far

$$\partial_\mu \boldsymbol{\mathcal{ADM}}^\mu = \partial_\mu \boldsymbol{\mathcal{EH}}^\mu - \partial_\mu \boldsymbol{\mathcal{D}}^\mu - \partial_r \boldsymbol{\mathcal{R}}^r - \partial_r \boldsymbol{\mathcal{S}}^r, \tag{4.160}$$

which can be used to obtain the transformations linking the $\Gamma\Gamma$, Dirac and ADM Lagrangian densities, since adding and subtracting these boundary terms to Einstein-Hilbert Lagrangian density we obtain the chain of equalities

$$\begin{aligned} \mathcal{L}_{EH} &= \mathcal{L}_{\Gamma\Gamma} + \partial_\mu \boldsymbol{\mathcal{EH}}^\mu \\ &= \mathcal{L}_{\mathcal{D}} - \partial_\mu \boldsymbol{\mathcal{D}}^\mu + \partial_\mu \boldsymbol{\mathcal{EH}}^\mu \\ &= \mathcal{L}_{ADM} - \partial_r \boldsymbol{\mathcal{R}}^r - \partial_r \boldsymbol{\mathcal{S}}^r - \partial_\mu \boldsymbol{\mathcal{D}}^\mu + \partial_\mu \boldsymbol{\mathcal{EH}}^\mu \\ &= \mathcal{L}_{ADM} + \partial_\mu \boldsymbol{\mathcal{ADM}}^\mu, \end{aligned} \tag{4.161}$$

allowing us to write the transformations linking the different Lagrangian densities

$$\begin{aligned} \mathcal{L}_{\mathcal{D}} &= \mathcal{L}_{\Gamma\Gamma} + \partial_\mu \boldsymbol{\mathcal{D}}^\mu, \\ \mathcal{L}_{ADM} &= \mathcal{L}_{\mathcal{D}} + \partial_r \boldsymbol{\mathcal{R}}^r + \partial_r \boldsymbol{\mathcal{S}}^r. \end{aligned} \tag{4.162}$$

Chapter 5

Quantization Methods

The tremendous experimental success of Quantum Mechanics (QM) makes it a cornerstone of our present knowledge of the physical world. The basic paradigm of QM is here revised. We will first introduce the notion of Hilbert space of states and then we discuss the standard procedure to quantize a dynamical system owing to the classical to quantum correspondence (Poisson brackets vs. commutators). We will discuss the dynamics in both the Schrödinger and Heisenberg representations and we provide the tools to realize semiclassical states. Then, we present the generalization of the canonical quantization procedure by introducing Weyl algebra and GNS construction. The former can be seen as the "exponentiated" version of the commutation relations, while the latter allows to construct the Hilbert space directly from the algebra. Under some hypotheses it is possible to probe how the GNS construction applied to the Weyl algebra leads to the Schrödinger representation (Stone-Von Neumann theorem). However, if some of these hypotheses are relaxed, one finds new inequivalent quantum representations of the Weyl algebra. We will discuss one of this new representations, namely the polymer representation, which has the property to set up naturally a minimum length scale. Finally we will return back to the canonical quantization procedure and we will outline the issues one encounters when quantizing some relevant constrained physical systems.

5.1 Classical and quantum dynamics

In this section we will revise some fundamental features of QM, in order to fix the notation and to give a clear insight of the basic formalism underlying the quantization procedure [99, 220, 262].

5.1.1 *Dirac observables*

Observables are described by Hermitian operators that act on vectors in Hilbert spaces: a *Hilbert space* is an abstract vector space, where the notions of distances and angles can be computed by means of the scalar product $\langle\beta|\alpha\rangle$, which is a map from two copies of the Hilbert space to $\mathbb{C}$ for which

$$\langle\alpha|\beta\rangle = (\langle\beta|\alpha\rangle)^*, \tag{5.1}$$

$$\langle\alpha|\alpha\rangle \geq 0 \quad \text{and} \quad \langle\alpha|\alpha\rangle = 0 \Leftrightarrow |\alpha\rangle = 0. \tag{5.2}$$

A (complete) set of basis vectors exists and the linear combinations of these vectors describe the set of all the possible states the system can be found in. An element (*ket*) $|\alpha\rangle$ in a Hilbert space H can be decomposed along a basis of the Hilbert space, $\{|a\rangle\}$, satisfying the identity $\sum_a |a\rangle\langle a| = I$, so that

$$|\alpha\rangle = \sum_a |a\rangle\langle a|\alpha\rangle. \tag{5.3}$$

The result of a measure does not depend on the characterization of the Hilbert space or on the choice of the basis, but only on the definition of observables. On the one hand, from a physical point of view, the equivalence of a description of an observable A_1 in a Hilbert space H_1 with that of A_2 in H_2 is given by the *unitary transformation* U $(U^\dagger U = 1)^1$, $U : H_1 \to H_2$, such that $A_1 = U^{-1}A_2U$. On the other hand, the crucial operation is the definition of *self adjoint operators*, *i.e.* $A = A^\dagger$, which are associated with classical quantities since their eigenvalues are real (see below).

When a measure of an observable A is performed, the system, which was previously described as a linear combination of the eigenstates of A, (5.3), falls into one of these states $|a'\rangle \in \{|a\rangle\}$, which corresponds to the eigenvalue a' of A, *i.e.* the measure operation changes the state of the system unless the system was already in an eigenstate of the observable. Since (5.3) describes the system before the measure, it is not possible a *priori* to know which state $|a'\rangle \in \{|a\rangle\}$ will be the result of the measure, but the probability $P_{a'}$ of a particular result $|a'\rangle$ is given by

$$P_{a'} = \frac{|\langle a'|\alpha\rangle|^2}{|\langle\alpha|\alpha\rangle|^2}. \tag{5.4}$$

The expectation value of an observable A with respect to $|\alpha\rangle$, $\langle A\rangle_\alpha$ is defined as

$$\langle A\rangle_\alpha \equiv \frac{\langle\alpha|A|\alpha\rangle}{|\langle\alpha|\alpha\rangle|^2} = \sum_{|a'\rangle\in\{|a\rangle\}} a'P_{a'}. \tag{5.5}$$

[1] in a finite dimension Hilbert space † coincides with Hermitian conjugation

From a phenomenological point of view, the probability (5.4) has to be determined after a lot of measures on systems of identical preparation, all characterized by the same ket $|\alpha\rangle$, *i.e.* a set of pure states. A mixture is a set of states $|\alpha_i\rangle$, each one with a weight w_i, such that $\sum_i w_i = 1$. The ensemble average of an operator A on a mixture is defined as

$$[A] = \sum_i w_i \frac{\langle\alpha_i|A|\alpha_i\rangle}{|\langle\alpha_i|\alpha_i\rangle|^2} = \sum_i w_i \langle A\rangle_{\alpha_i}, \tag{5.6}$$

where the expectation value (5.5) is weighted by the w_i's. If the density operator ρ is introduced,

$$\rho \equiv \sum_i w_i \frac{|\alpha_i\rangle\langle\alpha_i|}{|\langle\alpha_i|\alpha_i\rangle|^2}, \tag{5.7}$$

the ensemble average can be rewritten as

$$[A] = tr(\rho A). \tag{5.8}$$

The density operator is self adjoint, and $tr(\rho) = 1$. For a pure state, $|\alpha_j\rangle$, $w_j = 1$, the density operator $\rho = |\alpha_n\rangle\langle\alpha_n|$ is idempotent, $\rho^2 = \rho$, so that $tr(\rho^2) = 1$.

It's worth remarking that probabilities (5.4) are given by squared amplitudes, and the vector $|\alpha\rangle$ in H cannot be measured: the vector $|\alpha\rangle$ in H and the ray $\lambda|\alpha\rangle$, $\lambda \in \mathbb{C}$, $|\lambda|^2 = 1$ are a representation of the same physical state, as they give the same contribution to a measure. *It is rays, rather then vectors, that represent physical states in* H. From now on, we will work with normalized vectors $|\alpha\rangle$, *i.e.* $\langle\alpha|\alpha\rangle = 1$.

5.1.2 *Poisson brackets and commutators*

Quantum states are characterized by the canonical commutation relations, which hold between those operators that define the quantum system. Let x and p be the coordinates of the phase space. The canonical commutation relations

$$[x_i, x_j] = 0, \quad [p_i, p_j] = 0, \quad [x_i, p_j] = i\hbar\delta_{ij}, \tag{5.9}$$

can be inferred from the quantum properties of translations alone: such a derivation can also be performed for quantum objects that have no classical analogue. However, quantum-mechanical commutators $[\quad]$ can be obtained from the corresponding classical Poisson brackets $[\quad]_{cl}$

$$[A(p,x), B(p,x)]_{cl} \equiv \sum_s \left(\frac{\partial A}{\partial x_s}\frac{\partial B}{\partial p_s} - \frac{\partial A}{\partial p_s}\frac{\partial B}{\partial x_s}\right) \tag{5.10}$$

by the replacement

$$[\ \ ]_{cl} \rightarrow \frac{[\ \ ]}{i\hbar}. \tag{5.11}$$

This identification is possible because both Poisson brackets and commutators obey the algebraic properties

$$[A, A] = 0, \quad [A, B] = -[B, A], \quad [A, c] = 0, \quad c \in \mathbb{C}, \tag{5.12}$$

$$[A + B, C] = [A, C] + [B, C], \quad [A, BC] = [A, B]C + B[A, C], \tag{5.13}$$

$$[A, [B, C]] + [B, [A, C]] + [C, [A, B]] = 0, \tag{5.14}$$

the factor $1/i\hbar$ being necessary in (5.11), because, while the Poisson brackets of two real functions are real, the commutator of two Hermitian operator is anti-Hermitian. This way, the third of (5.9) can be derived from the classical $[x_i, p_j]_{cl} = \delta_{ij}$ through (5.11).

5.1.3 *Schrödinger representation*

After implementing the canonical commutation relations from a quantum point of view, it is possible to find out the representations of such operators.

In the coordinate representation, the *wave function* $\psi_\alpha(x)$ for the state $|\alpha\rangle$ is given by

$$\psi_\alpha(x) = \langle x|\alpha\rangle \tag{5.15}$$

while, in the momentum representation, we get

$$\psi_\alpha(p) = \langle p|\alpha\rangle. \tag{5.16}$$

These two representations are linked by Fourier duality, *i.e.* it is possible to pass from a representation to the other one by Fourier transform. The normalization of the state $|\alpha\rangle$ implies that

$$\int_D dx |\psi_{\alpha'}(x)|^2 = 1, \tag{5.17}$$

and the equivalent version in the momentum representation, so that wave functions belong to the Hilbert space of square-integrable functions $\mathrm{H} = L^2(D, dx)$, where the integration domain D (and therefore, the specification of L^2) depend on the physics to be described.

The probability of a transition of the system between two given states, say $|\alpha\rangle \to |\alpha'\rangle$, is given by (5.5)

$$P_{\alpha\to\alpha'} = \left| \int_D dx\, \psi^*_{\alpha'}(x)\psi_\alpha(x) \right|^2 \leq 1. \tag{5.18}$$

The momentum p can be shown to be the generator of translations in the coordinate representation, starting from the definition of infinitesimal translations in the coordinate representation, *i.e.* in the representation where the eigenstates of the coordinates are the basis vectors. This means that a physical state $|\alpha\rangle$ can be decomposed as

$$|\alpha\rangle = \int dx' |x'\rangle\langle x'|\alpha\rangle, \tag{5.19}$$

and the coordinate eigenstates satisfy the (multiplicative) relation[2]

$$\hat{x}|x'\rangle = x'|x'\rangle. \tag{5.20}$$

The definition of a translation $\hat{\mathcal{T}}(\Delta x')$ of an infinitesimal quantity $\Delta x'$ reads

$$\begin{aligned} \left(1 - \frac{i\hat{p}\Delta x'}{\hbar}\right)|\alpha\rangle &= \int dx' \hat{\mathcal{T}}(\Delta x')|x'\rangle\langle x'|\alpha\rangle \\ = \int dx'|x' + \Delta x'\rangle\langle x'|\alpha\rangle &= \int dx'|x'\rangle\langle x' - \Delta x'|\alpha\rangle \\ &= \int dx'|x'\rangle \left(\langle x'|\alpha\rangle - \Delta x' \frac{\partial}{\partial x}\langle x'|\alpha\rangle\right) \end{aligned} \tag{5.21}$$

such that

$$\hat{p}|\alpha\rangle = \int dx'|x'\rangle \left(-i\hbar\frac{\partial}{\partial x}\langle x'|\alpha\rangle\right); \tag{5.22}$$

the action of the operator $\hat{p}$ on wave functions is then $\hat{p} = -i\hbar\partial/\partial x$.
By the properties of translations and by the definition of the momentum operator as the generator of translations (up to an arbitrary constant with the dimensions of a length), the canonical commutation relations are defined for a quantum system without any direct comparison with their classical analogues: this derivation is particularly efficient, as it allows one to define both abstract quantum systems, and quantum systems whose observables have no classical counterpart.

Two main differences between the Poisson brackets in the classical case and the canonical commutation relations in the quantum case have to be stressed. Their dimensions are different, as they differ by the presence of differentiations with respect to x and p (which requires the introduction of $\hbar$). Furthermore, while, in the classical case, the Poisson brackets for two real functions are entirely real, the commutator of two Hermitian operators is anti-Hermitian.

[2] In what follows we will not put the $\hat{}$ on multiplicative operators: $\hat{x} = x$.

5.1.4 *Heisenberg representation*

The motion equation for an operator A^H in the Heisenberg picture, whose Schrödinger representation does not depend on time explicitly,

$$\frac{dA^H}{dt} = \frac{1}{i\hbar}[A^H, H] \tag{5.23}$$

can be derived, from a quantum-mechanical point of view, from the properties of the time-evolution operator and the definition of A^H alone. On the other hand, such a motion equation could be obtained, from a classical point of view, for a function $A(p, x)$ that does not depend on time explicitly,

$$\frac{dA(p,x)}{dt} = [A, H] \tag{5.24}$$

through the replacement (5.11). Again, the replacement (5.11) allows one to obtain quantum relations from classical ones, but (5.23) can be worked out from quantum properties even for operators that have no classical analogue, thus pointing out once more that classical mechanics can be derived as the limit of quantum mechanics, but the converse is not true.

5.1.5 *The Schrödinger equation*

The celebrated *Schrödinger equation* is obtained by assuming a Hamiltonian operator H in the quantum regime as consisting of a kinetic part $\frac{p^2}{2m}$ and a potential $V(x)$, which corresponds to a local Hermitian operator, *i.e.*

$$H = \frac{\hat{p}^2}{2m} + V(x), \tag{5.25}$$

and then by attributing to the quantum operator H the same physical meaning of its classical counterpart, that of the generator of time evolution.

The time-independent Schrödinger equation is the partial differential equation obeyed by the energy eigenfunctions $u_E(x)$, *i.e.*

$$-\left(\frac{\hbar^2}{2m}\right)\nabla^2 u_E(x) + V(x)u_E(x) = Eu_E(x), \tag{5.26}$$

while the time-dependent Schrödinger equation is given by the equation describing the time evolution of the wave function $\psi(x,t) = \langle x|\alpha, t\rangle$ for a ket $|\alpha, t\rangle$ in the coordinate representation, *i.e.*

$$i\hbar\frac{\partial}{\partial t}\psi(x,t) = -\left(\frac{\hbar^2}{2m}\right)\nabla^2\psi(x,t) + V(x)\psi(x,t). \tag{5.27}$$

5.1.6 *Quantum to classical correspondence: Hamilton-Jacobi equation*

5.1.6.1 *Hamilton-Jacobi formalism*

For a mechanical system whose Hamiltonian H depends explicitly on the time t, a canonical transformation [13] can be looked for, such that the coordinates and the momenta (x, p) evaluated at the time t can be written as constant quantities (x_0, p_0) at the time t, *i.e.*

$$x = x(x_0, p_0, t), \quad p = p(x_0, p_0, t), \tag{5.28}$$

if the new variables are constant in time, the transformed Hamiltonian $K = H + \partial F/\partial t$, where F is the generating function of the canonical transformation, vanishes identically, and the equations of motion read

$$\frac{\partial K}{\partial P_i} = \dot{Q}_i = 0, \quad -\frac{\partial K}{\partial Q_i} = \dot{P}_i = 0. \tag{5.29}$$

If $F \equiv F(x, P, t)$, then $p_i = \partial F/\partial x_i$, and K rewrites

$$K = H\left(x_1, ..., x_n, \partial F/\partial x_1, ..., \partial F/\partial x_n\right) + \frac{\partial F}{\partial t} = 0, \tag{5.30}$$

which is known as the *Hamilton-Jacobi equation*, and its solution, S, is the principal Hamilton function. S is the generating function of a canonical transformation that leads to constant coordinates and momenta, and defines an equivalence between the $2n$ first-order differential equations of motion and the first-order partial-derivative Hamilton-Jacobi equation. The Hamilton principal function is related to the Lagrangian L by

$$S = \int dt L + const, \tag{5.31}$$

since

$$\frac{dS}{dt} = \frac{\partial S}{\partial x_i}\dot{x}_i + \frac{\partial S}{\partial t} = p_i \dot{x}_i - H. \tag{5.32}$$

It is always possible to split the function S into two parts, one that depends on the $\{x_i\}$ only, and one that depends on time only, if the Hamiltonian H does not depend on time explicitly: in this case, the Hamilton-Jacobi equation reads

$$\frac{\partial S}{\partial t} + H\left(x_i, \frac{\partial S}{\partial x_i}\right) = 0, \tag{5.33}$$

so that it can be guessed that the expression of S as a function of the new momenta $\{\alpha_i\}$ should write

$$S(x_i, \alpha_i, t) = W(x_i, \alpha_i) - \alpha_1 t. \tag{5.34}$$

After direct substitution of (5.34) in (5.33), one finds

$$H\left(x_i, \frac{\partial W}{\partial x_i}\right) = 0, \tag{5.35}$$

which is independent of time, and the integration constant α_1 corresponds to the constant value of the Hamiltonian H. The function W, the characteristic Hamilton function, is defined by the condition $H(x_i, p_i) = \alpha_i$, and generates a canonical transformation where all the coordinates are cyclic (the Lagrangian does not depend on them).

5.1.6.2 *The WKB method*

From the Schrödinger equation in one dimension

$$\left[\frac{\hat{p}^2}{2m} + V(x)\right]\psi = H\psi = E\psi, \tag{5.36}$$

the probability density

$$\rho(x,t) = |\psi(x,t)|^2 \tag{5.37}$$

and the probability flux

$$j(x,t) = \left(\frac{\hbar}{m}\right)\mathrm{Im}(\psi^+ \frac{d\psi}{dx}) = \frac{\rho\nabla S}{m}, \tag{5.38}$$

which obey the continuity equation

$$\frac{\partial\rho}{\partial t} + \frac{dj}{dx} = 0, \tag{5.39}$$

lead to the expression of the wave function

$$\psi(x,t) = \sqrt{\rho(x,t)}\exp\frac{iS(x,t)}{\hbar}. \tag{5.40}$$

After direct substitution of (5.40) in the time dependent Schrödinger equation (5.36), it is easy to recognize that the solution of the classical Hamilton-Jacobi equation is

$$S(x,t) = W(x) - Et = \pm\int_x dx' \sqrt{2m[E - V(x')]} - Et : \tag{5.41}$$

we stress that the phase of the wave function can be interpreted as the solution of the Hamilton-Jacobi equation from a classical point of view. For a stationary state $\partial\rho/\partial t = 0$: by means of the continuity equation (5.39), an expression for ρ is found, which, substituted in the wave function (5.40), leads to the approximated solution

$$\psi(x,t) \simeq \left(\frac{const}{[E - V(x)]^{1/4}}\right)\exp\left[\pm\left(\frac{i}{\hbar}\right)\int_x dx' \sqrt{2m[e - V(x')]} - \frac{iEt}{\hbar}\right], \tag{5.42}$$

which is known as the WKB (Wemtzel, Kramers, Brillouin) solution.

5.1.7 *Semi-classical states*

The transition from a quantum to a classical picture is one of the most intriguing issue one deals with. This is usually achieved by some (semiclassical) states. These states $|\psi\rangle$ are characterized by the following properties

- the expectation value of observables A, *i.e.* $\frac{\langle\psi|A|\psi\rangle}{\langle\psi|\psi\rangle}$, follows a classical trajectory,
- the deviations $\delta A^2 = \frac{\langle\psi|A^2|\psi\rangle - \langle\psi|A|\psi\rangle^2}{\langle\psi|\psi\rangle^2}$ are negligible at macroscopic scales.

The second condition is quite arbitrary, since it depends on the choice of the "macroscopic" scale. Usually, phase-space coordinates are taken as observables and the semiclassical states are fixed such that the lower bound of the uncertainty relation holds for their deviations.

5.1.7.1 *Coherent states*

In order to fulfill these requirements let us introduce *coherent states* for a one-dimensional system, whose phase space is parametrized by the canonical variables (x, p). Given a length scale d, the following operator can be defined

$$\hat{a} = \frac{1}{\sqrt{2}d}(\hat{x} + i\frac{d^2}{\hbar}\hat{p}). \tag{5.43}$$

The coherent states are the eigenstates of $\hat{a}$, *i.e.*

$$\hat{a}|\psi\rangle_{coh} = a|\psi\rangle_{coh}, \tag{5.44}$$

and their wave-function reads as

$$\psi_{coh}(x) = Ne^{-\frac{(x-x_0)^2}{2d^2}} e^{ik_0(x-x_0)}, \tag{5.45}$$

x_0 and $k_0 d^2$ being the real and imaginary part of a, respectively, while N is a normalizing factor. From equation (5.45), it can be inferred that the expectation values of x and $\hat{p}$ are given by x_0 and $k_0/\hbar$, respectively, and the associated deviations are d^2 and $\hbar^2/d^2$. Therefore, the coherent states are peaked on the classical phase space point (x_0, p_0). In the case the dynamical system is a harmonic oscillator, coherent states remain coherent during the evolution and it can be shown that the dynamics of the expectation values coincide with the classical ones, while deviations remain bounded.

5.1.7.2 *Complexifier technique*

The extension of the procedure above to more complicated configuration spaces requires an improvement of the semiclassical technique. In particular, it is possible to generalize the operator $\hat{a}$, by introducing the "complexifier" [279], *i.e.* a smooth positive definite function C, which generates the classical analogous of a, which we denote by z, as follows

$$z = \Sigma_{n=0}^{\infty} \frac{i^n}{n!}[x, C]_n = e^{-C/\hbar} x e^{C/\hbar}, \tag{5.46}$$

where $[C, x]_0 = x$ and $[C, x]_{n+1} = [C, [C, x]_n]$. If we define the state $\psi_{x'}(x) = e^{-\hat{C}/\hbar}\delta_{x'}(x)$, $\delta_{x'}(x) = \delta(x - x')$ and we consider its analytic continuation $\psi_z(x)$, it can be demonstrated that ψ_z is an eigenstate of the operator $\hat{z}$. From this condition, ψ turns out to be peaked around a classical phase space point determined by the eigenvalue z.

The complexifier technique generalizes the definition of coherent states and allows to construct semiclassical states also in more complicated cases with respect to the harmonic oscillator. The standard case (discussed in the previous paragraph) can be inferred fixing $C = \frac{p^2}{2}$.

5.1.7.3 *The Ehrenfest theorem*

If motion equations are looked for, the proper Hamiltonian operator has to be found. In QM, for a physical system, which has a classical analogue, the variables x and p have to be replaced by the correspondent operators. The Heisenberg equations of motion can be evaluated for a free particle, whose Hamiltonian is taken of the form

$$H = \frac{\hat{p}^2}{2m} = \sum_i \frac{\hat{p}_i^2}{2m}, \tag{5.47}$$

where $i = 1, 2, 3$, and m is the mass of the particle. In the Heisenberg picture, one has

$$\frac{d\hat{p}_i}{dt} = \frac{1}{i\hbar}[\hat{p}_i, H] = 0, \tag{5.48}$$

$$\frac{dx_i}{dt} = \frac{1}{i\hbar}[x_i, H] = \frac{\hat{p}_i}{m} = \frac{\hat{p}_i(0)}{m}, \tag{5.49}$$

the momentum operator is a constant of the motion, $\hat{p}_i(0) = \hat{p}_i(t)$, while for the position operator, the solution

$$x_i(t) = x_i(0) + \left(\frac{\hat{p}_i(0)}{m}\right) t \tag{5.50}$$

is found.

If a potential $V(x)$ is added to the Hamiltonian (5.47), *i.e.*

$$H = \frac{\hat{p}^2}{2m} + V(x), \tag{5.51}$$

equations (5.48) and (5.49) can be rewritten as

$$\frac{d\hat{p}_i}{dt} = \frac{1}{i\hbar}[\hat{p}_i, V(x)] = -\frac{\partial}{\partial x_i}V(x), \tag{5.52}$$

$$\frac{dx_i}{dt} = \frac{1}{i\hbar}[x_i, H] = \frac{\hat{p}_i}{m}, \tag{5.53}$$

so that

$$\frac{dx_i^2}{dt^2} = \frac{1}{m}\frac{d\hat{p}_i}{dt}, \tag{5.54}$$

which, together with (5.52), gives the quantum analogue of the Newton law,

$$m\frac{d^2x}{dt^2} = -\nabla V. \tag{5.55}$$

Its expectation value reads

$$m\frac{d^2}{dt^2}\langle x\rangle = -\langle\nabla V\rangle, \tag{5.56}$$

and is known as the *Ehrenfest theorem*: expectation values are independent of the picture, and there is no relic of $\hbar$. In the case of the free Hamiltonian, the time evolution of a wave-packet describes a delocalization as time goes by and after a certain amount of time the state is no more semiclassical. In the case of the potential $V(x)$, the motion of center of the wave-packet is that of a classical particle with a potential $V(x)$ and the delocalization does not necessary occur. In this last case the semiclassical description does not drop down.

Therefore, the property of a state being semiclassical can be violated during the dynamic evolution. For an application of the Ehrenfest theorem in the framework of canonical quantum gravity, see [154].

5.2 Weyl quantization

Weyl quantization [296] consists of assuming canonical commutation relation for two operators $\hat{p}$, $\hat{x}$, as in (5.9), and in establishing a different (Weyl) representation of the operators. One can thereafter implement a quantization program, and then recover information about the standard quantization method via the so-called GNS construction.

5.2.1 *Weyl systems*

Given a symplectic vector space (E, σ), *i.e.* a vector space E endowed with a symplectic (non-degenerate, antisymmetric, bilinear) form σ, a *Weyl system* is the strongly-continuous[3] map W from E to unitary transformations on some Hilbert space H

$$W : E \to \mathcal{U}(\mathrm{H}), \tag{5.57}$$

and the Weyl form of the commutation relations reads

$$W(e_1)W(e_2) = e^{\frac{i}{\hbar}\sigma(e_1,e_2)}W(e_2)W(e_1), \tag{5.58}$$

where the cocycle of the representation is determined by the symplectic structure σ.

Complex coordinates, and the construction of a Fock space, with creation and annihilation operators, can be defined by the introduction of a complex form $J : E \to E$, $J^2 = -1$. An inner product on E can be defined by using J and σ.

It is possible to decompose the vector space E as $\mathcal{L} \oplus \mathcal{L}^*$, where $\mathcal{L} \subset E$ is a Lagrangian (both isotropic and coisotropic) subspace of E. The Hilbert space H is the space of square-integrable functions ϕ on $\mathcal{L}$ endowed with the translation-invariant Lebesgue measure $d\mu$, *i.e.* $\mathrm{H} = L^2(d\mu, \mathcal{L})$. In this decomposition, vectors on E can be defined as $e = (\alpha, \beta)$, $\beta \in \mathcal{L}$, $\alpha \in \mathcal{L}^*$, and the action of W on the functions ϕ reads

$$U(\alpha)\phi(x) \equiv W((\alpha, 0))\phi(x) = e^{\frac{i}{\hbar}\alpha x}\phi(x) \tag{5.59}$$

$$V(\beta)\phi(x) \equiv W((0, \beta))\phi(x) = \phi(x - \beta). \tag{5.60}$$

The vacuum expectation values of the operators U and V depends on the metric g constructed out of J, *i.e.* $g(e_1, e_2) = \sigma(e_1, Je_2)$.

5.2.2 *The Stone-von Neumann uniqueness theorem*

Within this scheme the Schrödinger representation is obtained via the following representation of W

$$W(\alpha, \beta)\psi(x) = e^{\frac{i}{2}\alpha\beta}e^{i\alpha x}\psi(x + \beta). \tag{5.61}$$

Such a representation is unitary, irreducible and weakly-continuous[4] in $\alpha + J\beta$. The last condition ensures that $U(\alpha)$ and $V(\beta)$ are weakly continuous in α and β, hence the corresponding generators $\hat{x}$ and $\hat{p}$ are self-adjoint and well-defined.

[3] A strongly continuous map is continuous in the strong operator topology.

[4] Weakly continuous means continuous in the weak operator topology.

The *Stone-von Neumann Uniqueness theorem* [235] states that any unitary, irreducible and weakly-continuous representation of the Weyl commutation relation on $\mathbb{C}^n$ is isomorphic to the Schrödinger representation. Furthermore, as corollary, it is also possible to show that any representation of the Weyl commutation relation on $\mathbb{C}^n$ is the direct sum of copies of the Schrödinger representation.

5.3 GNS construction

The *Gelfand Naimar Segal (GNS) construction* allows one to gain insight onto different representations of a given algebra [158][5]. From a mathematical point of view, it is based on defining a state as a normalized positive linear form.

Given a function ϕ and an algebra $\mathcal{A}$, $\phi : \mathcal{A} \to \mathbb{C}$ is called a linear form over $\mathcal{A}$ if

$$\phi(\alpha A + \beta B) = \alpha\phi(A) + \beta\phi(B), \quad \forall A, B \in \mathcal{A}, \ \forall \alpha, \beta \in \mathbb{C}\,. \tag{5.62}$$

We can endow $\mathcal{A}$ with a norm $\|..\|$ and if

$$\|AB\| \leq \|A\| \, \|B\| \,, \quad \forall A, B \in \mathcal{A}, \tag{5.63}$$

then $\mathcal{A}$ is a Banach algebra. If $\mathcal{A}$ is a Banach algebra, ϕ is bounded if

$$\mid \phi(A) \mid \leq c \, \|A\| \tag{5.64}$$

and the lowest bound for c is the norm of ϕ.

If

$$(Ax, y) = \overline{(y, Ax)} = (x, A^* y), \quad \forall A \in \mathcal{A}, \tag{5.65}$$

then $\mathcal{A}$ is a * algebra. If $\mathcal{A}$ is a * algebra, with unit, the linear form ϕ is a state if it is

$$\textit{real} \quad \phi(A^*) = \phi(A) \tag{5.66}$$

$$\textit{positive} \quad \phi(A^* A) \geq 0 \tag{5.67}$$

$$\textit{normalized} \quad \|\phi\| = 1. \tag{5.68}$$

As a result, a positive linear form over a Banach * algebra with unit is bounded, and $\|\phi\| = \phi(I)$. Furthermore, it satisfies the Schwarz inequality

$$\mid \phi(AB) \mid^2 \leq \phi(AA^*)\phi(BB^*). \tag{5.69}$$

[5] An (abstract) algebra $\mathcal{A}$ is a set of elements $A, B, \ldots \in \mathcal{A}$ which is closed under the action of a binary operation (the product), *i.e.* $AB = C \in \mathcal{A}$.

In fact, if we assume that the self-adjoint elements of $\mathcal{A}$ correspond to physical observables, and that the unit element I corresponds to the trivial observable, whose value is 1 for any physical state, then such a linear form can be interpreted as an expectation functional over physical observables.

Each positive linear form ω over a $*$ algebra $\mathcal{A}$ defines a Hilbert space H_ω and a representation π_ω of $\mathcal{A}$ by linear operators.

Since $\mathcal{A}$ is a linear space over the field $\mathbb{C}$, ω defines a Hermitian semi-definite product on $\mathcal{A}$, *i.e.*

$$\langle A|B\rangle = \omega(A,B), \quad \langle A|A\rangle \geq 0, \quad |\langle A|B\rangle|^2 \leq \langle B|B\rangle\langle A|A\rangle, \quad \forall A,B \in \mathcal{A}. \tag{5.70}$$

The set $\mathcal{J} \subset \mathcal{A}$, $\mathcal{J} = \{X \in \mathcal{A} : \omega(X^*X) = 0\}$, is a left ideal, and is called the Gelfand ideal of the state. Eliminating this set from $\mathcal{A}$, *i.e.* considering $\mathcal{A}/\mathcal{J}$ allows us to obtain linear space equipped with Hermitian positive-definite scalar product, and a vector ψ in this space corresponds to the equivalence class $[A]$ of the elements of $\mathcal{A}$ modulo $\mathcal{J}$, $\psi = \{\mathcal{A} + \mathcal{A}\}$, with

$$\langle\psi|\psi\rangle \equiv \|\psi\|^2 > 0. \tag{5.71}$$

It is worth remarking that the scalar product defined in (5.70) does not depend on $[A]$. Because of that, the action of the representation $\pi_\omega(A)$, defined on $\mathcal{A}/\mathcal{J} \subset \mathrm{H}_\omega$, is

$$\pi_\omega(A)\psi = [AB] \quad if \quad \psi = [B]. \tag{5.72}$$

A representation π is called cyclic if a cyclic vector $\Omega \in \mathrm{H}$ exists. A vector Ω is called cyclic if $\pi(\mathcal{A})\Omega$ is dense in H. If $\mathcal{A}$ has a unit, $\Omega = [I]$. Furthermore

$$\omega(A) = \langle\Omega|\pi_\omega(A)|\Omega\rangle, \tag{5.73}$$

and it is sometimes referred to as the vacuum state. Similarly, any vector $\psi \in \mathrm{H}_\omega$ defines a state

$$\omega_\psi(A) = \langle\psi|\pi_\omega(A)|\psi\rangle. \tag{5.74}$$

The GNS theorem states that, given a C^* algebra $\mathcal{A}$ endowed with unit, and a linear positive functional ω in a compact subset of $\mathcal{A}$, the triple $(\mathrm{H}_\omega, \pi_\omega, \Omega)$ *exists and is unique.*

If the algebra is commutative, the Hilbert space H_ω can be seen as the space of square-summable functions on the spectra of the C* algebra.

The Schrödinger representation is obtained by the GNS-construction fixing $\omega(W(\alpha,\beta)) = e^{-\frac{\alpha^2+\beta^2}{2}}$.

5.4 Polymer representation

The abstract techniques described in the previous section can be specified for the case of a physical system, where a cutoff has to be somehow introduced, and the removal of the cutoff can be analyzed within the framework of the proper GNS construction. To this end, it is possible to fix a physical meaning to the tools previously established.

5.4.1 *Difference operators versus differential operators*

The implementation of a cut-off in a physical system, suggested by the need to describe a discretized spacetime, is reflected by the introduction of difference operator (instead of differential operators) in the quantum version of a model. Before discussing in detail the features of the polymer representation of quantum mechanics, it is interesting to remark that there are in principle several possibilities to define difference operators [115], *i.e.*

$$D_{x_1,x_2}f(x) = \frac{f(x_2)-f(x_1)}{x_2-x_1}, \quad x_2 \neq x_1, \quad x_1, x_2 \in \mathbb{R}, \tag{5.75}$$

and the differential case should be recovered in the limiting process $x_2 \to x_1$. In particular, the derivative can be replaced by two kinds of operators

- additive operators D^a, *i.e.*

$$D^a f(x) = \frac{f(x+a)-f(x-a)}{(x+a)-(x-a)} = \frac{f(x+a)-f(x-a)}{2a}, \quad a \in \mathbb{R} \tag{5.76}$$

- multiplicative operators D^q, *i.e.*

$$D^q f(x) = \frac{f(qx)-f(q^{-1}x)}{qx-q^{-1}x} = \frac{1}{x}\frac{f(qx)-f(q^{-1}x)}{q-q^{-1}}, \quad q \in \mathbb{R}. \tag{5.77}$$

Considering difference operators instead of differential operators can be consistent for those cases, where differential operators do not exist, or where a discretized underlying structure is hypothesized, *i.e.* if a lattice is considered. According to the previous definitions, two kinds of lattices can be recognized, respectively

- uniform a-lattices, i.e. $\mathbb{L}_a = \{x_0 + ja | j \in \mathbb{Z}, x_0 \in \mathbb{R}\}$
- uniform q-lattices, i.e. $\mathbb{L}_q = \{x_0 q^j | j \in \mathbb{Z}, x_0 \in \mathbb{R}, x_0 \neq 0\}$.

The introduction of a cut-off is closely related to the definition of a scale, and the consequent continuum limit. The relevance in introducing a scale is the possibility to focus the attention from the points of the lattice to the intervals defined by the lattice, with the aim of approximating continuous functions on $\mathbb{R}$ with functions that are constants on such intervals. For any given scale, one can approximate functions on the lattice, and one can pass from one scale to the next one by a coarse-graining map [105].

5.4.2 *The polymer representation of Quantum Mechanics*

The polymer representation of quantum mechanics [27, 105, 106, 127] consists in defining abstract kets, labeled by a real number, and then considering a suitable finite subset of them, whose Hilbert space is defined by the corresponding inner product [106]. This procedure can be shown to be an inequivalent representation of the Weyl algebra with respect to the ordinary Schrödinger one. This representation helps one gain insight onto some particular features of quantum mechanics, when an underlying discrete structure is somehow hypothesized. The request that the Hamiltonian associated to the system be of direct physical interpretation defines the polymer phase space, and the continuum limit can be recovered by the introduction of the concept of scale [105].

In the simplest toy model, we can consider a particle moving on the real line $\mathbb{R}$, so that the symplectic vector space will be its phase space $\Gamma = \mathbb{R}^2$, which can be automatically decomposed in terms of the x space ($\mathcal{L}$) and the p space ($\mathcal{L}^*$). Accordingly, the Hilbert space H of the functions ϕ will be $\mathrm{H} = L^2(\mathbb{R}, d\mu)$. The exponentiated versions of the operators x and p,

$$U(\alpha) = e^{\frac{i}{\hbar}\alpha x}, \quad V(\beta) = e^{\frac{i}{\hbar}\beta p} \tag{5.78}$$

generates the Weyl form of the commutation relation

$$U(\alpha)V(\beta) = e^{\frac{i}{\hbar}\alpha\beta}V(\beta)U(\alpha). \tag{5.79}$$

The Weyl algebra associated to this system is the set of finite linear combinations of the generators.

The introduction of a cutoff can be related to the particular choice of the complex structure J_d

$$J_d = \begin{pmatrix} 0 & -d^2 \\ 1/d^2 & 0 \end{pmatrix}, \tag{5.80}$$

where the parameter d has the dimensions of a length. It is possible to show that the Hilbert space H of the functions ϕ for the system is one with

Gaussian measure, $\mathrm{H}_d = L^2(\mathbb{R}, d\mu_d)$, where

$$d\mu_d = K_d^2\, dx = \frac{1}{d\sqrt{\pi}} e^{-\frac{x^2}{d^2}} dx. \tag{5.81}$$

Furthermore, the action of (5.78) on the functions ϕ and their vacuum expectation values are

$$U(\alpha)\phi(x) = e^{\frac{i}{\hbar}\alpha x}\phi(x), \quad \langle U(\alpha)\rangle = e^{-\frac{1}{4}\frac{d^2\alpha^2}{\hbar^2}} \tag{5.82}$$

$$V(\beta)\phi(x) = e^{\frac{\beta}{d^2}(x-\beta/2)}\phi(x), \quad \langle V(\beta)\rangle = e^{-\frac{1}{4}\frac{\beta^2}{d^2}}, \tag{5.83}$$

and the value in the exponents of the vacuum expectation values depends on the metric $g_d(e_1, e_2) = \omega(e_1, Je_2)$, for $e_1 = e_2 = (0, \alpha)$ and $e_1 = e_2 = (\beta, 0)$, respectively. The states of H_d generated by the action of the operator $U(\alpha)$ on the vacuum $\phi_0(x) = 1$ are

$$\phi_\alpha(x) = U(\alpha)\phi_0(x) = e^{\frac{i}{\hbar}\alpha x}, \tag{5.84}$$

and the inner product, defined by (5.81), is

$$\langle \phi_\mu, \phi_\nu\rangle_d = \int \frac{1}{d\sqrt{\pi}} e^{-\frac{x^2}{d^2}} e^{-\frac{i}{\hbar}\mu x} e^{\frac{i}{\hbar}\nu x} dx. \tag{5.85}$$

Information about the Schrödinger representation of the model, $\phi(x) \in \mathrm{H}_S = L^2(\mathbb{R}, dx)$, can be recovered by the mapping

$$\psi(x) = K\phi(x) = \frac{e^{-\frac{x^2}{2d^2}}}{d^{1/2}\pi^{1/4}}\phi(x), \quad \psi_0(x) = \frac{e^{-\frac{x^2}{2d^2}}}{d^{1/2}\pi^{1/4}}\phi_0(x) = \frac{e^{-\frac{x^2}{2d^2}}}{d^{1/2}\pi^{1/4}}. \tag{5.86}$$

For finite values of d, the mapping is well-defined, so that all the d representations are unitarily equivalent to the Schrödinger one. Contrastingly, in both limits $d \to 0$ and $1/d \to 0$, the descriptions are not unitarily equivalent to the Schrödinger one anymore.

5.4.3 *Kinematics*

One can start by considering abstract kets $|\mu\rangle$, $\mu \in \mathbb{R}$, and a suitable subset defined by $\mu_i \in \mathbb{R}$, $i = 1, 2, ..., N$. These kets are assumed to be an orthonormal basis, *i.e.* $\langle\mu|\nu\rangle = \delta_{\mu\nu}$, along which any state ϕ can be projected. This defines a Hilbert space H_{pol}, on which two basic operators act, the symmetric "label" operator, $\hat{\epsilon}$, such that $\hat{\epsilon}|\mu\rangle = \mu|\mu\rangle$, and a one-parameter family of unitary operators, $\hat{s}(\lambda)$, such that $\hat{s}(\lambda)|\mu\rangle = |\mu + \lambda\rangle$. Because all kets are orthonormal, $\hat{s}(\lambda)$ is discontinuous, and cannot be obtained from any Hermitian operator by exponentiation. It is worth noting that this Hilbert space is not separable[6].

For the toy model of a 1-dimensional system, whose phase space is described by the variables p and x, the polymer representation techniques find interesting applications when one of the two variables is supposed to be discrete. This discreteness will affect both wave functions, obtained by projecting the physical state on the p or x basis (polarization), and the operators associated to the canonical variables, acting on them.

For later purposes, we will discuss only the case of a discrete position variable x, and the corresponding momentum polarization. In this case, wave functions are given by

$$\psi_\mu(p) = \langle p|\mu\rangle = e^{ip\mu}. \tag{5.87}$$

Accordingly, the "label" operator $\hat{\epsilon}$ is easily identified with $\hat{x}$, *i.e.*

$$\hat{x}\phi_\mu = -i\frac{\partial}{\partial p}\psi_\mu = \mu\psi_\mu, \tag{5.88}$$

while the "shift" operator does not exist, as discussed previously.

It can be shown that the corresponding Hilbert space is $\mathrm{H}_{pol} = L^2(\mathbb{R}_B, d\mu_H)$, *i.e.* the set of square-integrable functions defined on the Bohr compactification of the real line $\mathbb{R}_B$[7], with a Haar measure $d\mu_H$. Since the kets $|\mu\rangle$ are arbitrary but finite, the wave functions can be interpreted as quasi-periodic function, with the inner product

$$\langle\psi_\mu|\psi_\lambda\rangle = \int_{\mathbb{R}_B} d\mu_H \bar{\psi}\mu(p)\psi_\lambda(p) = \lim_{L\to\infty}\frac{1}{2L}\int_{-L}^{L} dp\ \bar{\psi}\mu(p)\psi_\lambda(p) = \delta_{\mu,\lambda}. \tag{5.89}$$

5.4.4 *Dynamics*

The Hamiltonian operator H describing a quantum-mechanical system is usually a function of both coordinate and momentum, *i.e.*

$$H = H(x,p) = \frac{p^2}{2m} + V(x) \tag{5.90}$$

while, in the particular case of a discrete position variable in the momentum polarization, p cannot be implemented as an operator. As a first step, a

[6]A Hilbert space is separable if and only if it admits a countable orthonormal basis.

[7]The Bohr compactification is a particular compactification, see [168] for the mathematical definition. $\mathbf{R}_{Bohr}$ is the dual group of $\mathbf{R}$, equipped with a discrete topology, in which any open set do not contain more than one point.

suitable approximation for the kinetic term has to be provided. For this purpose, it is useful to restrict the arbitrary kets $|\mu_i\rangle$, $i \in \mathbb{R}$ to $|\mu_i\rangle$, $i \in \mathbb{Z}$, *i.e.* to introduce the notion of regular graph γ_{μ_0}, defined as a numerable set of equidistant points, whose separation is given by the parameter μ_0, $\gamma_{\mu_0} = \{x \in \mathbb{R} | x = n\mu_0, \forall n \in \mathbb{Z}\}$. The associated Hilbert space $\mathrm{H}_{\gamma_{\mu_0}}$ is separable. Because of the regular graph γ_{μ_0}, the eigenfunctions of $\hat{p}_{\mu_0}$ must be of the form $e^{im\mu_0 p}$, $m \in \mathbb{Z}$, which are Fourier modes, of period $2\pi/\mu_0$. The inner product (5.89) is equivalent to the inner product on a circle S^1 with uniform measure, *i.e.*

$$\langle \phi(p)|\psi(p)\rangle_{\mu_0} = \frac{\mu_0}{2\pi}\int_{-\pi/\mu_0}^{\pi/\mu_0} \hat{\phi}(p)\psi(p), \tag{5.91}$$

with $p \in (-\pi\mu_0, \pi/\mu_0)$, so that $\mathrm{H}_{\gamma_{\mu_0}} = L^2(S^1, dp)$. Within this space, it is possible to construct an approximation for the "shift" operator, *i.e.* a regulated operator $\hat{p}_{\mu_0}$,

$$\hat{p}_{\mu_0}|\mu_n\rangle = -\frac{i}{2\mu_0}\left(|\mu_{n+1}\rangle - |\mu_{n-1}\rangle\right). \tag{5.92}$$

More precisely, the polymer paradigm can be understood as the formal substitution

$$p \to \frac{1}{\mu_0}\sin(\mu_0 p), \tag{5.93}$$

where the incremental ratio (5.92) has been evaluated for exponentiated operators. The Hamiltonian operator H_{μ_0}, which lives in $\mathrm{H}_{\gamma_{\mu_0}}$, reads $H_{\mu_0} = \frac{\hat{p}^2_{\mu_0}}{2m} + V(\hat{x})$, where the action of the new multiplication operator $\hat{p}^2_{\mu_0}$ on wave functions in the momentum polarization is

$$\hat{p}^2_{\mu_0}\psi(p) = \frac{1}{2\mu_0^2}\left[1 - \cos(2\mu_0 p)\right]\psi(p), \tag{5.94}$$

while the differential operator $\hat{x}$ (5.88) is well defined.

5.4.5 *Continuum limit*

The physical Hilbert space of such theories can be constructed as the continuum limit of effective theories at different scales, and can be illustrated to be unitarily isomorphic to the ordinary one, $\mathrm{H}_S = L^2(\mathbb{R}, dp)$.

To this end, it is useful to remark that it is impossible to obtain H_S starting from a given graph $\gamma_0 = \{x \in \mathbb{R} | x = n\mu_0, \forall n \in \mathbb{Z}\}$ by dividing each interval μ_0 into 2^k in new intervals of length $a_k = a_0/2^k$, because H_S cannot be embedded into H_{pol}.

It is however possible to go the other way round and to look for a continuous wave function that is approximated by a wave function over a graph, in the limit of the graph becoming finer. In fact, if one defines a scale C_k, *i.e.* a decomposition of $\mathbb{R}$ in terms of the union of closed-open intervals that have lattice points as end points and cover $\mathbb{R}$ without intersecting, one is then able to approximate continuous functions with functions that are constant on these intervals. As a result, at any given scale C_k, the kinetic term of the Hamiltonian operator can be approximated as in (5.94), and effective theories at given scales are related by coarse-graining maps. In particular, it is necessary to regularize the Hamiltonian, treated as a quadratic form, as a self-adjoint operator at each scale by introducing a normalization factor in the inner product. The convergence of microscopically-corrected Hamiltonians is based on the convergence of energy levels and on the existence of completely normalized eigencoverctors compatible with the coarse-graining operation.

5.5 Quantization of Hamiltonian constraints

First-class and second-class constraints must be treated in different ways for what concern quantization [167].

Second-class constraints χ_α can be safely solved classically, since the symplectic structure induced on the constraints hypersurface is not degenerate and there is no obstruction with quantizing coordinates in the reduced phase space. A different approach is to work with the coordinates of the full phase space and to quantize Dirac brackets [167] instead of Poisson ones. Since it is often difficult to find irreducible representations of the Dirac brackets, it can be useful in some cases to turn second-class constraints into first-class constraints by adding extra-variables.

First-class constraints can be quantized by means of the reduced-phase-space method and of the Dirac prescription. The reduced-phase-space method consists in eliminating the gauge degrees of freedom by fixing some gauge conditions, which make the total set of constraints second-class. Hence, one solves explicitly the constraints and perform the canonical quantization in the reduced phase space. The issue of this approach is how to reconcile it with gauge invariance. This can be done by a full BRST (Becchi-Rouet-Stora-Tyutin) analysis (see for instance [97]).

The *Dirac prescription* is the most powerful tool: it consists in canonically quantizing the full set of phase space coordinates in a certain space and in imposing the constraints as quantum operators annihilating physical states. If the resulting space can be endowed with a scalar product, it is called *physical Hilbert space*. In the next paragraphs, we will apply the Dirac prescription to some relevant examples. We will also discuss the *group averaging technique*, which is a powerful tool to equip the space of solutions of the constraints with a scalar product.

5.5.1 *Non-relativistic particle*

The action for a single non-relativistic particle in a potential V can be written as a function of the canonical variables $\{x^i, p_i\}$, $i = 1, 2, 3$, and the absolute time t:

$$S[x^i, p_i] = \int dt \left[p_i \frac{dx^i}{dt} - H\left(x^i, p_i, t\right) \right]. \tag{5.95}$$

An equivalent expression for (5.95) can be obtained by considering an arbitrary label time τ and the set of variables $\{x^\mu = \left(t, x^i\right), p_\mu = (p_0, p_i)\}$, $\mu = 0, 1, 2, 3$, through the introduction of the constraint $\mathcal{H} \equiv p_0 + H\left(x^\mu, p_i\right) = 0$ and a Lagrange multiplier N, which are needed in order to restore the right number of degrees of freedom,

$$S[x^\mu, p_\mu; N] = \int d\tau \left(p_\mu \dot{x}^\mu - N\mathcal{H}\right) \qquad \dot{x}^\mu = \frac{dx^\mu}{d\tau}. \tag{5.96}$$

The variation with respect to p_0 leads to the physical connotation of N, $N = \dot{t}$, while the variations with respect to x^i, p_i and t lead to the Hamilton equations, expressed for the parameter τ, and to the energy-balance equation $\dot{H} = (\partial H/\partial t)\dot{t}$, respectively.

The quantization of the constraints consists in the quantization of the variables involved in the constraints, and in the definition of the spaces where these operators live in.
After defining the operator analogues, obeying standard commutation relations, of the canonical variables, $\mathcal{H}$, which becomes an operator itself as an effect the substitution, $\mathcal{H} = 0 \rightarrow \hat{\mathcal{H}} = 0$, picks up the physical states, $\hat{\mathcal{H}}\psi = 0$, for which a suitable Hilbert space and appropriate self-adjoint operators have to be defined. If the coordinate representation is chosen for the operators, *i.e.* $x^\mu \rightarrow \hat{x^\mu} \equiv x^\mu$, and $p_\nu \rightarrow \hat{p}_\nu \equiv -i\hbar\partial_\nu$, the wave functions are $\psi(x^\mu)$. Since the operators $\hat{x}^\mu$ and $\hat{p}_\mu$ satisfy the commutation relation $[\hat{x}^\mu, \hat{p}_\nu] = i\hbar\delta^\mu_\nu$, the factor ordering of these operators becomes crucial in

the requirement of preserving the covariance and the algebra of constraints, as well as in the definition of the inner product.

As an example for the definition of the inner product, the dynamics of a non relativistic particle can be investigated, and the quantum constraint for (5.96) can be expressed as invariant under diffeomorphisms in the x^i-space (endowed with a metric h_{ij}): a factor ordering for $\hat{\mathcal{H}}$ can be chosen, such that a Laplacian on the field $\psi(x^i)$ appears,

$$\hat{\mathcal{H}} = \hat{p}_0 + \frac{1}{2}h^{-1/4}\hat{p}_i\sqrt{h}\, h^{ij}\hat{p}_j h^{-1/4} + V. \tag{5.97}$$

Accordingly, the inner product

$$\langle\psi_1, \psi_2\rangle \equiv \int_{t=const} d^3x \; \psi_1^*(x,t)\psi_2(x,t) \tag{5.98}$$

is formally consistent with the space of the constraints $\hat{\mathcal{H}}\psi = 0$ and defines the norm of the Hilbert space of the solutions; in fact, the two fields obey the constraint $\mathcal{H}\psi_1 = \mathcal{H}\psi_2 = 0$ and the continuity equation

$$\partial_t \rho_{12} + \partial_i j^i_{12} = 0, \tag{5.99}$$

where

$$\rho_{12} \equiv \psi_1^*\psi_2 \tag{5.100}$$

and

$$j^i_{12} \equiv \frac{i\hbar}{2}h^{ij}\left(\psi_1^*\overleftrightarrow{\partial}_j\psi_2\right) = \frac{i\hbar}{2}h^{ij}\left(\psi_1^*\partial_j\psi_2 - \partial_j(\psi_1^*)\psi_2\right). \tag{5.101}$$

It is worth noting that the constraint $\hat{\mathcal{H}}\psi = 0$ is the Schrödinger equation, and the inner product between two wave-function does not depend, by construction, on any particular time slice.

5.5.2 *Relativistic particle*

The definition of an inner product is not always straightforward, and one-particle states do not always find a precise physical meaning. This is the case of a relativistic particle (whose wave function we denote by $\phi(x^\mu)$), with a parametrized action of the form (5.96), in the x^μ representation in curved spacetime, whose constraint

$$\mathcal{H} \equiv \frac{1}{2m}\left(g^{\mu\nu}p_\mu p_\nu + m^2\right) = 0 \tag{5.102}$$

has to be quantized covariantly under spacetime diffeomorphisms.

A factor ordering of the super-Hamiltonian can be found, such that the D'Alembert operator acts on scalars, *i.e.*

$$\hat{\mathcal{H}}\phi(x^\mu) = \left[(g)^{-1/2}\hat{p}_\mu\sqrt{g}g^{\mu\nu}\hat{p}_\nu + m^2\right]\phi(x^\mu) = 0, \tag{5.103}$$

which is the Klein-Gordon equation in curved spacetime. As in the previous case, a continuity equation can be worked out,

$$\nabla_\mu j^\mu_{12} = 0, \quad j^\mu_{12} \equiv \frac{1}{2}g^{\mu\nu}\phi_1\overleftrightarrow{\partial}_\nu\phi_2 \tag{5.104}$$

for two fields ϕ_1 and ϕ_2. Nonetheless, the functional

$$\Omega\left[\phi_1,\phi_2\right] = \frac{1}{2}\int_\sigma d\sigma_\mu g^{\mu\nu}\phi_1\overleftrightarrow{\partial}_\nu\phi_2, \tag{5.105}$$

though independent of the spacetime hypersurface σ taken into account, does not define the Hilbert space of the solutions, because it is antisymmetric: it is, indeed, the symplectic form of such a space.

It is possible, however, to build a complex Hilbert space from the solutions of the real Klein-Gordon equation in stationary spacetime, endowed with a time-like hypersurface orthogonal Killing vector field t^μ

$$t^\mu = N^2 g^{\mu\nu}t_{,\nu}\,, \tag{5.106}$$

where $x^\mu = (t, x^i)$, N being the lapse function. In such coordinates, the Klein-Gordon equation reads

$$N^2\hat{\mathcal{H}}_\phi \equiv \ddot{\phi} + \hat{H}_N\phi = 0, \tag{5.107}$$

$$\hat{H}_N\phi \equiv N\left[-g^{-1/2}\partial_i\left(Ng^{1/2}g^{ij}\partial_j\phi\right) + Nm^2\phi\right]. \tag{5.108}$$

In the Hilbert space whose inner product reads

$$(\phi_1,\phi_2) \equiv \int g^{1/2}N^{-1}\phi_1\phi_2\, d^3x, \tag{5.109}$$

the operator $\hat{H}_N$ is symmetric and positive definite, and has a complete set of eigenfunctions $u_E(x^a)$ obeying the eigenvalue equation[8]

$$\hat{H}_N u_E(x^i) = E^2 u_E(x^i), \quad E \geq E_0 > 0\,, \tag{5.110}$$

the projection of the solutions $\phi(t, x^i)$ of the Klein-Gordon equation along this basis allows one to separate positive and negative frequencies,

$$\phi(t,x^i) = \phi^+(t,x^i) + \phi^-(t,x^i), \quad \phi^\pm(t,x^i) = \int_{E_0}^\infty dE\, u_E(x^i)e^{\mp iEt}. \tag{5.111}$$

[8]The existence of a minimum E_0 in the positive spectrum of the energy is guaranteed by the massive character of the particle.

The map J, *i.e.* the complex structure $J^2 = -1$ of the Hilbert space, sends any real solution of the Klein-Gordon equation in another real solution,

$$\phi \rightarrow J\phi \equiv i\phi^+ - i\phi^-, \tag{5.112}$$

does not depend on time and is compatible with Ω, *i.e.*

$$\Omega[\phi_1, J\phi_2] = \Omega[\phi_2, J\phi_1], \quad \Omega[\phi, J\phi] \geq 0\,. \tag{5.113}$$

Therefore, the space of the solution of the Klein-Gordon equation can be expressed as a complex vector space, where the antilinear scalar product reads

$$\langle\phi_1, \phi_2\rangle \equiv \Omega[\phi_1, J\phi_2] + i\Omega[\phi_1, J\phi_2], \tag{5.114}$$

$$\langle\phi_1, J\phi_2\rangle = i\langle\phi_1, \phi_2\rangle, \quad \langle J\phi_1, \phi_2\rangle = -i\langle\phi_1, \phi_2\rangle, \tag{5.115}$$

so that norms are positive definite, $\langle\phi, \phi\rangle \equiv \Omega[\phi, J\phi]$.

5.5.3 *Scalar field*

Let us consider a scalar field ϕ and perform the 3+1 splitting of the metric tensor described in section 4.2.1. The so-called kinematical action reads [193]

$$S^k(p_\mu, y^\mu) = \int_{\mathcal{M}} \{p_\mu \partial_0 y^\mu - b_0^\mu p_\mu\}\, d^4x, \tag{5.116}$$

where y^μ, p_μ denote spacetime coordinates and their momenta, while b_0^μ are the components of the vector $\vec{b}_0$ (4.89). The action of a self-interacting scalar field on a fixed background,

$$S^\phi(\pi_\phi, \phi) = \int_{\mathcal{M}} \left\{\pi_\phi \partial_0 \phi - N\mathcal{H}^\phi - N^i \mathcal{H}_i^\phi\right\} d^4x, \tag{5.117}$$

where π_ϕ is the momentum conjugate to ϕ, and the Hamiltonian terms

$$\mathcal{H}^\phi \equiv \frac{1}{2\sqrt{h}}{\pi_\phi}^2 + \frac{1}{2}\sqrt{h}\, h^{ij}\, \partial_i\phi\, \partial_j\phi + \sqrt{h}V(\phi) \quad \mathcal{H}_i^\phi \equiv \pi_\phi\, \partial_i\phi, \tag{5.118}$$

can be quantized by adding the kinematical action (5.116). In fact, without the kinematical action, the variation has to be performed with respect to ϕ and π_ϕ, but no precise role is assigned to N, N^i, and h_{ij}, since the background is fixed. If the kinematical action S^k is taken into account, so that

$$S^{\phi k} \equiv S^\phi + S^k = \int_{\mathcal{M}} \left\{\pi_\phi \partial_0 \phi + p_\mu \partial_0 y^\mu - N(\mathcal{H}^\phi + \mathcal{H}^k) - N^i(\mathcal{H}_i^\phi + \mathcal{H}_i^k)\right\} d^4x, \tag{5.119}$$

where $\mathcal{H}^k \equiv p_\mu \eta^\mu$ and $\mathcal{H}_i^k \equiv p_\mu b_i^\mu$, the field equations for ϕ remain unchanged, but the Hamiltonian constraints $\mathcal{H}^\phi = -p_\mu \eta^\mu$ and $\mathcal{H}_i^\phi = -p_\mu b_i^\mu$ are obtained.

The canonical quantization is accomplished by the assumption that the states of the system are functionals of the variables, y^μ and ϕ, $\Psi[y^\mu(x^i), \phi(x^i)]$, and by the implementation of the canonical variables to operators, $\{y^\mu, p_\mu, \phi, \pi_\phi\} \to \{\hat{y}^\mu,\ \hat{p}_\mu = -i\hbar\delta(\ )/\delta y^\mu,\ \hat{\phi},\ \hat{\pi}_\phi = -i\hbar\delta(\ )/\delta\phi\}$, the quantum dynamics being described by the equations

$$i\hbar n^\mu \frac{\delta\Psi}{\delta y^\mu} = \hat{\mathcal{H}}^\phi \Psi = \left[-\frac{\hbar^2}{2\sqrt{h}} \frac{\delta}{\delta\phi}\frac{\delta}{\delta\phi} + \frac{1}{2}\sqrt{h} h^{ij}\partial_i\phi\partial_j\phi + \sqrt{h}V(\phi) \right] \Psi \tag{5.120}$$

$$i\hbar e_i^\mu \frac{\delta\Psi}{\delta y^\mu} = \hat{\mathcal{H}}_i^\phi \Psi = -i\hbar\partial_i\phi \frac{\delta\Psi}{\delta\phi}. \tag{5.121}$$

The space of the solutions of (5.121) can be cast into a Hilbert space by defining the inner product

$$\langle \Psi_1 \mid \Psi_2 \rangle \equiv \int_{y^\mu = y^\mu(x^i)} \Psi_1^* \Psi_2 D\phi\,, \tag{5.122}$$

with

$$\frac{\delta \langle \Psi_1 \mid \Psi_2 \rangle}{\delta y^\mu} = 0\,, \tag{5.123}$$

which implies the conserved functional probability distribution $\varrho \equiv \langle \Psi \mid \Psi \rangle$. The semiclassical limit, $\psi = \exp iS^{\phi k}$ is recovered by addressing the WKB approximation (see section 5.1.6.2), *i.e.* by taking the wave functional $\Psi = \exp\left\{\frac{i}{\hbar}\Sigma(y^\mu, \phi)\right\}$ in (5.121) and retaining the leading order as $\hbar \to 0$. It is worth remarking that (5.121) has $5 \times \infty^3$ degrees of freedom, given by the scalar field ϕ and the four components of y^μ.

5.5.4 *The group averaging technique*

The group averaging [32,207] is a tool to find the physical Hilbert space. Given a constraint $\hat{C}$, acting on some quantum states defined in a Hilbert space H, one can formally define a solution of the condition $\hat{C}|\psi\rangle = 0$ as

$$|\psi_{\hat{C}}\rangle = \frac{1}{\Lambda}\int d\lambda\, \hat{U}(\lambda)|\psi\rangle, \tag{5.124}$$

where $\hat{U}(\lambda) = e^{i\lambda\hat{C}}$ is a one-parameter group of transformations generated by $\hat{C}$, while Λ is a normalizing factor. The scalar product between "averaged" physical states can be defined in terms of the original one as follows [32]

$$\langle \psi_{\hat{C}} | \phi_{\hat{C}} \rangle = \frac{1}{\Lambda} \int d\lambda \langle e^{-i\lambda\hat{C}} \psi | \phi \rangle. \tag{5.125}$$

However, if the one-parameter group $\hat{U}(\lambda)$ is not compact the bra $\langle \psi_{\hat{C}}|$ does not generically belong to H (it is possible that it be non-normalizable). In this case, a well-definite physical Hilbert space can be found as follows: one restricts to a certain dense subspace S of H and constructs $\langle \psi_{\hat{C}}| \in S^*$, S^* being the topological dual of S. Since $\mathrm{H} \subset S^*$ (see next paragraph) it may be the case that the expression (5.125) gives a well-definite scalar product in S^*, such that S^* can be identified with the physical Hilbert space. This procedure is used in LQG in order to solve the vector constraint (see section 8.3.2).

The main difficulty within this framework consists in finding the subspace S. Furthermore, there could be different possible choices, giving inequivalent quantum descriptions.

Rigged Hilbert space The formulation of quantum mechanics in terms of Hilbert spaces encounters problems when dealing with continuous spectra and with unbounded operators (see [108] and references therein). For instance, for a one-dimensional system with coordinate on the full real line the Hilbert space in the Schrödinger representation can be taken as the collection of square integrable functions $f(x) \in \mathrm{H} = L^2(\mathbb{R}, \mu)$. The action of the position operator on such a space does not leave it invariant, since, for instance, there are some functions $f(x) \in \mathrm{H}$ for which $xf(x)$ is not square integrable. Therefore, one restrict the full Hilbert space to a dense subset of it Φ on which all powers of observables and products among them (x, p, the Hamiltonian, ...) are well defined. Bras and kets can be seen as antilinear and linear functionals on Φ, respectively. This way, one introduces distributions within the space of states and a rigorous mathematical meaning can be given to expressions like

$$\langle x | x' \rangle = \delta(x - x'). \tag{5.126}$$

The mathematical structure underlying such a picture is that of Gelfand triples, *i.e.* $\Phi \subset \mathcal{H} \subset \Phi^*$, Φ^* being the dual or anti-dual space of Φ in which kets and bras live, respectively.

Therefore, bras and kets are rigorously defined as operators acting on the Hilbert space.

Chapter 6

Quantum Geometrodynamics

The standard Hamiltonian analysis of General Relativity is performed in ADM variables according with the tools discussed in Chapters 3 and 4. The resulting dynamic system is completely constrained and the total Hamiltonian weakly vanishes. The canonical quantization can be performed either after (ADM reduction) or before (Wheeler-DeWitt formulation) the solution of the constraints. In the former case, one achieve a well grounded quantum framework for suitable choices of the reference frame, thus posing the issues of the physical meaning of such frames and of the fate of General Covariance on a quantum level. In the latter, the constraints can be solved by the Dirac prescription for the quantization of constrained systems. In both cases, a proper Hilbert space for the functionals of the metric has not been found yet. This is due to the complicated structure of the configuration space and to the lack of an explicit expression for the solution of the constraints. Other relevant issues are the functional nature of the theory, which requires the definition of a proper regularization scheme, and the definition of a time parameter for the associated Schrödinger dynamics.

6.1 The Hamiltonian structure of gravity

6.1.1 *ADM Hamiltonian density*

As a first step we derive now the Hamiltonian formulation of General Relativity starting from the ADM action functional that we derived in Chapter 4. The action functional reads

$$S_{ADM} = \frac{c^3}{16\pi G} \int dx^0 d^3x N\sqrt{h} \left[\bar{R} + \left(h^{ir}h^{js} - h^{ij}h^{rs}\right) K_{ij}K_{rs}\right], \qquad (6.1)$$

and since it does not contain time derivatives of both N and N^k we easily obtain the primary constraints of the theory

$$\Pi = \frac{\partial \mathcal{L}_{ADM}}{\partial \partial_0 N} \approx 0, \qquad \Pi_k = \frac{\partial \mathcal{L}_{ADM}}{\partial \partial_0 N^k} \approx 0, \tag{6.2}$$

and the conjugate momenta to the three-metric h_{ij} follow[1]

$$\Pi^{ij} = \frac{\partial \mathcal{L}_{ADM}}{\partial \partial_0 h_{ij}} = \frac{c^3}{16\pi G}\sqrt{h}\left(h^{ri}h^{sj} - h^{ij}h^{rs}\right)K_{rs}. \tag{6.3}$$

Given the definition of the extrinsic curvature (4.111) we can write the velocities $\partial_0 h_{ij}$ as functions of the conjugate momenta. The only difficulty is to find the inverse of the tensor $h^{ri}h^{sj} - h^{ij}h^{rs}$ which is given by

$$G_{ijrs} = G_{rsij} = h_{ir}h_{js} - \frac{1}{2}h_{ij}h_{rs}, \qquad G_{ijrs}\left(h^{rm}h^{sn} - h^{mn}h^{rs}\right) = \delta_i^m \delta_j^n, \tag{6.4}$$

allowing us to invert the transformation

$$K_{rs} = \frac{16\pi G}{c^3}\frac{1}{\sqrt{h}}G_{ijrs}\Pi^{ij}, \qquad \partial_0 h_{rs} = \frac{16\pi G}{c^3}\frac{2N}{\sqrt{h}}G_{ijrs}\Pi^{ij} + 2D_{(r}N_{s)}. \tag{6.5}$$

Hence, one obtains for the Lagrangian density

$$\mathcal{L}_{ADM} = \frac{c^3}{16\pi G}N\sqrt{h}\bar{R} + \frac{16\pi G}{c^3}\frac{N}{\sqrt{h}}G_{ijrs}\Pi^{rs}\Pi^{ij}, \tag{6.6}$$

and thus for the Hamiltonian density

$$\begin{aligned} H_{ADM} =& \Pi^{ij}\partial_0 h_{ij} + \lambda\Pi + \lambda^i\Pi_i - \mathcal{L}_{ADM} \\ =& \lambda\Pi + \lambda^i\Pi_i + N^k\mathcal{H}_k + N\mathcal{H} + 2D_i\left(\Pi^{ij}N^k h_{kj}\right), \end{aligned} \tag{6.7}$$

where we have defined respectively the so-called *super-Hamiltonian*, $\mathcal{H}$, and *supermomentum* $\mathcal{H}_k$ as

$$\mathcal{H} = \frac{16\pi G}{c^3}\frac{1}{\sqrt{h}}G_{ijrs}\Pi^{rs}\Pi^{ij} - \frac{c^3}{16\pi G}\sqrt{h}\bar{R}, \tag{6.8}$$

$$\mathcal{H}_k = -2h_{kj}D_i\Pi^{ij}. \tag{6.9}$$

Since G_{ijrs} is contracted with two symmetric tensors in the pairs of indices (i, j) and (r, s) it is common to use its symmetric part which is called *supermetric*

$$G_{ijrs} = \frac{1}{2}\left(h_{ir}h_{js} + h_{is}h_{jr} - h_{ij}h_{rs}\right), \tag{6.10}$$

[1] Note that the conjugate momenta are tensorial densities.

that we will use from now on with the same symbol. With suitable boundary conditions the last term in (6.7), which is a spatial boundary term, vanishes and one can easily read the secondary constraints of the theory to be

$$[\Pi, H_{ADM}] = -\mathcal{H} \approx 0, \qquad [\Pi_k, H_{ADM}] = -\mathcal{H}_k \approx 0, \tag{6.11}$$

hence, *super-Hamiltonian and supermomentum coincide with the secondary constraints.* Thus, with vanishing spatial boundary terms, the whole Hamiltonian density in ADM variables is a linear combination of primary and secondary constraints, hence it weakly vanishes

$$H_{ADM} \approx 0. \tag{6.12}$$

Moreover, the equations of motion of the lapse function and the shifts vector read

$$\partial_0 N = [N, H_{ADM}] = \lambda, \qquad \partial_0 N^k = \left[N^k, H_{ADM}\right] = \lambda^k, \tag{6.13}$$

implying that the dynamics of N and N^k is completely arbitrary, even though the condition $N > 0$ must always hold. The symplectic structure of the phase-space is given by the equal-time fundamental Poisson brackets

$$\begin{aligned} &\left[N\left(x, x^0\right), \Pi\left(y, x^0\right)\right] = \delta^{(3)}\left(x - y\right), \\ &\left[N^i\left(x, x^0\right), \Pi_k\left(y, x^0\right)\right] = \delta^i_k \delta^{(3)}\left(x - y\right), \\ &\left[h_{ij}\left(x, x^0\right), \Pi^{rs}\left(y, x^0\right)\right] = \delta_i^{(r}\delta_j^{s)}\delta^{(3)}\left(x - y\right), \end{aligned} \tag{6.14}$$

through which the evolution of any regular function of the phase-space variables can be calculated. Indeed, there are no tertiary constraints as we will see in the next section.

6.1.2 *Constraints in the 3+1 representation*

It is interesting to note that the secondary constraints (6.11) are related with Einstein's equations $G_{0\mu} = 0$, in fact one has

$$G_{\mu\nu}\eta^\mu\eta^\nu = -\frac{\mathcal{H}}{2\sqrt{h}}, \qquad G_{\mu\nu}b_i^\mu b_i^\nu = \frac{\mathcal{H}_i}{2\sqrt{h}}. \tag{6.15}$$

This relationship between the secondary constraints and the Einstein's equations $G_{0\mu} = 0$ is not surprising, since the latter can actually be seen as constraint in the initial value formulation of General Relativity: they restrict the set of admissible initial conditions (metric and its first derivatives on a Cauchy hypersurface) and they are automatically preserved by the evolution. We define the smeared form of any regular function of the phase-space variables $M^i(x) = M\left(N(x), N^k(x), h_{ij}(x), \Pi(x), \Pi_k(x), \Pi^{ij}(x)\right)$ as

$$\vec{M}[\vec{m}] = \int_\Sigma d^3x M^i(x) m_i(x), \tag{6.16}$$

where $m_i = m_i(x)$ are test-functions. Let us now consider the algebra[2] generated by the secondary constraints in the smeared form, *i.e.*

$$\begin{aligned} \left[\vec{\mathcal{H}}[\vec{m}], \vec{\mathcal{H}}[\vec{n}]\right] &= \vec{\mathcal{H}}[\mathcal{L}_{\vec{n}}\vec{m}] \\ \left[\mathcal{H}[m], \vec{\mathcal{H}}[\vec{n}]\right] &= \mathcal{H}[\mathcal{L}_{\vec{n}}m] \\ [\mathcal{H}[m], \mathcal{H}[n]] &= \vec{\mathcal{H}}[\vec{q}], \end{aligned} \tag{6.17}$$

$\mathcal{L}$ being the Lie derivative, whose explicit form reads

$$\mathcal{L}_{\vec{n}}m = -n^i\partial_i m, \qquad \mathcal{L}_{\vec{n}}\vec{m} = -n^j\partial_j\vec{m} + m^j\partial_j\vec{n},$$

while the function $q^i = q^i\left(m, n, h^{ij}\right)$ is given by

$$q^i = h^{ij}\left(m\partial_j n - n\partial_j m\right). \tag{6.18}$$

From the relations (6.17) one concludes that the algebra of constraints is closed, *i.e.* all constraints are first-class. This fact means that other constraints are not required and that each constraint generates a transformation inside the constraint hypersurface. As a consequence, the number of degrees of freedom is reduced twice for each constraint restoring the right number of degrees of freedom for the gravitational field, which is two[3]. In the presence of first-class constraints, one must impose both the vanishing of constraints and of their Poisson action in order to find observables. In this respect, the imposition of the primary constraints reduce to deal with phase-space functions not depending on N, Π, N^i and Π_i, thus $F = F(h_{ij}, \Pi^{ij})$. From now on we will consider only the reduced phase space parametrized by $\{h_{ij}, \Pi^{ij}\}$, on which only the secondary constraints act.

Furthermore, it is worth noting that the algebra of the constraints is not a Lie one. In fact, in the expression (6.17) the right-hand side is a linear combination of constraints via some non constant (structure) functions.

[2]It is interesting to notice that one can render any field theory covariant under $3+1$ diffeomorphisms and the resulting constraints algebra is exactly the same as it is possible to see in Kuchař treatment in [179].

[3]The ADM canonical variables are 20, 8 first-class constraints, thus $20 - 8 \times 2 = 4$ independent variables in the phase space, which means 2 degrees of freedom.

The vanishing of the supermomentum constraints is related to the arbitrariness in the choice of the coordinate system on each Σ_{x^0}: under an infinitesimal diffeomorphism $x'^i = x^i - \xi^i$, the transformation induced on the three-metric at the first order in the parameters ξ^i is given by,

$$\delta_{\vec{\xi}} h_{ij} = h'_{ij}(x) - h_{ij}(x) \simeq -2D_{(i}\xi_{j)}, \tag{6.19}$$

in complete analogy with the four-dimensional case discussed in Chapter 4 (4.21). This result is precisely the one obtained via the Poisson action of $\vec{H}[\vec{\xi}]$,

$$[\vec{H}[\vec{\xi}], h_{ij}] = -2D_{(i}\xi_{j)} \simeq \delta_{\vec{\xi}} h_{ij}. \tag{6.20}$$

Similarly, it can be shown that the Poisson action of $\vec{H}[\vec{\xi}]$ on Π^{ij} gives the variation of the conjugate momenta under a three-dimensional diffeomorphism with parameters ξ^i, *i.e.*

$$[\vec{H}[\vec{\xi}], \Pi^{ij}] \simeq \delta_{\vec{\xi}} \Pi^{ij}, \tag{6.21}$$

whose explicit expression can be inferred from (6.3). Therefore, given a generic function $F = F(h_{ij}, \pi^{ij})$, one has at the first order

$$[\vec{H}[\vec{\xi}], F] \simeq F(\delta_{\vec{\xi}} h_{ij}, \delta_{\vec{\xi}} \Pi^{ij}). \tag{6.22}$$

Hence, the full imposition of the condition $\mathcal{H}_i \approx 0$ is equivalent to the restriction to the diffeomorphisms-invariant subspace of the phase-space. The associated configuration space is known as *superspace* (see [152] for a mathematical analysis) and it is parametrized by three-geometries (which we indicate with $\{h_{ij}\}$), *i.e.* the equivalence class of metrics related via three-diffeomorphisms.

In superspace the only dynamic information is encoded in the super-Hamiltonian constraint. After long calculations, one can show (the demonstration can be found in [283]) that on-shell this constraint generates the infinitesimal spacetime transformations parallel to the normal vector $\vec{\eta}$, *i.e.*

$$[H[\xi], F(h_{ij}, \Pi^{ij})] \simeq \frac{8\pi G}{c^4} F(\mathcal{L}_{\xi\vec{\eta}}\, h_{ij}, \mathcal{L}_{\xi\vec{\eta}}\, \Pi^{ij}), \tag{6.23}$$

in which $\mathcal{L}_{\xi\vec{\eta}}$ denotes the Lie derivative under a spacetime diffeomorphism along the direction $\vec{\eta}$ with parameter ξ. For instance, for h_{ij} one has

$$[H[\xi], h_{ij}] = -\frac{16\pi G}{c^4\sqrt{h}} 2\,\xi\, G_{ijkl}\Pi^{kl} = -2\,\xi\, K_{ij}, \tag{6.24}$$

which coincides with

$$\begin{aligned}
\mathcal{L}_{\xi\vec{\eta}}\, h_{ij} &= -\xi\eta^{\mu}\partial_{\mu} h_{ij} - g_{\mu i}\,\partial_j(\xi\eta^{\mu}) - g_{\mu i}\,\partial_j(\xi\eta^{\mu}) \\
= -\xi\frac{1}{N}\partial_0 h_{ij} &+ \xi\frac{N^k}{N}\,\partial_k h_{ij} + \frac{\xi}{N} h_{ik}\,\partial_j N^k + \frac{\xi}{N} h_{jk}\,\partial_i N^k \\
&= -\xi\frac{1}{N}\partial_0 h_{ij} + \frac{\xi}{N}\, D_j N_i + \frac{\xi}{N}\, D_i N_j = -2\,\xi\, K_{ij}\,,
\end{aligned}$$

where we used the expressions of η^μ (4.109), of K_{ij} (4.110) and that $g_{\mu i}\eta^\mu = 0$.

Hence, the constraint $\mathcal{H} \approx 0$ implies the invariance under those diffeomorphisms which are orthogonal to Σ_{x^0}. Therefore, the secondary constraints implement in the phase-space the invariance under the action of diffeomorphisms on and orthogonal to the spatial hypersurfaces.

6.1.3 *The Hamilton-Jacobi equation for the gravitational field*

A Hamilton-Jacobi formulation for the gravitational field is possible and it can be used as a first step towards quantization (see paragraph 5.1.6.2). Let us introduce the action functional $S = S[N, N_i, h_{ij}]$, such that momenta can be rewritten as functional derivatives of S with respect to the corresponding variables, *i.e.*

$$\Pi^{ij} = \frac{\delta S}{\delta h_{ij}}, \qquad \Pi = \frac{\delta S}{\delta N}, \qquad \Pi^i = \frac{\delta S}{\delta N_i}. \tag{6.25}$$

Primary constraints $\Pi \approx \Pi_k \approx 0$ ensure that S does not depend on N and N^i, while the vanishing of the supermomentum implies defining S in superspace.

In superspace, the full dynamics is described by the super-Hamiltonian constraint, which can be rewritten as the Einstein-Hamilton-Jacobi equation [246]

$$\frac{16\pi G}{c^3}\frac{1}{\sqrt{h}}G_{ijrs}\frac{\delta S}{\delta h_{ij}}\frac{\delta S}{\delta h_{rs}} - \frac{c^3}{16\pi G}\sqrt{h}\bar{R} = 0, \tag{6.26}$$

thus reducing the problem to that of ∞^3 particles moving in a spacetime manifold, with the metric tensor given by the supermetric itself, subjected to a potential $\sqrt{h}\,\bar{R}$.

In particular, the identification of a direction in which the supermetric is negative definite provides a way to introduce a time-like variable. An example can be given by rewriting the three-metric as

$$h_{ij} = u^{4/3}u_{ij}, \qquad \det(u_{ij}) = 1, \tag{6.27}$$

and taking u and u_{ij} as configuration variables; in this case, the Hamiltonian reads as follows

$$H = -\frac{3}{2c^3}\pi G\, p_u^2 + \frac{16\pi G}{c^3\, u^2}\, u_{ik}\, u_{jl}\, p^{ij}\, p^{kl} - \frac{c^3}{16\pi G}\, u^2\, \bar{R}(u, u_{ij}, \nabla u, \nabla u_{ij}), \tag{6.28}$$

p_u and p^{ij} being the conjugate momenta to u and u^{ij}, respectively. It is clear from the expression (6.28) that the variable u, linked to the determinant of the metric, is time-like.

This result is well-known in cosmological settings (see section 9.2); for instance in Robertson-Walker spacetimes the scale factor is as an appropriate time variable [111].

6.2 ADM reduction of the Hamiltonian dynamics

In view of giving a better physical characterization of the gravitational field dynamics, one can identify in the metric tensor the physical degrees of freedom ϕ_A. Hence, the Lagrangian density can be written in the so-called *canonical form, i.e.*

$$L = \int d^3x[\pi^A \partial_0 \phi_A - H_{true}(\phi_A, \pi^B)]. \tag{6.29}$$

This reduction has been performed in [17] in the framework of a Palatini-like formulation. The metric tensor can be split into the transverse traceless component h_{ij}^{TT}, plus the trace of the transverse part h^T and the longitudinal part h_i. The imposition of the constraints and the choice of a system of coordinates allow one to fix completely h^T, h_i and their conjugate momenta, thus h_{ij}^{TT} describe physical degrees of freedom. Finally, the full Lagrangian density reads

$$\mathcal{L} = \pi^{TTij} \partial_t h_{ij}^{TT} + T_0^0, \tag{6.30}$$

where

$$T_0^0 = \nabla^2 h^T(\pi^{TTij}, h_{ij}^{TT}, h_i(\pi^{TTkl}, h_{kl}^{TT}), \pi^T(\pi^{TTkl}, h_{kl}^{TT})). \tag{6.31}$$

The change of the system of coordinates implies different relations fixing h_i and π^T, such that the Hamiltonian density T_0^0 takes a new expression as a function of $(\pi^{TTij}, h_{ij}^{TT})$. This feature outlines that the Hamiltonian density depends on the reference frame.

The canonical form makes General Relativity similar to a field theory formulation, such that it is possible to define the energy and the momenta of the gravitation field in terms of the generators of translations T_0^μ, as follows

$$P^\mu = -\int_T d^3x T_0^\mu. \tag{6.32}$$

In an asymptotically-flat spacetime in which an average over oscillatory terms is performed, P^μ turns out to be invariant under coordinate transformations preserving the flatness at infinity (*i.e.* $g'_{\mu\nu} - \eta_{\mu\nu} \propto 1/r$ for $r \to \infty$, r being the radial coordinate).

As for phase-space transformations, one must restrict to transformations between Heisenberg frames, which means that the full metric $g_{\mu\nu}$ can be expressed in terms of canonical variables only, without any explicit coordinate dependence.

Therefore, within this scheme, P^μ fulfills the requirement for a well-defined energy-momentum for the gravitational field in an asymptotically flat spacetime. Furthermore, proper conditions can be fixed such that a wave-like behavior comes out for canonical variables in a certain spacetime region, where non-linearities can be neglected.

However, in general these non-linear terms into the Hamiltonian cannot be neglected and in this regime no proper quantum description has been achieved.

Therefore, the attempts towards quantization of gravity has been mainly based on solving constraints on a quantum level, following the Dirac procedure for the quantization of constrained systems, rather than on reducing the Hamiltonian to the canonical form.

6.3 Quantization of the gravitational field

The quantization of the gravitational field dynamics can be formally performed as for the relativistic particle (discussed in section 5.5.2), because the supermetric is hyperbolic. The space of states is that of proper functionals of configuration variables (in particular they are required to be differentiable in order to define the action of constraints),

$$\Psi = \Psi[N, N^i, h_{ij}]. \tag{6.33}$$

Hence, the configuration variables and the conjugate momenta are promoted to multiplicative and derivative operators as follows[4]

$$h_{ij}(x) \to \hat{h}_{ij}(x) \equiv h_{ij}(x), \qquad \Pi^{ij}(x) \to \hat{\Pi}^{ij}(x) \equiv -i\hbar \frac{\delta}{\delta h_{ij}(x)}, \tag{6.34}$$

$$N(x) \to \hat{N}(x) \equiv N(x), \qquad \Pi(x) \to \hat{\Pi}(x) \equiv -i\hbar \frac{\delta}{\delta N(x)}, \tag{6.35}$$

[4]It is worth noting that the functional derivatives $\delta/\delta..$ provides a dimensionality $1/(\text{length})^3$.

$$N^i(x) \to \hat{N}^i(x) \equiv N^i(x), \qquad \Pi_i(x) \to \hat{\Pi}_i(x) \equiv -i\hbar \frac{\delta}{\delta N^i(x)}. \tag{6.36}$$

This way, canonical commutation relations are established, in fact one has

$$[\hat{h}_{ij}(x), \hat{\Pi}^{kl}(y)] = i\hbar \frac{1}{2}(\delta_i^k \delta_l^j + \delta_j^k \delta_i^l)\, \delta^{(3)}(x-y), \tag{6.37}$$

$$[\hat{N}(x), \hat{\Pi}(y)] = i\hbar\, \delta^{(3)}(x-y), \tag{6.38}$$

$$[\hat{N}^i(x), \hat{\Pi}_j(y)] = i\hbar \delta_j^i\, \delta^{(3)}(x-y), \tag{6.39}$$

while the other commutators vanish. The space of states is not a Hilbert space, but this is not necessarily an issue, since the constraints are not solved yet.

Let us now implement the Dirac prescription for the quantization of constrained systems and define the physical states as those annihilated by the operators associated with the constraints.

6.3.1 *Quantization of the primary constraints*

The vanishing of Π and Π_i implies the following conditions on a quantum level

$$-i\hbar \frac{\delta}{\delta N} \Psi[N, N^i, h_{ij}] = 0, \qquad -i\hbar \frac{\delta}{\delta N^i} \Psi[N, N^i, h_{ij}] = 0. \tag{6.40}$$

These conditions can be easily solved by restricting to the wave functionals which do not depend on the lapse function and the shift vector, *i.e.*

$$\Psi = \Psi[h_{ij}]. \tag{6.41}$$

Hence, physical states are independent from the variables defining the spacetime slicing and they are functionals of the three-metric only.

6.3.2 *Quantization of the supermomentum constraint*

The supermomentum constraint (6.9) must be imposed as follows

$$\hat{\mathcal{H}}_i \Psi \equiv D_j \left[\frac{\delta \Psi}{\delta h_{ij}(x)} \right] = 0, \tag{6.42}$$

and this condition can be solved by requiring that the wave functionals depend on the three-geometry $\{h_{ij}\}$ only, rather than on any specific representation. An explicit solution can be given by requiring Ψ to be a function of scalar (smeared) quantities, as for instance $\int \sqrt{h}\, d^3x$, $\int \sqrt{h}\, \bar{R}\, d^3x$, $\int \sqrt{h} \bar{R}_{ij}\, \bar{R}^{ij}\, d^3x$, etc. The number of scalar quantities one can construct in this way is unbounded, thus one cannot write an expression for the generic solution of the supermomentum constraint.

6.3.3 *The Wheeler-DeWitt equation*

The super-Hamiltonian (6.8) must be turned into an operator as well and its action on the wave function has to be fixed, *i.e.*

$$\frac{16\pi G}{c^3}\hat{\mathcal{H}}\Psi \equiv -\left(16\pi l_P^2\right)^2 {}''\left(G_{ijkl}(x)[h_{mn}]\frac{\delta^2\Psi}{\delta h_{ij}(x)\delta h_{kl}(x)}\right)'' - \sqrt{h}\bar{R}(x)\Psi = 0, \tag{6.43}$$

where we introduced the fundamental length scale in Quantum Gravity, the Planck length

$$l_P = \sqrt{\frac{G\hbar}{c^3}}\,. \tag{6.44}$$

Equation (6.43) is the Wheeler-DeWitt equation, which is the fundamental equation giving the quantum dynamics for the gravitational fields. The symbol "(...)" denotes a proper factor ordering, which can be chosen such that the Dirac algebra (*i.e.* the relations (6.17)) is preserved (see also [90]). Indeed, the definition of the commutators is well defined only by adopting a proper regularization scheme (see section 6.4.2).

It is interesting to notice that if one had started from the ΓΓ Lagrangian density (4.35) and performed in a purely algebraic way the ADM transformation (4.91) one would have ended up with a Lagrangian density differing mainly for Dirac's boundary term (4.162) which is the interesting one as for the definition of the conjugate momenta. Indeed, the whole set of constraints would be different but since the transformation induced by Dirac's boundary term is strongly canonical one can easily write the new algebra as a linear combination of the ADM one (6.17). We report here the classical expressions for the set of constraints of the ΓΓ formulation in ADM variables

$$\phi = \Pi - \frac{\sqrt{h}}{N^2}\partial_k N^k \approx 0, \qquad \phi_k = \Pi_k - \partial_k\left(\frac{\sqrt{h}}{N}\right) \approx 0, \tag{6.45}$$

$$\chi_i = \mathcal{H}_i + \sqrt{h}\partial_i\left(\frac{1}{N}\partial_k N^k\right) \approx 0, \tag{6.46}$$

$$\chi = -\mathcal{H} + \frac{3}{8}\frac{\sqrt{h}}{N^2}\partial_i N^i \partial_k N^k + \frac{1}{2N}\Pi^{ij}h_{ij}\partial_k N^k \approx 0. \tag{6.47}$$

While on the classical side there is no difficulty arising in linking these two formulations, imposing the equivalence of the quantization procedures is a bit more subtle. Indeed, imposing on some functional Ψ the constraints ϕ, ϕ_k and χ_i one finds (via a heat-kernel regularization, see section 6.4.2)

$$\Psi[N, N^i, h_{ij}] = \exp\left(-\frac{i}{16\pi l_P^2}\int d^3x \frac{\sqrt{h}}{N}\partial_k N^k\right)\Psi[\{h_{ij}\}]. \qquad (6.48)$$

For the imposition of $\chi\Psi = 0$ to be equivalent to $\mathcal{H}\Psi = 0$ it is necessary to impose some restrictions on the possible operator orderings, which corresponds to the same prescription for the quantum algebra to close. Among other ordering prescriptions the symmetric one satisfies these requirements for any choice of the regulator which is a desirable property[5]. At the end one has

$$\chi\Psi[N, N^i, h_{ij}] = \exp\left(-\frac{i}{16\pi l_P^2}\int d^3x \frac{\sqrt{h}}{N}\partial_k N^k\right)\mathcal{H}\Psi[\{h_{ij}\}] = 0, \qquad (6.49)$$

enforcing the quantum equivalence through the selection of an operator ordering. Thus, the use of a strongly canonical transformation due to a boundary term in the classical theory may reduce the possible operator orderings (for the details see [90]).

6.4 Shortcomings of the Wheeler-DeWitt approach

In this section we discuss the relevant shortcomings of the metric approach to the quantization of General Relativity based on the Dirac prescription. We will clarify why the quantization of the Wheeler-DeWitt equation is still an open issue and a proper quantum formulation works only in specific models (mini-superspace), where some degrees of freedom are frozen (see Chapter 9).

6.4.1 *The definition of the Hilbert space*

A Hilbert space structure in the space of the solutions of the constraints (6.42), (6.43) (physical Hilbert space) has not been found. The difficulties are both related with finding a basis in the physical Hilbert space (in fact the Wheeler-DeWitt equation is not linear) and with defining a proper scalar product on it.

Indeed, in view of the analogy between the Wheeler-DeWitt equation and the Klein-Gordon one, DeWitt [111] proposed a formal definition for a

[5]By symmetric ordering we mean $\hat{\Pi}^{ij}\, G_{ijrs}\, \hat{\Pi}^{rs}$.

suitable scalar product in superspace. In fact, a time-like direction exists within superspace and it is associated with a "dilatation" u for the metric (see (6.27)). Furthermore, the following current is conserved for the solutions of the Wheeler-DeWitt equation

$$S_{12ij}(x)[\Psi_1, \Psi_2] = \frac{1}{2} G_{ijkl}(x) \Psi_1 \frac{\overleftrightarrow{\delta}}{\delta h_{kl}(x)} \Psi_2, \qquad \frac{\delta}{\delta h_{ij}(x)} S_{12ij}(x) = 0 \tag{6.50}$$

where

$$\Psi_1 \frac{\overleftrightarrow{\delta}}{\delta h_{kl}(x)} \Psi_2 = \Psi_1 \frac{\delta \Psi_2}{\delta h_{kl}(x)} - \frac{\delta \Psi_1}{\delta h_{kl}(x)} \Psi_2. \tag{6.51}$$

Hence, an inner product can be defined, *i.e.*

$$\Omega[\Psi_1, \Psi_2] \equiv \prod_x \int D\Sigma^{ij}(x) S_{12ij}(x)[\Psi_1, \Psi_2], \tag{6.52}$$

and it does not depend on which space-like hypersurface Σ^{ij} is chosen in superspace.

However, the scalar product (6.52) is not positive-defined and, unlike the case of the Klein-Gordon field, the separation of frequencies cannot be performed, since superspace is singular in $\eta = 0$.

There has been no real significant progress with respect to DeWitt's paper [111] as long as the definition of the scalar product in the physical Hilbert space is concerned. Therefore, *a proper Hilbert space has not been found for the gravitational field in the ADM representation.*

A further issue is related to the term $\sqrt{h}\,\bar{R}(x)$, which in this scheme behaves as a potential term. In fact, there is no restriction on the sign of the scalar curvature, thus tachyon-like objects (with respect to the metric in superspace) are expected to appear in a rigorous quantum framework.

6.4.2 *The functional nature of the theory*

The expression (6.43) retains a formal meaning only without a proper regularization for the product of operators on the same point x. This feature can be tamed by defining a triangulation of the spatial manifold and by extending the product to the vertices of the triangulation only. Such a regularization can be realized by applying to a canonical setting the tools of Regge calculus [251]. This framework also provides a regularized expression for the Wheeler-DeWitt equation (6.43) (see [159, 160] for the application to 2+1 and 3+1 gravity, respectively) and it simplifies the treatment of the potential term $\sqrt{h}\,\bar{R}(x)$ containing spatial derivatives of the metric (whose behavior in a continuous formulation is elusive). This approach is promising, since it gives a much more solid mathematical ground with respect to the standard paradigm. However, it is not clear yet if all the issues of the canonical quantization of General Relativity can be addressed in this way.

A different regularization procedure for the super-Hamiltonian operator can be realized in terms of a heat-kernel expansion [60, 191]. The idea is to replace (6.43) with the following condition

$$-\left(16\pi l_P^2\right)^2 \int dx' G_{ijkl}(x')K(x,x',t)\frac{\delta^2\Psi}{\delta h_{ij}(x)\delta h_{kl}(x')} - \sqrt{h}\,\bar{R}(x)\,\Psi = 0, \tag{6.53}$$

where the function K is the heat kernel, *i.e.*

$$\partial_t K(x,x',t) = \nabla^2_{(x)}K(x,x',t), \qquad \lim_{t\to 0} K(x,x',t) = \frac{\delta^{(3)}(x-x')}{\sqrt{h}}. \tag{6.54}$$

An explicit solution for K can be given via an expansion in powers of the parameter t (see [98] for some recent results). Finally, the Wheeler-DeWitt equation (6.53) can be rewritten as an expansion in t. Only a finite number of (diverging) contributions remain after taking the limit $t \to 0$ (those multiplied by non-positive powers of t), which can be associated with some renormalized constants. Within this scheme, one has full control over the divergences related to the functional nature of the theory, but the definition of a proper scalar product remains an open issue.

6.4.3 *The frozen formalism: the problem of time*

The Hamiltonian is the generator of time displacements in phase-space and this leads on a quantum level to the Schrödinger equation. The Hamiltonian of the gravitation field is a linear combination of the supermomentum and super-Hamiltonian constraints (6.7). Hence, according to the Dirac prescription the Hamiltonian annihilates physical states

$$\hat{H}\Psi = 0. \tag{6.55}$$

The equation above can be seen as the Schrödinger equation for a quantum state not depending on time. This means that quantum states do not evolve (*frozen formalism*), thus it suggests that there is no quantum dynamics.

The problem here is that we are identifying two notions of time which are distinct. On one hand, time in Quantum Mechanics is a fixed external parameter (see section 6.4.3.1), which lies at the basis of equal-time canonical commutation relations, as well as of the notion of Hilbert space, in which the scalar product is conserved. On the other hand, time in General Relativity is merely a coordinate, since the theory is invariant under generic spacetime transformations (time is not a physical observable).

A physical time can be chosen by fixing a particular foliation as fundamental and by labeling events on a manifold according to some physical clock.

This can be done either before or after the quantization, or by some phenomenological considerations, in a model where time plays no precise role.

If time is regarded to be fundamental, the canonical constraints have to be solved before the quantization of the system, where the internal time is not expressed as a functional of the canonical variables, but picked up from the set of all the other dynamic variables by suitable canonical transformations. Alternatively, matter fields interacting with the spacetime geometry can be used as physical clocks. The arbitrariness in the choice of the internal or external time variable is reflected in the non-uniqueness of the associated Schrödinger dynamics, which poses the problem of the fate of the invariance under the choice of the time variable on a quantum level or of the physical meaning of a preferred time direction.

If one wants to discover the role of time after the quantization, one should look at physical states, written as functionals of fixed background geometries, that carry such an information in the Wheeler-DeWitt approach: these states can be treated as operator, in the third-quantization scheme (a second quantization for the wave function of the gravitational field) [88, 122, 144, 244], or interpreted in their semi-classical limit via a kind of Born-Oppenheimer approximation [194] (see also section 6.4.3.4). However, both these approaches provide more interpretative and conceptual issues than hints towards a satisfying solution of the problem of time (see for instance [175]).

In a third approach, one can choose as physical clocks some observables, *i.e.* operators which commute with all the constraints. The issue is precisely to find such observables (functions which commute with all constraints). In this respect, there is a well-settled procedure to get an observable out of a gauge variant phase-space function f (*partial observable*) [114]: let us consider for simplicity the case with just one constraint H and choose a clock-like function T (in general there are as many clock-like functions as constraints); the gauge orbit generated by H can be parametrized by the

Lagrangian multiplier N and described mathematically via

$$\alpha_{H[N]}(f) = \sum_{n=0}^{\infty} \frac{1}{n!}[f, H[N]]_n, \tag{6.56}$$

in which $[f, H[N]]_0 = f$ and $[f, H[N]]_{n+1} = [[f, H[N]]_n, H[N]]$. The (complete) observable can be defined as that function $F_{[f,T]}(\tau)$ which tells the value of f when $T = \tau$ on the gauge orbit, *i.e.*

$$F_{[f,T]}(\tau) \approx \alpha_{H[N]}(f)\bigg|_{N=\beta}, \tag{6.57}$$

with β such that

$$\alpha_{H[\beta]}(T) \approx \tau. \tag{6.58}$$

The notions of complete and partial observables give an operational definition for observables, which can be extended on a quantum level too. However, they are still subjected to the arbitrariness in the definition of clock-like functions.

Also, many efforts have been made in order to define the history of a quantum system, without involving the notion of time. Two research lines in this direction are the consistent-history approach [161] and topos theory [176].

In the following, before analyzing some mechanisms for identifying time in Quantum Gravity (the Brown-Kuchař model, the multi-time approach and the Vilenkin's proposal), we briefly recall some basic features about the notion of time in quantum physics.

6.4.3.1 *Time in Quantum Physics*

The definition of time is not unique, despite its feature of commonly-experienced quantity. The measure of time does not shed light on its physical nature, both from a classical and from a quantum-mechanical perspective. In General Relativity, time is generalized as a coordinate, which is not, however, the physical time. The puzzle is far from being solved [53, 255], and the inadequacy of every a priori characterization leads to different possibilities of treating time in canonical quantum gravity [175]. Even from a classical point of view, in Newtonian physics, time t is a parameter, and can be measured by means of clocks. Clocks, however, do not measure time directly, but they display its representation $T(t)$, which might have a linear functional dependence on t, *i.e.* $T(t) = \alpha t$. The correspondence is nevertheless not perfect, as $-\infty < t < +\infty$, while $T(t)$ is defined within a finite interval, and is always affected by an experimental uncertainty. Furthermore, the measure of a time-dependent physical quantity Q depends on T rather than on t, *i.e.* $Q = Q(T)$.

The same idea of external parameter that describes the dynamics of a system is present in QM too. In fact, the Schrödinger equation treats the time variable t as disconnected from other physical coordinates.

6.4.3.2 *The Brown-Kuchař model*

The identification of time after quantization in General Relativity is based on the following procedure:

- the constraints are solved classically,
- a functional T of the configuration variables is identified with the time parameter,
- the Hamiltonian h associated with T is quantized.

A crucial feature of these models is the choice of the time-functional. Such functional must be a monotonically increasing function, at least locally, and, after the quantization, has to provide a well-defined and conserved probability density. Moreover, the separation between time and other variables can be performed in different ways, which provide us with different scenarios.

Among these approaches, we point our attention on the *Brown-Kuchař* one. In their work [83] (see also [264, 265]) a dust field is present and the bundle of world-lines identifies a preferred time-like direction. This direction plays the role of time. For what concerns the constraints, the super-Hamiltonian and the supermomentum are modified by the addition of the terms $\mathcal{H}^D$ and $\mathcal{H}_i^D$ due to the matter field, *i.e.*

$$\mathcal{H}' = \mathcal{H} + \mathcal{H}^D \approx 0\,, \tag{6.59}$$

$$\mathcal{H}_i' = \mathcal{H}_i + \mathcal{H}_i^D \approx 0\,. \tag{6.60}$$

In particular, in the case of a dust-like clock field comoving with the space-time slicing, $\mathcal{H}^D$ and $\mathcal{H}_i^D$ read

$$\mathcal{H}_i^D = P\,\partial_i T, \qquad \mathcal{H}^D = \sqrt{P^2 + h^{ij}\,\mathcal{H}_i^D\,\mathcal{H}_j^D}\,, \tag{6.61}$$

T being the proper time along the dust field, while P denotes its conjugate momentum. Hence, we can solve the supermomentum constraint $\mathcal{H}_i^D \approx$

$-\mathcal{H}_i$, substitute it into $\mathcal{H}' \approx 0$ and solve the latter with respect to P, getting the following condition

$$P + h(h_{ij}\,,\,\Pi^{ij}) \approx 0, \tag{6.62}$$

with

$$h = -\sqrt{G} \qquad G = \mathcal{H}^2 - h^{ij}\,\mathcal{H}_i\,\mathcal{H}_j. \tag{6.63}$$

Therefore, by taking T as a time parameter, the evolution is described by the Hamiltonian h. This Hamiltonian turns out to be positive-definite. The dynamic system is thus completely *deparametrised*, *i.e.* the constraints split into the sum of two terms, one for the gravitational degrees of freedom and the other for the non-gravitational ones. Furthermore, we stress that P appears linearly in the non-gravitational part and the constraints (6.62) and (6.60) strongly commute among themselves.

Hence, the quantization of gravity with a dust-like field can be performed and a Schrödinger-like equation is obtained from the quantization of the constraint (6.62) with the proper time along the dust field playing the role of time. A quantum description for the system can be given, together with a definition of an inner product (which in general is only formal), with a proper time parameter T.

The Brown-Kuchař approach relies on the generic dualism existing between time and matter [228], [218] in General Relativity. Indeed, the definition of the properties for a matter field to be a proper clock is not completely well settled. This is the issue of any approach based on the use of an external time, since there is no compelling reason to choose a certain clock field and different choices provide inequivalent physical Hamiltonians.

Actually, the request to deal with a deparametrised system appears more a technical than a physical requirement, and it also does not solve all the issues (the Hamiltonian h is a nonlocal operator, because of the square root). Nonetheless, the Brown-Kuchař model can be extended to other fields different from the simple dust fluid we discussed above. The request that the spatial gradients of such matter fields must be removed from the dynamic constraint could be regarded as a basic quality for a time variable.

In this respect, an interesting proposal has been developed in [280]: this work consists in the application of the Brown-Kuchař formulations in the presence of a K-essence, which comes out as a relic of the quantum description of the spacetime geometry, such that its presence can be physically motivated.

6.4.3.3 *Multi-time approach*

The multi-time approach realizes a generic, even though rather formal, procedure to derive an internal time parameter for the gravitational field [175]. This procedure is based on the reduction of the Lagrangian to a canonical form (see section 6.2). For gravity, one identifies among the components of the metric tensor h_{ij}, some variables H_r $(r = 1, 2)$ describing the two degrees of freedom. Hence, by performing a canonical transformation from $\{h_{ij}\,, \Pi^{ij}\}$ to $\{H_r\,, P^r\}$ plus a set $\{\xi^\mu\,, \pi_\mu\}$ $(\mu = 0, \dots, 3)$ of embedding variables, *i.e.* with no physical meaning, the action can be rewritten as

$$S = -\frac{c^3}{16\pi G}\int d^4x(\pi_\mu\,\partial_0\xi^\mu + P^r\,\partial_0 H_r - N\,\mathcal{H} - N^i\,\mathcal{H}_i)\,, \tag{6.64}$$

in which $\mathcal{H} = \mathcal{H}(\xi^\mu\,, \pi_\mu\,, H_r\,, P^r\,)$ and $\mathcal{H}_i = \mathcal{H}_i(\xi^\mu\,, \pi_\mu\,, H_r\,, P^r\,)$. One can solve the constraints $\mathcal{H} \approx 0$ and $\mathcal{H}_i \approx 0$ in order to get an expression for the momenta π_μ in terms of other phase-space coordinates, and then substitute into the action (6.64), so having

$$S = -\frac{c^3}{16\pi G}\int d^4x(P^r\partial_0 H_r - \pi_\mu(\xi^\mu, H_r, P^r)\partial_0\xi^\mu). \tag{6.65}$$

Finally, the multi-time idea consists in a canonical quantization of ξ^μ and H_r variables, taking wave functionals $\Psi = \Psi(\xi^\mu, H_r)$ whose evolution is provided by the following set of Schrödinger-like equations

$$i\hbar\frac{\delta\Psi}{\delta\xi^\mu} = \pi_\mu\left(\xi^\mu, H_r, \frac{\delta}{\delta H_r}\right)\Psi. \tag{6.66}$$

This formalism finds some applications in cosmological settings [51].

6.4.3.4 *Vilenkin proposal*

Even though a time parameter could be introduced (and eventually a minisuperspace description could be addressed, see section 9.1), nevertheless the implementation of the quantum description is highly non-trivial. For instance, the wave function of the Universe [239, 291], as a solution of the Wheeler-DeWitt equation, on one hand, should maintain its role of defining the probability density of events happening in the Universe, and, on the other hand, should be able to reproduce the physical features of the Universe itself, both quantum and classical.

One of the most striking differences between the ordinary quantum and the cosmological wave function is that, while the former depends on time explicitly and allows, under very reasonable hypotheses, for the definition of a positive-definite probability distribution dp, *i.e.*

$$\Psi = \Psi(q_i, t) \to dp = |\Psi(q_i, t)|^2 \, d^N q, \tag{6.67}$$

the latter does not. In fact, it depends only on the three-dimensional metrics and eventually on matter fields ϕ_A, *i.e.* $\Psi = \Psi\left[h_{ij}, \phi_A\right]$, while the dependence on "time" would make little sense, because of the invariance under arbitrary parametrizations of the label time.

As far as the probabilistic interpretation is concerned, two main cases can be distinguished, when the superspace variables are all semiclassical, or when a quantum subsystem is taken into account.

If we restrict the investigation to (homogeneous) superspace models, whose action reads

$$S = \int dt \left(P_\alpha \partial_0 h^\alpha - N\left[g^{\alpha\beta} P_\alpha P_\beta + U(h)\right]\right), \tag{6.68}$$

where h^α labels some generalized superspace variables $\alpha = 1, \ldots, n$[6], and P_α their conjugate momenta, the potential $U = h^{1/2}[V(\phi) - {}^3R]$ (here we specify matter as a self-interacting scalar field ϕ, having potential $V(\phi)$) defines the Wheeler-DeWitt equation

$$\left(\nabla^2 - U\right)\Psi = 0. \tag{6.69}$$

In the case of semiclassical variables, the wave function can be written as the superposition of functions of the action $S(h)$, as [291]

$$\Psi = A(h) e^{iS(h)}, \tag{6.70}$$

which admits a WKB expansion and leads to the conserved current

$$j^\alpha = |A|^2 \nabla^\alpha S, \quad \nabla_\alpha j^\alpha = 0. \tag{6.71}$$

Here, the classical action S is an equivalence class of classical trajectories, and a family of $n-1$ hypersurfaces Σ_α crossed once by the trajectories, $\dot{h}_\alpha d\Sigma^\alpha > 0$. If we take $h_n = t$, equation (6.71) rewrites

$$\frac{\partial \rho}{\partial t} + \partial_a j^a = 0, \quad a = 1, ..., n-1, \tag{6.72}$$

where $j^a = \rho \partial_0 h^a$, and describes ρ as the "distribution function for an ensemble of classical universes".

The previous discussion can be easily generalized to a superposition of (6.70), $\Psi = \sum_k \Psi_k = \sum_k A_k e^{iS_k}$. If a family of equal-time hypersurfaces

[6] Here α is not a vierbein index.

can be found for all the possible S_k's, then the total distribution function can be expressed as

$$\rho = \sum_k \rho_k + \text{cross terms}, \tag{6.73}$$

where the cross terms can be shown to produce no physically-relevant interference.

The possibility of including small quantum subsystems, which do not modify significantly the dynamics, among the superspace variables can be taken into account. Equation (6.69) rewrites

$$\left(\nabla^2_{cl} - U_{cl} - H_q\right) \Psi = 0, \tag{6.74}$$

where the index $_{cl}$ refers to classical variables only, and $_q$ to quantum effects only. The pertinent wave function is

$$\Psi(h,q) = \sum_k \Psi_k(h)\chi_k(h,q), \tag{6.75}$$

which leads to the definition of the currents

$$j^\alpha = |\chi|^2 |A|^2 \nabla^\alpha_{cl} S \equiv j^\alpha_{cl} \rho_\chi, \tag{6.76}$$

$$j^\nu = -\frac{i}{2}|A|^2 \left(\chi^* \nabla^\nu \chi - \chi \nabla^\nu \chi^*\right) \equiv \frac{1}{2}|A|^2 j^\nu_\chi, \tag{6.77}$$

where α labels semiclassical variables, while ν refers to the quantum subsystem: the currents are related by the continuity conditions

$$\nabla_\alpha j^\alpha + \nabla_\nu j^\nu = 0, \quad \nabla_{cl\,\alpha} j^\alpha_{cl} = 0, \tag{6.78}$$

which lead to the probability distribution

$$\rho(h,q,t) = \rho_{cl}(h,t)|\chi|^2. \tag{6.79}$$

The normalization of probabilities can be easily checked, if one considers the volume element $d\Sigma = d\Sigma_0 d\Sigma_q$, such that

$$\int \rho_{cl} d\Sigma_{cl} = 1, \tag{6.80}$$

$$\int |\chi|^2 (\det[g_{\mu\nu}])^{1/2} d^m q = 1. \tag{6.81}$$

For an implementation of these Vilenkin's ideas in the minisuperspace of a Bianchi IX, see section 9.4.2.

Chapter 7

Gravity as a Gauge Theory

All interactions but gravity are described by Yang-Mills gauge theories, whose basic property is the invariance under the action of an internal Lie group. This invariance is a key point for the quantum formulation, (for instance it is crucial for renormalization), and it implies that Yang-Mills models fulfill the requirements for a predictive Quantum Field Theory. A Yang-Mills-like formulation for gravity is expected to simplify the quantization issue, since one would have at disposal the tools developed for other interactions. In this section, we will review the formulation of Yang-Mills gauge theory both in continuum and on a lattice. Then, we will push forward the analogies between Yang-Mills models and General Relativity on a Lagrangian level, but we will also stress the differences. We will discuss a first order Lagrangian formulation for gravity in terms of vierbein vectors and spin connections, by pointing out all the shortcomings which prevent the identification with a gauge theory of the Lorentz group. We will present Poincaré gauge theory as an example of a Yang-Mills model for gravity not equivalent to General Relativity. We will then analyze the Holst formulation, which is equivalent to General Relativity (at least in vacuum), but whose phase space structure resembles that of $SU(2)$ gauge theory.

7.1 Gauge theories

7.1.1 *The Yang-Mills formulation*

Gauge theories are a mathematical tool that describes interactions through the invariance of the action $S(\phi, \partial_\mu \phi)$ for the field $\phi(x) \equiv \{\phi_r(x)\}$ under Lie groups of transformations whose elements we denote by $U(\varepsilon)$

[139, 231]. The invariance of the action is expressed by

$$0 = \delta S = \delta \int d^4x \mathcal{L}(\phi(x), \partial_\mu \phi(x)) = \int d^4x \delta \mathcal{L}, \tag{7.1}$$

which implies,

$$\delta \mathcal{L} = 0 \Rightarrow \frac{\partial \mathcal{L}}{\partial \phi_r} \delta \phi_r + \frac{\partial \mathcal{L}}{\partial \partial_\mu \phi_r} \delta \partial_\mu \phi_r = 0. \tag{7.2}$$

It is therefore crucial to know the expressions for $\delta \phi_r$ and $\delta \partial_\mu \phi_r$.

Let's consider the action of the operator $U(\varepsilon)$ on ϕ[1]

$$\phi \rightarrow \phi' = U(\varepsilon)\phi, \quad U(\varepsilon) = e^{ig\varepsilon^a \tau_a} \simeq I + i\varepsilon^a \tau_a, \quad \varepsilon^a \ll 1\,, \tag{7.3}$$

which induces on each component the transformation

$$\begin{aligned} &\phi_r(x) \rightarrow \phi'_r(x) = \phi_r(x) + \delta \phi_r(x), \\ &\delta \phi_r(x) = \phi'_r(x) - \phi_r(x) = \epsilon^a (\delta \phi_r)_a = i\varepsilon^a \tau_a^{rs} \phi_s, \end{aligned} \tag{7.4}$$

where ε^a are a set of infinitesimal parameters, and τ_a the generators, that obey the commutation rule

$$[\tau_a, \tau_b] = i\, C_{ab}^c\, \tau_c. \tag{7.5}$$

Vanishing structure constants C_{ab}^c define *Abelian* groups, while *non-Abelian* groups have non-vanishing structure constants.

If the parameters ε^a are constant, the transformation is called global, and $\partial_\mu \delta \phi_r = \delta \partial_\mu \phi_r$. This way, after substitution of the Euler-Lagrange equation in (7.2), the conserved current j^μ, $\partial_\mu j^\mu = 0$, is found

$$j_a^\mu \equiv \frac{\partial \mathcal{L}}{\partial \partial_\mu \phi_r} (\delta \phi_r)_a = \frac{\partial \mathcal{L}}{\partial \partial_\mu \phi_r} \tau_a^{rs} \phi_s, \tag{7.6}$$

which allows one to define the conserved charges Q_a

$$Q_a = \int d^3x j_a^0 = \int d^3x \frac{\partial \mathcal{L}}{\partial \partial_\mu \phi_r} \tau_a^{rs} \phi_s, \tag{7.7}$$

according to the Noether theorem.

If the parameters ε^a are not constant, $\varepsilon^a = \varepsilon^a(x)$, the transformation is called local, *i.e.* a gauge transformation, and $\partial_\mu \delta \phi_r \neq \delta \partial_\mu \phi_r$. To restore the invariance of the Lagrangian density, it is therefore necessary to define a new derivative, the *gauge covariant derivative* D_μ, that commutes with the variation operation. In fact, the ordinary derivative

$$d\phi = \phi(x + dx) - \phi(x) = dx^\mu \partial_\mu \phi \tag{7.8}$$

[1]In this section the indices $a, b, c, \ldots$ denotes the internal indices of the gauge group, while in the following we will use them for the spatial indices of the local Lorentz frame.

is ill-defined under a local transformation (7.3) because the fields in two different points x and $x + dx$ transform according to different laws. The transport operator $T(x, y)$, when applied to a field,

$$T(x,y)\phi(y) \to T'(x,y)\phi'(y) = U(x)T(x,y)\phi(y), \tag{7.9}$$

generates an object with the same transformation properties of the field itself, so that

$$T(x,y) \to T'(x,y) = U(x)T(x,y)U^{-1}(y). \tag{7.10}$$

Since the transport operator is an element of the transformation group, it can be expressed as a function of the generators, and, for infinitesimal transformations one has

$$T(x, x+dx) = I + i\, dx^\mu A^a_\mu(x)\tau_a, \tag{7.11}$$

where the *connections* A^a_μ, which are the gauge fields, in the combination $dx^\mu A^a_\mu$, play the role of $\varepsilon^a(x)$. The covariant derivative D_μ,

$$D_\mu\phi(x) = (\partial_\mu + i\, A^a_\mu\tau_a)\phi(x), \tag{7.12}$$

as $T(x, y)$, transforms as (7.10) under a local transformation (7.3). The transformation law for the gauge fields can be obtained from the definition (7.12), (7.9) and (7.11), and reads

$$A_\mu(x) = A^a_\mu(x)\tau_a \to A'_\mu(x) = U(x)A_\mu(x)U^{-1}(x) - iU(x)\partial_\mu U(x)^{-1} \tag{7.13}$$

and, for infinitesimal ε^a

$$A^a_\mu(x) \to A'^a_\mu(x) = A^a_\mu + \delta A^a_\mu = A^a_\mu + C^a_{bc}A^b_\mu\varepsilon^c - \partial_\mu\varepsilon^a. \tag{7.14}$$

The properties of the transport operator along a closed loop allow one to verify that the Lagrangian density for the gauge fields,[2]

$$\mathcal{L} = -\frac{1}{4}G^a_{\mu\nu}G^{\mu\nu}_a, \tag{7.15}$$

where the *field strength* is given by

$$G^a_{\mu\nu}(x) = \partial_\nu A^a_\mu - \partial_\mu A^a_\nu + A^b_\mu A^c_\nu C^a_{bc}, \tag{7.16}$$

is invariant under the transformation (7.13).

The same result can be achieved by defining a covariant derivative that commutes with the variation operation via the introduction of compensating fields, the gauge fields, *i.e.*

$$\delta D_\mu\phi = i\varepsilon^a(x)\tau_a D_\mu\phi. \tag{7.17}$$

The transformation law (7.14) follows from direct calculation, while the expression for $G^a_{\mu\nu}$ is given by the commutator of the covariant derivatives,

$$[D_\nu, D_\mu]\phi = iG^a_{\mu\nu}\tau_a\phi. \tag{7.18}$$

Applying the Jacobi identity to (7.18), it is easy to verify that $G^i_{\mu\nu}$ obeys the Bianchi identity,

$$D_\rho G^a_{\mu\nu} + D_\nu G^a_{\rho\mu} + D_\mu G^a_{\nu\rho} = 0. \tag{7.19}$$

[2]The internal indices are raised and lowered by the Euclidean internal metric δ_{ab}.

7.1.2 *Hamiltonian formulation*

We now perform the standard Hamiltonian analysis for the system described by the action $S = \int d^4x\sqrt{-g}\mathcal{L}$, $\mathcal{L}$ being Lagrangian density (7.15), in a generic curved manifold using the $3+1$ decomposition of the metric tensor discussed in section 4.2.1. The configuration variables are the gauge fields A^a_μ and their conjugate momenta are

$$\pi^0_a = 0 \qquad \pi^i_a = \sqrt{-g}\, G^{a0i}, \tag{7.20}$$

where spacetime indexes have been raised with the metric tensor. The first set of conditions in (7.20) are primary constraints. Via a Legendre transformation one gets

$$H = \int \bigg[\Lambda^a \pi^0_a + \pi^i_a(\partial_i A^a_0 + C^a_{bc}\, A^b_0\, A^c_i) - N^i\, \pi^j_a\, G^a_{ij} \\ + \frac{N}{2}\left(\frac{1}{\sqrt{h}}\pi^i_a\, \pi^j_a\, h_{ij} + \frac{1}{2}\sqrt{h}\, G^a_{ij}\, G^a_{kl}\, h^{ik}\, h^{jl}\right)\bigg] d^3x \tag{7.21}$$

Λ^a being Lagrangian multipliers. Hence we get as secondary constraints the conditions

$$G_a = [\pi^0_a, H] = \partial_i \pi^i_a - C^c_{ab}\, A^b_i\, \pi^i_c \approx 0. \tag{7.22}$$

This set of constraints is known as the *Gauss constraint*, since for a $U(1)$ gauge theory they reduce to the Gauss equation for electrodynamics (3.103). By the procedure described in section 3.4.3, it can be shown that G_a generates Yang-Mills gauge transformations and in this respect the constraint $G_a \approx 0$ enforces gauge invariance in phase-space. Therefore, the full extended Hamiltonian reads

$$H_E = \int \bigg[\Lambda^a \pi^0_a + l^a G_a - N^i\, \pi^j_a\, G^a_{ij} \\ + \frac{N}{2}\left(\frac{1}{\sqrt{h}}\pi^i_a\, \pi^j_a\, h_{ij} + \frac{1}{2}\sqrt{h}\, G^a_{ij}\, G^a_{kl}\, h^{ik}\, h^{jl}\right)\bigg] d^3x \tag{7.23}$$

l^a being Lagrangian multipliers, and it provides the constrained dynamics of a Yang-Mills model.

7.1.3 *Lattice gauge theories*

The proper framework for the quantization of non-Abelian gauge theories is the path-integral formulation. A proper regularization of the partition function, from which the elements of the S-matrix can be evaluated, has been given via a lattice formulation, in which the gauge connections are defined only along edges of a fixed lattice.

A lattice k consists of links and plaquettes; if an orientation is chosen for them, edges e and faces f are defined, respectively, as oriented links and plaquettes, but physical quantities are independent on the choice of the (arbitrary) orientation. In particular, the lattice k is composed of the edges on the boundary ∂k and the edges in the interior, k^0, so that

$$k = k^0 \cup \partial k, \tag{7.24}$$

and $E_k = \{e\}$ denotes the set of all the edges of k.

The notion of connection g as restricted to a lattice is an application that maps the edges into elements of a (compact Lie) gauge group G,

$$g : E_k \to G, \tag{7.25}$$

$$e \mapsto g_e, \tag{7.26}$$

g_e being an element of the gauge group G, *i.e.* the holonomy of the connection along e (see section 8.1 for the definition of holonomies). We shall call the configuration space of the connections on k $\mathcal{A}_k$.

Path integrals[3], such as

$$W[\phi] = \int \left(\prod_{e \in k} dg_e \right) e^{iS(g)} \phi[g], \tag{7.27}$$

are the quantities that describe physical information. In (7.27), $\left(\prod_{e\in k} dg_e\right)$ is the Haar measure on G, $S(g)$ is the action, and the function $\phi[g] \in L^2_0(\mathcal{A}_k)$ denotes the physical states of the system. The action $S[g]$ can be written as the sum of terms referred to each face $f \in k^0$ (face action):

$$S[g] = \sum_{f \in k^0} S_f. \tag{7.28}$$

Each term S_f is required to be gauge invariant and to depend on the edges of the face itself only.

Throughout this discussion, we will be interested in boundary-state amplitudes $\Omega[\phi]$, whose weighting functionals $\phi[\partial g] \in L^2_0(\mathcal{A}_{\partial k})$ depend on the group elements carried by boundary edges,

$$\Omega[\phi] := \int \left(\prod_{e \in k} dg_e \right) e^{iS(g)} \phi^*(\partial g). \tag{7.29}$$

[3]Although in [102] the whole description is developed without specifying the choice of a Minkowskian or a Euclidean background, for our purposes it will be more convenient to depict the model in a Minkowskian frame.

The expression above can be seen as the wave function of a physical state, that is the probability amplitude of obtaining a physical state from vacuum. The dynamics of any lattice gauge model can be turned in a physically equivalent description by means of the spin-foam formalism which will be described in section 7.1.3.2.

7.1.3.1 *Spin-network States*

In order to properly define a quantum theory on $L^2(\mathcal{A}_k)$, we need some basis states. These are given by spin networks. A *spin network* is an oriented graph, whose edges e are labeled by irreducible representations of the considered gauge group ρ_e (colors), and whose vertices v are labeled by an *intertwiner* I_v, *i.e.* an element of the tensor product

$$\otimes_i V_{\rho_{i\,out}} \otimes_j V^*_{\rho_{j\,in}}, \tag{7.30}$$

where V_ρ is the space of the irreps of ρ and *in/out* denotes the edges incoming/outgoing in v. In this picture, intertwiners form a map between $\otimes_i V_{\rho_{i\,out}}$ and $\otimes_j V_{\rho_{j\,in}}$. If one decomposes the space (7.30) in the sum of irreps, a subspace can be found, which transforms according to the trivial representation (*invariant intertwiners*), as it is composed of invariant vectors. A spin-network state or functional $\psi_S[g]$ can be associated to a spin network, such that

$$\psi_S[g] : g_e \to \rho_{g_e}, \tag{7.31}$$

ρ_{g_e} being the irreps ρ evaluated on g_e, and reads

$$\psi_S(g) = \left(\prod_{v\in k} I_v\right) \cdot \left(\prod_{e\in k} D_{\rho_e}\rho_e(g_e)\right), \tag{7.32}$$

where D_{ρ_e} is a normalizing factor equal to the square root of the dimension of V_{ρ_e}, while the second product extends over all the edges emanating from v and $\cdot$ denotes index contraction. The correspondence (7.31) is not one-to-one, so that spin networks can be defined to be equivalent if they lead to the same spin-network functional. Spin-network states with invariant intertwiners define the space $L_0^2(\mathcal{A}_k)$ of gauge-invariant functional of the connections A_k: according to the Peter-Weyl theorem, the matrix elements of the irreps of a group form a basis for the functions of the Hilbert space of the L^2 functions of the group. If an orthonormal basis B_k for the invariant intertwiners I_v is chosen, for instance for two-valence vertices the intertwiner reads

$$I^a{}_b = \frac{1}{D_{\rho_e}}\delta^a{}_b, \tag{7.33}$$

so that the basis for $L_0^2(\mathcal{A}_k)$ will be orthonormal too; in this case, the spin-network functional simply rewrites

$$\psi_S[g] = tr\left[\rho\left(\prod_i g_{e_i}\right)\right]. \tag{7.34}$$

If only the edges on the boundary ∂k are taken into account, $B_{\partial k}$ is the orthonormal basis for $L_0^2(A_{\partial k})$. Loops are the spin networks on the edges that surround a face (the smallest graphs possible), and induce a basis for the face action (7.28), so that the exponential in (7.27) rewrites

$$e^{iS_f} = \sum_{S_f \in B_f} c_{S_f} \psi_{S_f}, \tag{7.35}$$

where c_{S_f} are suitable coefficients and the sum extends over all spin networks S_f whose states Ψ_{S_f} are in the basis B_f.

7.1.3.2 *Spin Foams*

Spin foams are 2-dimensional branched surfaces that carry irreps and intertwiners: the definition is analogous to that of spin networks, but one dimension has to be added. Each branched surface F is composed of its unbranched components F_i, so that $F = \bigcup_i F_i$.

Given a spin network ψ (we adopt here the same notation as for spin-network states), a spin foam F is an application such that

$$\forall \psi, F : 0 \to \psi, \tag{7.36}$$

in which 0 denotes the "zero" spin network with no edges and vertices. Given any two disjoint spin networks ψ and ψ', a spin foam F is an application that maps the former into the latter, or, equivalently,

$$\forall \psi, \psi', \;\; F : \psi \to \psi'; \qquad F : 0 \to \psi^* \circ \psi', \tag{7.37}$$

where $\circ$ denotes disjoint union.

A spin foam is non-degenerate if and only if each vertex is the end-point of at least one edge, each edge of at least one face and each face carries an irrep of the group G.

Equivalence classes can be established for spin foams: spin foams are equivalent if one can be obtained from the other by affine transformation, subdivision or orientation reversal of the lattice.

Let's analyze in some detail how to express the path integral (7.29) in terms of spin-foam amplitudes. The integration can be performed in two steps.

The path integrals can be integrated over k^0, as $\phi^*(\partial g)$ is not affected by the integration, *i.e.*

$$\Omega[g] = \int_{\partial g'=g} \left(\prod_{e\in k^0} dg'_e\right) e^{iS(g')}, \tag{7.38}$$

and then inserted into (7.29),

$$\Omega[\phi] = \int \left(\prod_{e\in\partial k} dg_e\right) \Omega(g)\phi^*(g), \tag{7.39}$$

so that (7.38) can be expanded into spin-network states. In fact, the exponential of the action can be expanded as

$$e^{iS(g)} = \prod_{f\in k^0} e^{iS_f(g)} = \prod_{f\in k^0} \sum_{S_f\in B_f} c_{S_f}\psi_{S_f(g)}, \tag{7.40}$$

and, when substituted in (7.38), it brings the result

$$\begin{aligned}\Omega[g] = \int_{\partial g'=g} \prod_{e\in k^0} dg'_e e^{iS(g')} &= \int_{\partial g'=g} \prod_{e\in k^0} dg'_e \prod_{f\in k^0} \sum_{S_f\in B_f} c_{S_f}\psi_{S_f}(g') \\ &= \sum_{\{f\}\to\{S_f\}} \int_{\partial g'=g} \prod_{e\in k^0} dg'_e \prod_{f\in k^0} c_{S_f}\psi_{S_f}(g),\end{aligned}$$

where, in the last step, the sum has been drawn out of the integral, and all the possible configurations S_f for each f have been taken into account. The introduction of spin foams is suggested by the need to evaluate each term of the sum,

$$\int_{\partial g'=g} \prod_{e\in k^0} dg'_e \prod_{f\in k^0} c_{S_f}\psi_{S_f}(g') \tag{7.41}$$

where spin networks are better organized into surfaces. In fact, two spin networks belong to the same unbranched surface F_i if they share only one edge, and if this edge is not shared with any other spin network. The unbranched surfaces F_i are either disconnected, or match other unbranched surfaces. In the latter case, the spin foam is defined as the branched surface $F = \cup_i F_i$.

In order to evaluate (7.41), two non-trivial cases can be distinguished, *i.e.*

(1) two loops match on one edge, and they carry the same label: the unbranched surface is defined as single-colored;
(2) more than two loops match on one edge, and Haar intertwiners (a generalization of intertwiner defined formerly) have to be introduced.

As a result, all the elements contribute to the sum as follows

(1) for each vertex, a factor A_v ;
(2) for each single-colored component, a factor $\prod_i A_{F_i}$, where $A_{F_i} \propto \prod_{f\in f_i} C_{f_\rho}$ and $C_{f_\rho} \propto C_{s_f}$, the proportionality factor being a suitable power of $dimV_\rho$;
(3) for each branching graph Γ_F, the projection properties of the Haar intertwiners have to be taken into account: as a result, for each vertex of the branching graph, one has to sum over all the possible ways to assign an intertwiner to the links of Γ_F.

Collecting all the terms together, one obtains

$$\Omega[g] = \sum_{F\subset k} \left(\prod_{v\in\Gamma_F} A_v\right)\left(\prod_i A_{F_i}\right)\psi_{S_f(g)}. \tag{7.42}$$

The product

$$\prod_{f\in F_i} C_{f_\rho} \tag{7.43}$$

in general depends on the discretization, and only in particular cases a geometrical interpretation is possible.

The insertion of (7.42) in (7.39) gives the final expression of the path integral $\Omega[\phi]$. If one expands $\phi(g)$ in terms of the orthonormal basis of spin networks (7.34),

$$\phi(g) = \sum_{S\in B(\partial k)} \phi_S \psi_S(g), \tag{7.44}$$

(7.39) reads

$$\Omega[\phi] = \sum_{F\subset k} \left(\prod_{v\in\Gamma_F} A_v\right)\left(\prod_i A_{F_i}\right)\phi^*_{S_F}, \tag{7.45}$$

i.e. the only non-vanishing contributions are brought by boundary spin networks, and each spin-foam amplitude is weighted by the coefficient of the corresponding boundary state. The comparison between the path-integral formulation (7.29) and (7.45) is eventually accomplished by noticing that the integration over connections is replaced by the sum over spin foams, and the spin-foam amplitudes weighted by the boundary functional ϕ_{S_F} play the role of the invariant measure times the exponential of the action, with a boundary weighting coefficient $\phi(g)$.

7.1.3.3 *Background independence*

The mismatch between the idea of background independence and the geometrical interpretation of spin-foam models can be analyzed by considering two possibilities:

(1) spin foams can be identified with the entire lattice, which plays the role of a discrete spacetime;
(2) spin foams can be interpreted as lattice-independent geometrical objects, which live on the lattice itself. The lattice, in this case, is considered as an auxiliary field, which has to be removed in the definitive model.

In the second case, the amplitudes described in the initial model must depend on the geometry of the spin foams only, *i.e.*, in the sum

$$\Omega_k[\phi] = \sum_{F \in k} \left(\prod_{v \in \Gamma_F} A_v \right) \left(\prod_i A_{F_i} \right) \phi^*_{S_f}, \tag{7.46}$$

each factor A depends on the branching graph only. The sum in (7.46) can be extended to a background-independent sum over all the equivalence classes of spin foams F on a given manifold M, *i.e.* $\sum_{F \in k} \to \sum_{F \in M}$, so that

$$\Omega_k[\phi] = \sum_{F \in M} \left(\prod_{v \in \Gamma_F} A_v \right) \left(\prod_i A_{F_i} \right) \phi^*_{S_f}, \tag{7.47}$$

where abstract (or topological) spin foams are defined by means of abstract spin-network states, the equivalence class of spin-networks states, invariant under homeomorphisms of the boundaries. The extension (7.47) is possible only by the modification of the Hilbert space, as spin networks are not defined on the boundaries ∂k. The new space of boundary states H is defined as

$$\mathrm{H}_{\partial M} = \left\{ \sum_{i=1}^{n} a_i S_i : \; a_i \in \mathbb{C}, S_i \subset M, n \in \mathbb{N} \right\}, \tag{7.48}$$

i.e. a finite combination of spin-network states, endowed with the structure of scalar product

$$\langle S, S' \rangle = \delta_{SS'}, \tag{7.49}$$

for which the dual space $\mathrm{H}^*_{\partial M}$ is defined, as usual, as

$$\mathrm{H}^*_{\partial M} = \{\phi\}, \qquad \phi : \mathrm{H}_{\partial M} \to \mathbb{C}. \tag{7.50}$$

The definition of such a Hilbert space is followed by the problem of over-counting, due to the homomorphisms h of the manifold M,

$$h : M \to M, \qquad A(h^* M) = A(M), \qquad \phi_{S_{h^*F}} = \phi_{S_F}, \tag{7.51}$$

which can be gauged away *à la* Faddeev-Popov. The result is the definition of abstract or topological spin foams, and the corresponding spin-network states are the equivalence class of spin-network states invariant under homeomorphisms of the boundaries.

Spin foams can be applied to covariant quantum gravity, where they play the role of the path integral [162], as the tool that connects different gravity states (geometries) in time. In particular, an equivalence class of three-geometries $h = \{h_{ij}\}$ on a three-dimensional hypersurface Σ is represented by a spin-network state, and the history between two different states is the spin-foam amplitude, *i.e.*

$$\langle h_2, \Sigma_2 | h_1, \Sigma_1 \rangle = \int_{g/g(\Sigma_1 = h_1), g/g(\Sigma_2) = h_2} Dg \, e^{iI_{EH}(g)}, \tag{7.52}$$

where I_{EH} denotes the Einstein-Hilbert action. The measure Dg is aimed at outlining the conceptual analogy with Feynman's approach [10, 128] rather than at defining any specific integration measure, which will be explicitly given, when needed, throughout the calculations.

Of course, the composition of spin foams must be defined, such that the transition between two states is independent of the intermediate states among which the transition is decomposed, *i.e.* in the sum

$$\langle h_3, \Sigma_3 | h_1, \Sigma_1 \rangle = \sum_{h_2} \langle h_3, \Sigma_3 | h_2, \Sigma_2 \rangle \, \langle h_2, \Sigma_2 | h_1, \Sigma_1 \rangle, \tag{7.53}$$

the intermediate states 2 must carry a trivial representation, in the sense specified in the previous paragraphs. As the probability of creating a state from the vacuum, a spin foam is defined as

$$|h_1, \Sigma_1\rangle = \int_{g/g(\Sigma_1) = h_1} Dg \, e^{iI_{EH}(g)}. \tag{7.54}$$

Spin-network states can be defined following a procedure which is slightly different from the previous one, in order to realize how the geometrical properties of the state fit the constraints of the ADM formulation [283]. This is discussed in section 8.5.1.

7.2 Gravity as a gauge theory of the Lorentz group?

The formulation of spinors in curved spacetime outlines that $\omega_{\mu}^{\alpha\beta}$ behaves as the connection of the local Lorentz group in the vierbein representation. Let us briefly revise it.

7.2.1 *Spinors in curved spacetime*

The mathematical description of *spinors* requires the definition of the *Dirac algebra*, whose main properties in a flat Minkowski spacetime can be summarized as follows:

- the existence of a Clifford algebra, *i.e.* an algebra generated by four matrices γ^{μ} (Dirac matrices) satisfying $\{\gamma^{\mu}, \gamma^{\nu}\} = 2\eta^{\mu\nu}$, where $\{..\}$ denotes the anticommutator $\{A, B\} = AB + BA$;
- the independence of Dirac matrices by coordinates, *i.e.* $\partial_{\mu}\gamma^{\nu} = 0$;
- the conjugation relations $\gamma^{\mu\dagger} = \gamma^{0}\gamma^{\mu}\gamma^{0}$.

The extension to a curved spacetime is non-trivial [80]. For instance, the analogous of the first condition reads

$$\{\gamma^{\mu}, \gamma^{\nu}\} = 2g^{\mu\nu}, \tag{7.55}$$

and since the right-hand side of the expression above generically depends on the coordinates, Dirac matrices are not constant anymore. Hence, the second property cannot hold but we can generalize it by defining a new covariant derivative D^{ψ}_{μ} for spinors for which[4]

$$D^{\psi}_{\mu}\gamma^{\nu} = \nabla_{\mu}\gamma^{\nu} - [\Gamma_{\mu}, \gamma^{\nu}] = 0. \tag{7.56}$$

Hence, the connection Γ_{μ} turns out to be

$$\Gamma_{\mu} = -\frac{1}{4}\gamma_{\nu}\nabla_{\mu}\gamma^{\nu}. \tag{7.57}$$

It is illuminating to see what is going on in terms of the vierbein representation. The vierbein projection of Dirac matrices $\gamma^{\alpha} = e^{\alpha}_{\mu}\gamma^{\mu}$ coincide with the flat ones, since

$$\{\gamma^{\alpha}, \gamma^{\beta}\} = e^{\alpha}_{\mu}\, e^{\beta}_{\nu}\, \{\gamma^{\mu}, \gamma^{\nu}\} = 2\, e^{\alpha}_{\mu}\, e^{\beta}_{\nu}\, g^{\mu\nu} = 2\eta^{\alpha\beta}, \tag{7.58}$$

and we can use the derivatives D^{γ}_{μ} with Ricci coefficients (1.120) to relate spacetime derivatives with those in flat space, *i.e.*

$$\begin{aligned}\nabla_{\mu}\gamma^{\nu} &= e^{\alpha}_{\mu}\, e^{\nu}_{\beta}\, D^{(\gamma)}_{\alpha}\gamma^{\beta} = e^{\alpha}_{\mu}\, e^{\nu}_{\beta}\, (\partial^{(\gamma)}_{\alpha}\gamma^{\beta} + \gamma^{\beta}{}_{\delta\alpha}\gamma^{\delta}) \\ &= e^{\alpha}_{\mu}\, e^{\nu}_{\beta}\, \gamma^{\beta}{}_{\delta\alpha}\gamma^{\delta} = -e^{\nu}_{\beta}\, \omega^{\beta\alpha}_{\mu}\, \gamma_{\alpha}.\end{aligned} \tag{7.59}$$

[4]The additional term acts as a commutator on matrices since a commutator is a derivation for matrices (the Leibniz rule holds).

Hence, one finds

$$\Gamma_\mu = -\frac{i}{2}\omega_\mu^{\alpha\beta}\Sigma_{\alpha\beta}\,, \tag{7.60}$$

where $\Sigma_{\alpha\beta}$ are the generators of the Lorentz group, *i.e.*

$$\Sigma_{\alpha\beta} = \frac{i}{4}[\gamma_\alpha, \gamma_\beta]. \tag{7.61}$$

Hence the full derivative reads

$$D_\mu^\psi \psi = \partial_\mu \psi - \frac{i}{2}\omega_\mu^{\alpha\beta}\Sigma_{\alpha\beta}\psi\,, \tag{7.62}$$

and from the comparison with the expression (7.12) one sees how $\omega_\mu^{\alpha\beta}$ behaves as the gauge connection associated with the group of local Lorentz transformations. The Dirac Lagrangian density for spinors on a curved spacetime invariant under local Lorentz transformation can be written from the standard one by replacing ordinary derivatives with D_μ^ψ, so finding

$$\Lambda_\psi = \frac{i\hbar c}{2}\left[\left(\partial_\mu\bar{\psi} + \frac{i}{2}\omega_\mu^{\alpha\beta}\bar{\psi}\Sigma_{\alpha\beta}\right)\gamma^\mu\psi - \bar{\psi}\gamma^\mu\left(\partial_\mu\psi - \frac{i}{2}\omega_\mu^{\alpha\beta}\Sigma_{\alpha\beta}\psi\right)\right]. \tag{7.63}$$

The invariance under local Lorentz transformations is now manifest. The main difference with respect to the flat case is that in a curved spacetime one can choose a different local Lorentz frame at each point. Therefore, some new connections have to be introduced in order to ensure the invariance under the local Lorentz symmetry. On the contrary, in a Minkoswkian spacetime one usually identifies the tangent space with the manifold itself (in other words one identifies all the tangent spaces), thus the only Lorentz symmetry is a global one.

Torsion arises as a consequence of the back-reaction of spinors on the spacetime in a first order formulation. In fact, the Lagrangian (7.63) contains the spin connections, thus in the presence of spinors the structure equation (4.85) is modified as follows

$$\partial_{[\mu}e^\alpha_{\nu]} - \omega^\alpha_{[\mu|\beta|}e^\beta_{\nu]} = \frac{1}{4}\epsilon^\alpha{}_{\beta\gamma\delta}e^\beta_\nu e^\gamma_\mu J^\delta_A \qquad J^\alpha_A = \bar{\psi}\gamma^\alpha\gamma_5\psi. \tag{7.64}$$

Hence, the spin connections do not coincide with the expression (1.124), but they contain an additional torsion-like term. In particular, the axial current acts as a source for torsion, which reads

$$T^\rho_{\mu\nu} = \frac{1}{4}\varepsilon^\rho{}_{\nu\mu\sigma}e^\sigma_\delta J^\delta_A. \tag{7.65}$$

The modified expression for $\omega_\mu^{\alpha\beta}$ can be substituted into the total Einstein-Dirac action so finding

$$S = \int \Big[-\frac{c^3}{16\pi G} R + \frac{i\hbar}{2} \left({}^{(0)}D_\mu^\psi \bar{\psi}\,\gamma^\mu\,\psi - \bar{\psi}\,\gamma^\mu\,{}^{(0)}D_\mu^\psi \psi \right) - \frac{3\pi G\hbar^2}{c^3} \eta_{\alpha\beta}\, J_A^\alpha\, J_A^\beta \Big]\, e\, d^4x\,, \tag{7.66}$$

where the Einstein-Hilbert term is constructed for a spacetime with torsion, while the derivatives ${}^{(0)}D_\mu^\psi$ contain only the torsion-less spin connections (1.124). It is worth noting the presence of four fermions interaction terms, which make the effective spinor action non-renormalizable on a perturbative level.

7.2.2 *Comparison between gravity and Yang-Mills theories*

The analogy between a Yang-Mills gauge theory and gravity is restricted to the coupling between Dirac spinors and spin connections. First of all, the gravitational Lagrangian density differs from the one of gauge bosons, because instead of a term of the kind G^2 (7.15), which here means $g^{\mu\rho}\, g^{\nu\sigma}\, \eta_{\alpha\gamma}\, \eta_{\beta\delta}\, R_{\mu\nu}^{\alpha\beta} R_{\rho\sigma}^{\gamma\delta}$, we have $e\, e_\alpha^\mu\, e_\beta^\nu\, R_{\mu\nu}^{\alpha\beta}$. Then, the spin connections are not the only variables, but there are additional fields, vierbein vectors. The attempts to interpret e_μ^α also as connections (associated with translations) point to Poincaré gauge theory (see section 7.3).

In this respect, let us now give an argument explaining why the spin connections are entirely determined by vierbein vectors. An infinitesimal Lorentz rotation can be written as

$$\delta x^\mu = \varepsilon^{\mu\nu}(x) x_\nu = \theta^\mu(x) \qquad \varepsilon^{\mu\nu} = -\varepsilon^{\nu\mu}, \tag{7.67}$$

and it cannot be distinguished from an infinitesimal translation

$$\delta x^\mu = \varepsilon^\mu(x). \tag{7.68}$$

Since $\varepsilon^{\mu\nu}$ are arbitrary functions of spacetime coordinates, then θ^μ and also ε^μ are arbitrary. This reduction of local Lorentz transformations to translations explains why, in the standard treatment, the connections associated with the former, $\omega_\mu^{\alpha\beta}$, can be obtained from those associated with the latter, e_μ^α.

Finally, it is remarkable to point out that in order to give a real physical meaning to spin connections, they should be actual independent fields and so they should not be entirely determined by gravitational or matter fields or both. The reason for this request is clear in the Einstein-Cartan theory

[165]: in this framework, torsion is non-vanishing only in spacetime points where a spin density is present. This means that torsion is not a propagating field and there is no way to detect it (one can roughly think that torsion is "overwhelmed" by the spin density in any place and any time it is). Such a kind of field has no real physical meaning.

Therefore, the interpretation of spin connections as real independent physical fields requires the introduction of a fundamental torsion, which must be present also in vacuum [197]. This can be realized only by extending General Relativity.

7.3 Poincaré gauge theory

Poincaré Gauge Theory (PGT) [58, 59] is aimed at describing local Poincaré transformations within the framework of the gauge formalism, *i.e.* by the introduction of covariant derivatives and conserved currents.

Let us consider an infinitesimal global Poincaré transformation in Minkowski space, formed by global Lorentz transformations and translations,

$$x^{\mu} \to x'^{\mu} = x^{\mu} + \tilde{\varepsilon}^{\mu}{}_{\nu} x^{\nu} + \tilde{\varepsilon}^{\mu}, \tag{7.69}$$

and the consequent transformation law for spinor fields

$$\psi(x) \to \psi'(x) = \left(1 + \frac{1}{2}\tilde{\varepsilon}^{\mu\nu} M_{\mu\nu} + \tilde{\varepsilon}^{\mu} P_{\mu}\right)\psi(x), \tag{7.70}$$

where the generators $M_{\mu\nu} = L_{\mu\nu} + \Sigma_{\mu\nu}$[5] and P_{μ} obey Lie-algebra commutation relations. If the matter Lagrangian density is assumed to depend on the spinor field and on its derivatives only, $L = L(\psi, \partial_a \psi)$, and if the equations of motion are assumed to hold, the conservation law $\partial_{\mu} J^{\mu} = 0$ is found, where

$$J^{\mu} = \frac{1}{2}\tilde{\varepsilon}^{\nu\lambda} M^{\mu}{}_{\nu\lambda} - \tilde{\varepsilon}^{\nu} T^{\mu}_{\ \nu}, \tag{7.71}$$

and the canonical energy-momentum and angular-momentum tensors are defined, respectively, as

$$T^{\mu}_{\ \nu} = \frac{\delta \mathcal{L}}{\delta \psi_{,\mu}} \partial_{\nu}\psi - \delta^{\mu}{}_{\nu}\mathcal{L}, \tag{7.72}$$

$$M^{\mu}{}_{\nu\lambda} = \left(x_{\nu} T^{\mu}_{\ \lambda} - x^{\lambda} T^{\mu}_{\ \nu}\right) - S^{\mu}{}_{\nu\lambda} \equiv \left(x_{\nu} T^{\mu}_{\ \lambda} - x^{\lambda} T^{\mu}_{\ \nu}\right) + \frac{\partial \mathcal{L}}{\partial(\partial_{\mu}\psi)}\Sigma_{\nu\lambda}\psi. \tag{7.73}$$

[5] $L_{\mu\nu}$ denotes the orbital angular momentum operator.

As the parameters in (7.71) are constant, according to Noether's theorem, the conservation laws for the energy-momentum current and for the angular-momentum currents, together with the related charges, are established:

$$\partial_\mu T^\mu_{\ \nu} = 0 \rightarrow \partial_0 P^\nu = \partial_0 \int d^3x T^{0\nu} = 0 \tag{7.74}$$

$$\partial_\mu M^\mu_{\ \nu\lambda} = 0 \rightarrow \partial_0 M^{\nu\lambda} = \partial_0 \int d^3x M^{0\nu\lambda} = 0. \tag{7.75}$$

When the theory is locally implemented, equations (7.71)-(7.75) do not hold any more, and compensating gauge fields have to be introduced in order to restore local invariance. As a first step, a covariant derivative $D_\alpha\psi$ is defined as

$$D_\alpha\psi = e^\mu_\alpha D_\mu\psi = e^\mu_\alpha \left(\partial_\mu + A_\mu\right)\psi = e^\mu_\alpha \left(\partial_\mu + \frac{1}{2}A^{\beta\gamma}_\mu \Sigma_{\beta\gamma}\right)\psi, \tag{7.76}$$

where the compensating fields e^μ_α and $A^{\beta\gamma}_\mu$, and the generator $\Sigma_{\beta\gamma}$ have been taken into account. This way, the Lagrangian density depends on the covariant derivative of the fields, instead of the ordinary one, $\mathcal{L} = \mathcal{L}(\psi, D_\alpha\psi)$; covariant derivatives (7.76) don't commute, but satisfy the commutation relation

$$[D_\mu, D_\nu]\psi = \frac{1}{2}F^{\alpha\beta}_{\ \ \mu\nu}\Sigma_{\alpha\beta}\psi, \quad [D_\gamma, D_\delta]\psi = \frac{1}{2}F^{\alpha\beta}_{\ \ \gamma\delta}\Sigma_{\alpha\beta}\psi - F^\epsilon_{\ \gamma\delta}D_\epsilon\psi, \tag{7.77}$$

where $F^{\alpha\beta}_{\ \ \mu\nu}$ and $F^\epsilon_{\ \gamma\delta}$ are the Lorentz field strength and the translation field strength, respectively.

Covariant energy-momentum and spin currents, T'^μ_ν and $S'^\mu_{\ \alpha\beta}$, can be found, in analogy with the global case, after the substitution $\partial_\mu \rightarrow D_\mu$, and are found to be equivalent to the dynamic currents $\tau^\mu_{\ \nu}$ and $\sigma^\mu_{\ \alpha\beta}$,

$$T'^\mu_\nu = \tau^\mu_{\ \nu} = e_\alpha^{\ \mu}\frac{\partial\mathcal{L}}{\partial e_\alpha^{\ \nu}}, \tag{7.78}$$

$$S^\mu_{\ \alpha\beta} = \sigma^\mu_{\ \alpha\beta} = -\frac{\partial\mathcal{L}}{\partial A^{\alpha\beta}_{\ \ \mu}}, \tag{7.79}$$

whose meaning will be outlined throughout the rest of this section.

A simple and illuminating example by Hehl *et al.* [165] illustrates the inadequacy of special relativity to describe the behavior of matter fields under global Poincaré transformations. Global Poincaré transformations preserve distances between events and the metric properties of neighboring matter fields: comparing field amplitudes in nearby points before performing the transformation, and then transforming the result, or comparing the transformed amplitudes of the fields is equivalent. This property is known as rigidity condition, as matter fields behave as rigid bodies under this kind of transformations. On the contrary, it can be shown that the action of local Poincaré transformations can be interpreted as an irregular deformation of matter fields, thus predicting different phenomenological evidences for the field and for the transformed field. The compensating gauge fields e^{α}_{μ} and $A^{\alpha\beta}_{\mu}$, introduced to restore local invariance, describe geometrical properties of the spacetime: it can be demonstrated that PGT has the geometrical structure of a Riemann-Cartan spacetime.

The geometrical approach to PGT can be carried out by considering the most general metric-compatible linear connections, with 24 independent components, which can be written as a function of the torsion field $T^{\mu}{}_{\nu\rho}$ (4.83).

Geometric covariant derivatives are defined as

$$D^{\psi}_{\mu}\psi = (\partial_{\mu} + \omega_{\mu})\,\psi = \left(\partial_{\mu} + \frac{1}{2}\omega^{\alpha\beta}{}_{\mu}\Sigma_{\alpha\beta}\right)\psi, \tag{7.80}$$

where spin connections $\omega^{\alpha\beta}{}_{\mu}$ consist of the bein projection of the Ricci rotation coefficients and the contortion field, respectively: $\omega_{\alpha\beta\mu} = -\gamma_{\alpha\beta\mu} + K_{\alpha\beta\mu}$.

The gauge potentials $e_{\alpha}{}^{\mu}$ are generally interpreted as the connection between the orthonormal frames and the coordinate frames, while the introduction of the gauge potentials $\omega^{\beta\gamma}_{\alpha}$ is connected with the relative rotations of the orthonormal basis at neighboring points: this induces a change in the derivative operator, *i.e.*

$$\partial_{\alpha} \to D^{\psi}_{\alpha} \equiv \partial_{\alpha} + \frac{1}{2}\omega^{\beta\gamma}_{\alpha}\Sigma_{\beta\gamma}. \tag{7.81}$$

The comparison between the gauge approach and the geometrical approach leads to the identification of the gauge field $A^{\alpha\beta}_{\mu}$ with spin connections $\omega^{\alpha\beta}{}_{\mu}$, and the field $e_{\alpha}{}^{\mu}$ with the components of the vierbein. This way, the identification of the Lorentz field strength with the curvature, and that of the translation field strength with torsion are straightforward.

Torsion contributes to the gravitational dynamics, according to its gravitational action: it has been illustrated that [164] the most general form for a Lagrangian $\mathcal{L}_T$, which allows for equations of motion that are at most of second order in the field derivatives, is

$$\mathcal{L}_T = AT_{\alpha\beta\gamma}T^{\alpha\beta\gamma} + BT_{\alpha\beta\gamma}T^{\beta\alpha\gamma} + CT_{\alpha}T^{\alpha}, \tag{7.82}$$

where $T_\alpha = T^\beta{}_{\beta\alpha}$. The values of the parameters A, B, C are to be determined according to the Physics that has to be described, and some relevant examples are discussed in [163, 166, 237, 293].

For later purposes, it will be convenient to restate the description of PGT in a slightly different formalism, which allows for a better explication of the role of spin.

Equation (7.70) can be written as

$$\psi(x) \rightarrow \psi'(x) = \left(1 + \frac{1}{2}\varepsilon^{\mu\nu}\Sigma_{\mu\nu} + \varepsilon^\mu P_\mu\right)\psi(x), \tag{7.83}$$

$$\varepsilon^\gamma \equiv \widetilde{\varepsilon}^\gamma + \widetilde{\varepsilon}_\alpha{}^\beta \delta^\alpha_i x^i, \qquad \varepsilon^{\alpha\beta} = \widetilde{\varepsilon}^{\alpha\beta}, \tag{7.84}$$

the generators of translations and spin rotations satisfying the relations:

$$[\Sigma_{\mu\nu}, \Sigma_{\rho\sigma}] = \eta_{\rho[\mu}\Sigma_{\nu]\sigma} - \eta_{\sigma[\mu}\Sigma_{\nu]\rho}, \quad [\Sigma_{\mu\nu}, P_\rho] = -\eta_{\rho[\mu}\partial_{\nu]}, \quad [P_\mu, P_\nu] = 0. \tag{7.85}$$

The advantage of (7.83) consists in keeping pure rotations distinguished from translations. The orbital angular momentum is this way kept independent of the spin angular momentum: the former is strictly related with the energy-momentum, thus with the rotation-dependent part of ε^μ, while the latter is connected with the pure-rotation parameter $\varepsilon^{\mu\nu}$. In fact, if the analogy is drawn between a generic diffeomorphism and a global Poincaré transformation, it is impossible to perform translations and rotations independently, but, when a local symmetry is considered, this becomes possible, because the parameters defining the transformation are allowed to vary freely.

It is worth noting that, after the geometrical identification of the covariant gauge derivative, equation (7.79) becomes an algebraic relation between spin and torsion: since the relation is not differential, torsion is not predicted to propagate, but its existence is bound to the presence of spin-$\frac{1}{2}$ matter fields. Finally, the field equations read:

$$\frac{1}{e}D^{(\omega)}_\beta\left(ee^\alpha_\mu e^\beta_\nu\right) = \mathcal{S}^\alpha{}_{\mu\nu}, \tag{7.86a}$$

$$R^\alpha_\mu - \frac{1}{2}e^\alpha_\mu R = \frac{8\pi G}{c^4}T^\alpha_\mu. \tag{7.86b}$$

The first equation is the first Cartan structure equation, which provides one with the expression of connections of the group of rotations as a function of connections of the group of translation and matter fields, while the second equation is the Einstein's dynamic equation: tensor fields involved in the equations above must satisfy the identities (1.96) and (1.95), thus predicting a non-propagative behavior for torsion.

Thus we see that while the gravitational field in General Relativity is a fictitious gauge field of the Lorentz group (the only independent fields are the vierbein and the spinors), as far as torsion is included we can construct a gauge theory of the Poincaré group. In such a theory, the vierbein field accounts for the local character of translations (infinitesimal diffeomorphisms), while the torsion field has the proper independent behavior to be a gauge field of the Lorentz group (a contribution to the corresponding connections is provided also by the vierbein via the Ricci coefficients).

7.4 Holst action

The difficulties in quantizing General Relativity in the metric representation suggest to look for a different formulation of gravity. If one wants to modify General Relativity, one cannot contradict its well-tested predictions. An easy way to do that is to give a reformulation of gravity which does not modify the equations of motion, such that the classical dynamics is unchanged while some new features may arise on a quantum level. This is the case for the Holst reformulation of gravity.

7.4.1 *Lagrangian formulation*

Holst action reads [170]

$$S_{H} = -\frac{c^3}{16\pi G}\int d^4x\; e\, e^{\mu}_{\alpha}\, e^{\nu}_{\beta}\left(R^{\alpha\beta}_{\mu\nu} - \frac{1}{2\gamma}\varepsilon^{\alpha\beta}{}_{\delta\varepsilon}\, R^{\delta\varepsilon}_{\mu\nu}\right), \tag{7.87}$$

γ being the Immirzi parameter [173], which we assume to be real, while $R^{\alpha\beta}_{\mu\nu}$ is given in (1.133). It is convenient to introduce the following quantity

$$^{(\gamma)}p^{\delta\varepsilon}{}_{\alpha\beta} = \left(\delta^{\delta\varepsilon}_{\alpha\beta} - \frac{1}{2\gamma}\varepsilon_{\alpha\beta}{}^{\delta\varepsilon}\right), \qquad \delta^{\delta\varepsilon}_{\alpha\beta} = \delta^{[\delta}_{\alpha}\delta^{\varepsilon]}_{\beta}\,, \tag{7.88}$$

which can be seen as a linear operator acting on the direct product of two internal Lorentz vectors (bi-vectors). This operator can be inverted for $\gamma \neq \pm i$ and the explicit expression of the inverse is

$$^{(\gamma)}p^{-1\alpha\beta}_{\delta\varepsilon} = \frac{\gamma^2}{\gamma^2+1}\left(\delta^{\alpha\beta}_{\delta\varepsilon} + \frac{1}{2\gamma}\varepsilon^{\alpha\beta}{}_{\delta\varepsilon}\right). \tag{7.89}$$

In order to evaluate the equations of motion, let us consider a first order formulation and perform independent variations with respect to ω^{ab}_{μ} and

e^{μ}_{α}. In particular, the variation of S_H with respect to the spin connections gives

$$\begin{aligned}\delta_\omega S_H &= -\frac{c^3}{16\pi G}\int d^4x\; e\, e^{\mu}_{\delta}\, e^{\nu}_{\varepsilon}\, {}^{(\gamma)}p^{\delta\varepsilon}{}_{\alpha\beta}\delta\, R^{\alpha\beta}_{\mu\nu} \\ &= \frac{c^3}{16\pi G}\int d^4x\, \delta\omega^{\alpha\beta}_{\nu}\, D^{(\omega)}_{\mu}\left({}^{(\gamma)}p^{\delta\varepsilon}{}_{\alpha\beta}\, e\, e^{\mu}_{\delta}\, e^{\nu}_{\varepsilon}\right) \\ &= \frac{c^3}{16\pi G}\int d^4x\, \delta^{(\gamma)}A^{\alpha\beta}_{\nu}\, D^{(\omega)}_{\mu}(e\, e^{\mu}_{\alpha}\, e^{\nu}_{\beta}),\end{aligned}$$

where in the first line we made a partial integration and in the second we used $D^{(\omega)}_{\mu}({}^{(\gamma)}p^{\delta\varepsilon}{}_{\alpha\beta}) = 0$ (see section 1.12), while we defined

$$ {}^{(\gamma)}A^{\alpha\beta}_{\mu} = {}^{(\gamma)}p^{\alpha\beta}{}_{\delta\varepsilon}\, \omega^{\delta\varepsilon}_{\mu}. \tag{7.90}$$

Since ${}^{(\gamma)}A^{\alpha\beta}_{\mu}$ and $\omega^{\alpha\beta}_{\mu}$ are mapped onto each other via invertible operators, the conditions of stationarity with respect to these variables are equivalent

$$\delta_\omega S_H = 0 \Leftrightarrow \delta_A S_H = 0. \tag{7.91}$$

Therefore, one gets the following equations of motion from $\delta_\omega S_H = 0$

$$D^{(\omega)}_{\mu}(e\, e^{\mu}_{\alpha}\, e^{\nu}_{\beta}) = 0. \tag{7.92}$$

This equation can be solved by fixing ω as in (1.124). If one substitutes the expression for $\omega = \omega(e)$ inside the action (7.87), the term $-e\, e^{\mu}_{\alpha}\, e^{\nu}_{\beta}\, R^{\alpha\beta}_{\mu\nu}$ coincides with the Einstein-Hilbert Lagrangian $\sqrt{g}R$, while the second term vanishes as a consequence of the cyclic identity for the Riemann tensor (1.95), since it can be shown from (1.132) that the following relation holds

$$\varepsilon^{\alpha\beta}{}_{\gamma\delta}\, e\, e^{\mu}_{\alpha}\, e^{\nu}_{\beta}\, R^{\gamma\delta}_{\mu\nu}(e) = -\varepsilon^{\mu\nu\rho\sigma}\, R_{\mu\nu\rho\sigma}(g) = 0. \tag{7.93}$$

Therefore, the Holst action differs from the Einstein-Hilbert one by a term vanishing "on-shell" (*i.e.* on the equations of motion) and the two formulations are classically equivalent. This is not the case anymore in the presence of matter fields coupled with spin connections, as for spinor fields [248].

7.4.2 *Hamiltonian formulation*

We can perform the Hamiltonian analysis as in section 8.1 by applying the ADM splitting procedure to vierbein vectors. The easiest choice is to partially fix the local Lorentz symmetry such that the vector e^{0}_{μ} is orthogonal to the spatial hypersurfaces (time-gauge condition), *i.e.* $e^{0}_{\mu} = \eta_\mu$ and generically

$$e_{\mu}{}^{\alpha} = \begin{pmatrix} N & N^i e^a_i \\ 0 & e^a_i \end{pmatrix}, \tag{7.94}$$

whose associated inverse vectors are given by

$$e_\alpha{}^\mu = \begin{pmatrix} 1/N & -N^i/N \\ 0 & e_a^i \end{pmatrix}. \tag{7.95}$$

The determinant of the metric can be rewritten as

$$\sqrt{-g} = e = N\,\bar{e}, \tag{7.96}$$

$\bar{e}$ being the determinant of e_i^a. By substituting the expressions above in the action (7.87) one finds

$$S_H = -\frac{c^3}{16\pi G}\int dx^0 d^3x \Big[2\,E_a^i\,\Big(\partial_0{}^{(\gamma)}A_i^a - \partial_i\omega_0^{0a} + \frac{1}{2\gamma}\varepsilon^a{}_{bc}\,\partial_i\omega_0^{bc} - 2\eta_{bc}\,\omega_{[0}^{0b}\,\omega_{i]}^{ca}$$

$$-\frac{1}{\gamma}\varepsilon^a{}_{bc}\,\omega_{[0}^{b0}\,\omega_{i]}^{0c} + \frac{1}{\gamma}\varepsilon^a{}_{bc}\,\eta_{df}\,\omega_{[0}^{bd}\,\omega_{i]}^{fc}\Big) - 2N^i\,E_a^j\,\Big(R_{ij}^{0a} - \frac{1}{2\gamma}\varepsilon^a{}_{bc}\,R_{ij}^{bc}\Big)$$

$$+N\,\bar{e}\,e_a^i\,e_b^j\Big(R_{ij}^{ab} + \frac{1}{\gamma}\varepsilon^{ab}{}_c\,R_{ij}^{0c}\Big)\Big] \tag{7.97}$$

in which we fixed ${}^{(\gamma)}A_i^a = {}^{(\gamma)}A_i^{0a}$, while E_a^i denotes densitized dreibein inverse vectors (inverse densitized triads)

$$E_a^i = \bar{e}\,e_a^i. \tag{7.98}$$

${}^{(\gamma)}A_i^a$ are Ashtekar-Barbero variables [43] and they are the only variables whose time derivatives appear in the action. The expression above can be written as follows performing an integration by parts

$$S_H = -\frac{c^3}{16\pi G}\int dx^0 d^3x \Big[2E_a^i\,\partial_0{}^{(\gamma)}A_i^a + 2{}^{(\gamma)}A_0^a(\partial_i E_a^i - \gamma\varepsilon_{ab}{}^c\,{}^{(\gamma)}A_i^b\,E_c^i)$$

$$+\frac{2}{\gamma}(\gamma^2+1)\varepsilon^a{}_{bc}\,\omega_0^{b0}\,\omega_i^{0c}\,E_a^i - 2N^i\,E_a^j\,\Big({}^{(\gamma)}F_{ij}^a + \frac{\gamma^2+1}{\gamma}\varepsilon^a{}_{bc}\,\omega_i^{0b}\,\omega_j^{0c}\Big)$$

$$+N\frac{E_a^i\,E_b^j}{\gamma\,\bar{e}}\Big(\varepsilon^{ab}{}_c\,{}^{(\gamma)}F_{ij}^c + \frac{\gamma^2+1}{\gamma}\,\bar{R}_{ij}^{ab}\Big)\Big], \tag{7.99}$$

where

$${}^{(\gamma)}F_{ij}^a = \partial_i{}^{(\gamma)}A_j^a - \partial_j{}^{(\gamma)}A_i^a - \gamma\varepsilon^a{}_{bc}\,{}^{(\gamma)}A_i^b\,{}^{(\gamma)}A_j^c \tag{7.100}$$

and

$$\bar{R}_{ij}^{ab} = \partial_i\omega_j^{ab} - \partial_j\omega_i^{ab} - \eta_{cd}\,\omega_i^{ac}\,\omega_j^{db} + \eta_{cd}\,\omega_j^{ac}\,\omega_i^{db}. \tag{7.101}$$

The vanishing of the variations with respect to ${}^{(\gamma)}A_0^a$ and ω_0^{b0} provides the following constraints

$$G_a = \partial_i E_a^i - \gamma \varepsilon_{ab}{}^c\, {}^{(\gamma)}A_i^b\, E_c^i = 0, \qquad \varepsilon^a{}_{bc}\, \omega_i^{0c}\, E_a^i = 0. \tag{7.102}$$

By inserting the second condition into the first one, we get

$$\partial_i E_a^i - \omega_a{}^b{}_i\, E_b^i = 0, \tag{7.103}$$

which can be solved by requiring ω_i^{ab} to be the spin connections associated with dreibein e_i^a, *i.e.*

$$\omega_i^{ab} = \bar{\omega}_i^{ab}(E) = \eta^{bc}\, e_c^j\, D_i e_j^a. \tag{7.104}$$

As soon as equation (7.104) holds, the conditions (7.102) are equivalent and we retain only the first one. Hence, the whole action can be written as[6]

$$S = -\frac{c^3}{8\pi G} \int dx^0 d^3x \left(E_a^i\, \partial_t {}^{(\gamma)}A_i^a + \Lambda^a\, G_a - N^i\, \mathcal{V}_i - \frac{N}{2}\, \mathcal{S} \right), \tag{7.105}$$

and since Λ^a[7], N and N^i are Lagrangian multipliers, one gets the following constraints

$$\begin{aligned} G_a &= \partial_i E_a^i - \gamma \varepsilon_{ab}{}^c\, {}^{(\gamma)}A_i^b\, E_c^i \approx 0 \\ \mathcal{V}_i &= E_a^j\, {}^{(\gamma)}F_{ij}^a \approx 0 \\ \mathcal{S} &= -\frac{E_a^i\, E_b^j}{\gamma\, \bar{e}} \left(\varepsilon^{ab}{}_c\, {}^{(\gamma)}F_{ij}^c + \frac{\gamma^2+1}{\gamma}\, \bar{R}_{ij}^{ab} \right) \approx 0. \end{aligned} \tag{7.106}$$

Therefore, the connections ${}^{(\gamma)}A_i^a$ are the only relevant configuration variables, while inverse densitized triads give the conjugate momenta. The Poisson brackets read

$$[{}^{(\gamma)}A_i^a(x^0, x^i)\,,\, E_b^j(x^0, y^i)] = -\frac{8\pi G}{c^3}\, \delta_b^a\, \delta_j^i\, \delta^3(x-y), \tag{7.107}$$

the other vanishing.

The Hamiltonian is a linear combination of constraints:

$$H = \frac{c^3}{8\pi G} \int d^3x \left(\Lambda^a\, G_a - N^i\, \mathcal{V}_i - \frac{N}{2}\, \mathcal{S} \right), \tag{7.108}$$

where for clarity we skipped the terms containing primary constraints. Among them there is $G_a = 0$ which is equivalent to the Gauss constraints of

[6] An alternative procedure to derive the constraints is presented in section 7.4.4.

[7] The explicit expression of Λ^a is

$$\Lambda^a = {}^{(\gamma)}A_0^a + \frac{\gamma^2+1}{\gamma^2}\omega_0^{a0} - \frac{\gamma^2+1}{\gamma^2} N^i\, \omega_i^{0a}.$$

a $SU(2)$ gauge theory. *The emergence of a $SU(2)$ gauge symmetry is the key point of this formulation*, since it will allow us to adopt some quantization techniques used for Yang-Mills models (see next chapter). In particular, the $SU(2)$ connection is obtained from ${}^{(\gamma)}A^a_i$ via a rescaling: $A^a_i = \gamma {}^{(\gamma)}A^a_i$, such that G_a becomes exactly the Gauss constraint of the $SU(2)$ group (7.22), while the Immirzi parameter enters the Poisson brackets as follows

$$[A^a_i(x^0, x^i)\,,\, E^j_b(x^0, y^i)] = -\frac{8\pi G}{c^3}\,\gamma\,\delta^a_b\,\delta^i_j\,\delta^3(x-y). \tag{7.109}$$

The other constraints are the vector constraint $\mathcal{V}_i = 0$, the scalar one $\mathcal{S} = 0$, which are equal modulo the Gauss constraint to the super-momentum and super-Hamiltonian constraints [283], respectively

$$\mathcal{V}_i|_{G_a=0} = \mathcal{H}_i, \qquad \mathcal{S}|_{G_a=0} = \mathcal{H}\,. \tag{7.110}$$

The constraints (7.106) are first-class. There is a convenient way to rewrite A^a_i:

$$A^a_i = \gamma\,\omega^{0a}_i - \frac{1}{2}\varepsilon^a{}_{bc}\,\bar{\omega}^{bc}_i = -\gamma K^a_i + \Gamma^a_i, \tag{7.111}$$

where K^a_i is related with the extrinsic curvature, *i.e.*

$$K^a_i = K_{ij}\,e^{ja}, \tag{7.112}$$

while Γ^a_i is a three-dimensional connection, defined as

$$\Gamma^a_i = -\frac{1}{2}\varepsilon^a{}_{bc}\,e^b_j\,D_i e^{jc}. \tag{7.113}$$

In order to demonstrate the relation (7.111), let us first note that

$$\omega^{0a}_i = e^{a\mu}\,\nabla_i e^0_\mu = e^{a\mu}\,(\partial_i e^0_\mu - \Gamma^\nu_{i\mu}\,e^0_\nu), \tag{7.114}$$

and because of the form of the vierbein vectors (7.94) and (7.95), $e^0_i = e^0_a = 0$ and

$$\omega^{0a}_i = -N\Gamma^0_{ij}\,e^{aj} = -K_{ij}\,e^{aj}, \tag{7.115}$$

where we used the relation (4.137).

For further applications, it will be useful to rewrite $\mathcal{S}$ in terms of K^a_i and A^a_i. This can be done by using $\bar{\omega}^{ab}_i = \varepsilon^{ab}{}_c\,(\gamma\,\omega^{0c}_i - A^c_i)$, which leads to

$$\begin{aligned}
\frac{E^i_a\,E^j_b}{\bar{e}}\,\bar{R}^{ab}_{ij} &= 2\left[\partial_{[i}\left(\varepsilon^{ab}{}_c\,(\gamma\,\omega^{0c}_{j]} - A^c_{j]})\right) - \bar{\omega}^a_{[i\ c}\,\varepsilon^{cb}{}_d\,(\gamma\,\omega^{0d}_{j]} - A^d_{j]})\right] \\
&= 2\frac{E^i_a\,E^j_b}{\bar{e}}\left[\gamma\,D^{\bar{\omega}}_{[i}\left(\varepsilon^{ab}{}_c\,\omega^{0c}_{j]}\right) - \varepsilon^{ab}{}_c\,\partial_{[i}A^c_{j]} + \bar{\omega}^a_{[i\ c}\,\varepsilon^{cb}{}_d\,(\gamma\,\omega^{0d}_{j]} + A^d_{j]})\right] \\
&= 2\frac{E^i_a\,E^j_b}{\bar{e}}\left[\gamma\,D^{\bar{\omega}}_{[i}\left(\varepsilon^{ab}{}_c\,\omega^{0c}_{j]}\right) - \varepsilon^{ab}{}_c\,\partial_{[i}A^c_{j]} + \varepsilon^a{}_{cf}\,(\gamma\,\omega^{0f}_{[i} - A^f_{[i})\,\varepsilon^{cb}{}_d\,(\gamma\,\omega^{0d}_{j]} + A^d_{j]})\right] \\
&= 2\frac{E^i_a\,E^j_b}{\bar{e}}\left[\gamma\,D^{\bar{\omega}}_{[i}\left(\varepsilon^{ab}{}_c\,\omega^{0c}_{j]}\right) - \frac{1}{2}\varepsilon^{ab}{}_c\,F^c_{ij} - \gamma^2\,\omega^{0a}_{[i}\,\omega^{0b}_{j]}\right],
\end{aligned}$$

in which $F^a_{ij} = \gamma\, {}^{(\gamma)}F^a_{ij}$ and in the last line we avoid all the symmetric terms under the exchange of the indexes a and b inside the square brackets. Since $D^{\bar{\omega}}_i e^j_a = 0$, the first term in the expression above can be written as

$$\frac{E^i_a\,E^j_b}{\bar{e}}\gamma\, D^{\bar{\omega}}_i\left(\varepsilon^{ab}{}_c\,\omega^{0c}_j\right) = E^i_a\,e^j_b\,\gamma\, D^{\bar{\omega}}_i\left(\varepsilon^{ab}{}_c\,\omega^{0c}_j\right) = E^i_a\,\gamma\, D^{\bar{\omega}}_i\left(\varepsilon^{ab}{}_c\,\omega^{0c}_j\,e^j_b\right), \tag{7.116}$$

thus it is proportional to the derivatives of the first condition in (7.102) and it vanishes on the constraints hypersurface. Therefore, the scalar constraint can be rewritten as

$$\mathcal{S} \approx \varepsilon^{ab}{}_c\,\frac{E^i_a\,E^j_b}{\bar{e}}\,F^c_{ij} - 2(\gamma^2+1)\frac{E^{[i}_a\,E^{j]}_b}{\bar{e}}K^a_i\,K^b_j\,. \tag{7.117}$$

7.4.3 *Ashtekar variables*

The cases $\gamma = i$ and $\gamma = -i$ (*Ashtekar connections* [19,20,37]) are very peculiar, since ${}^{(\pm i)}A^{\alpha\beta}_\mu$ turns out to be the self-dual and the anti-self-dual part of $\omega^{\alpha\beta}_\mu$, respectively, *i.e.* they satisfy

$$\pm\frac{i}{2}\,\varepsilon^{\alpha\beta}{}_{\gamma\delta}\,{}^{(\pm i)}A^{\gamma\delta}_\mu = {}^{(\pm i)}A^{\alpha\beta}_\mu. \tag{7.118}$$

This way, by assigning the components ${}^{(\pm i)}A^{0a}_\mu$, the full connections $\omega^{\alpha\beta}_\mu$ are determined (${}^{(\pm i)}A^{\alpha\beta}_\mu$ are complex quantities, so ${}^{(\pm i)}A^{0a}_\mu$ and $\omega^{\alpha\beta}_\mu$ have the same number of degrees of freedom).
Moreover, in the Holst Lagrangian the curvature term $R^{\alpha\beta}_{\mu\nu} \pm \frac{i}{2}\varepsilon^{\alpha\beta}{}_{\gamma\delta}\,R^{\gamma\delta}_{\mu\nu}$ appears and it contains only ${}^{(\pm i)}A^{\alpha\beta}_\mu$. Thus, being ${}^{(\pm i)}A^{\alpha\beta}_\mu$ the only variables appearing in the Lagrangian density, one makes the variations with respect to them. This way the condition (7.92) still comes out as an equation of motion, even if the relation between ${}^{(\pm i)}A^{\alpha\beta}_\mu$ and $\omega^{\alpha\beta}_\mu$ cannot be inverted. In particular, the full action can be written as follows in terms of ${}^{(\pm i)}A^{0a}_\mu$

$$S_H = -\frac{c^3}{16\pi G}\int d^4x\; e\,(e^\mu_0\,e^\nu_a \pm \frac{i}{2}\varepsilon_a{}^{bc}\,e^\mu_b\,e^\nu_c)\,{}^{(\pm i)}F^a_{\mu\nu} \tag{7.119}$$

where ${}^{(\pm i)}F^a_{\mu\nu} = \partial_{[\mu}\,{}^{(\pm i)}A^{0a}_{\nu]} \pm \frac{i}{2}\varepsilon^a{}_{bc}\,{}^{(\pm i)}A^{0b}_{[\mu}\,{}^{(\pm i)}A^{0c}_{\nu]}$ is the field strength associated with the $SU(2)$ connections ${}^{(\pm i)}A^{0a}_\nu$.

All these properties are due to the fact that the Ashtekar connections are associated with the $SU(2)$ subgroups in which we can decompose the whole Lorentz group. The Lorentz group is isomorphic to the direct product of two complex $SU(2)$ groups (in terms of which one defines left-handed and right-handed representations). The self-dual and the anti-self-dual parts of $SO(1,3)$ correspond precisely to the projection on these two $SU(2)$ groups, which are related by complex conjugation. The Holst action for $\gamma = i, -i$ is simply the self-dual and the anti-self-dual projection of the Einstein-Hilbert one, respectively. It is the above mentioned possibility to split the Lorentz group into $SU(2) \otimes SU(2)$, which provides an explanation for the fact that only ${}^{(i)}A^{\alpha\beta}_{\mu}$ and ${}^{(-i)}A^{\alpha\beta}_{\mu}$ are present in the action for $\gamma = i$ and $\gamma = -i$, respectively. On the contrary, for $\gamma \neq \pm i$, the full action cannot be rewritten in terms of ${}^{(\gamma)}A^{\alpha\beta}_{\mu}$ only, but also ${}^{(-\gamma)}A^{\alpha\beta}_{\mu}$ is present in (7.99) (ω^{0a}_{μ} and ω^{ab}_{μ} are linear combinations of ${}^{(\gamma)}A^{\alpha\beta}_{\mu}$ and ${}^{(-\gamma)}A^{\alpha\beta}_{\mu}$).

Hence, Ashtekar connections can be seen as the basic configuration variables in which we can split $\omega^{\alpha\beta}_{\mu}$ and for this reason, historically, they were considered first for a quantum formulation for gravity in terms of some $SU(2)$ connections.

A further property of Ashtekar connections is the simplification of the scalar constraint: the second term in (7.117) vanishes. This way, the full set of constraints, except for a square root of the three-metric determinant in the scalar constraint, is polynomial in configuration variables. This is a promising result in view of quantization.

This simplification is due to the fact that ${}^{(\pm i)}A^{a}_{i}$ is the pull-back on Σ_{x^0} of a spacetime connection [263], *i.e.* ${}^{(\pm i)}A^{ab}_{\mu}$. On the contrary, ${}^{(\gamma)}A^{\alpha\beta}_{\mu}$ for $\gamma \neq \pm i$ does not behave as a connection, because it is associated with a group of transformations which is not a subgroup inside the Lorentz group (there have been some attempts [123] to reproduce Ashtekar-Barbero connections via the pull back of a spacetime one related with a $SU(2)$ group, but the physical interpretation of such a group is not clear). As a consequence, the behavior of real connections under diffeomorphisms is much more complicated and this is reflected into the structure of the scalar constraint.

Hence, Ashtekar connections have a deeper mathematical meaning than the real ones. The reason why real connections instead of Ashtekar ones have been adopted in LQG resides in the difficulties one encounters when quantizing complex connections. In particular, since the momenta determine the geometry, one expects them to be real. Hence, some reality conditions must be implemented [215, 254], for which no solution is known on a quantum level. Moreover, the complexification of the $SU(2)$ group is not compact and the whole machinery of LQG (in particular the definition of a positive-definite scalar product) works only for compact groups.

Although Ashtekar variables are not anymore the basic configuration variables of LQG, nevertheless they are still intriguing, mainly with respect to understand the coupling of the gravitational degrees of freedom with fundamental chiral particles (see for instance [234] in which a gauge theory of the complexified Lorentz group has been used to recover the Lagrangian density of the $SU(2)$ sector of the electro-weak model and the one proper of Ashtekar reformulation of gravity).

7.4.4 *Removing the time-gauge condition*

We have seen how the 3+1 splitting procedure of Holst action is based on a partial gauge fixing of the local Lorentz symmetry, realized via the time-gauge condition $e^0_i = 0$ (7.94). This way, the Gauss constraint proper of a $SU(2)$ gauge theory is obtained. We will discuss now the Hamiltonian formulation without any gauge fixing, by pointing out that the emergence of the $SU(2)$ Gauss constraint is a generic property regardless of whether the local Lorentz frame is fixed or not.

In an arbitrary local Lorentz frame the vierbein vectors can be written as [45]

$$e_\mu{}^\alpha = \begin{pmatrix} N & N^i e^a_i \\ \chi_a e^a_i & e^a_i \end{pmatrix}. \tag{7.120}$$

The new variables χ_a are the parameters of the internal boost mapping the vectors (7.120) into a set of vectors comoving with respect to the 3+1 splitting (of the kind (7.94)). The Hamiltonian analysis of the Holst action (7.87) in a generic Lorentz frame gives new constraints, which results to be second-class (see section 3.2.2). Let us parametrize the phase-space via the variables $\omega^{\alpha\beta}_\mu$ and the associated momenta ${}^{(\gamma)}\pi^\mu_{\alpha\beta}$, the whole system of constraints reads

$$\begin{aligned}
\mathcal{S} &= \frac{1}{\sqrt{h}}\,\pi^i_{\delta\zeta}\,\pi^{j\zeta}_{\ \ \varepsilon}\,{}^{(\gamma)}p^{\delta\varepsilon}{}_{\alpha\beta}\,R^{\alpha\beta}_{ij} \approx 0 \\
\mathcal{V}_i &= {}^{(\gamma)}p_{\alpha\beta}{}^{\delta\varepsilon}\,\pi^j_{\delta\varepsilon}\,R^{\alpha\beta}_{ij} \approx 0 \\
G_{\alpha\beta} &= D^{(\omega)}_i\pi^i_{\alpha\beta} = \partial_i\pi^i_{\alpha\beta} - 2\omega_{i[\alpha}{}^{\delta}\,\pi^i_{|\delta|\beta]} \approx 0 \\
C^{ij} &= \varepsilon^{\alpha\beta\delta\varepsilon}\,\pi^{(i}_{\alpha\beta}\,\pi^{j)}_{\delta\varepsilon} \approx 0 \\
D^{ij} &= \varepsilon^{\alpha\beta\delta\varepsilon}\,\pi^k_{\alpha\zeta}\,\pi^{(i\zeta}{}_{\beta}\,D^{(\omega)}_k\pi^{j)}_{\delta\varepsilon} \approx 0
\end{aligned} \tag{7.121}$$

where

$$^{(\gamma)}\pi^{\mu}_{\alpha\beta} = {}^{(\gamma)}p^{\delta\varepsilon}{}_{\alpha\beta}\,\pi^{\mu}_{\delta\varepsilon}. \tag{7.122}$$

$\mathcal{S}$ and $\mathcal{V}_i$ denote the scalar and vector constraints, while the six conditions $G_{\alpha\beta}$ enforce local Lorentz invariance (the associated Lagrangian multipliers are $\omega_t^{\alpha\beta}$). The additional constraints $C^{ij} \approx 0$ arise because the momenta $\pi^i_{\alpha\beta}$ are not all independent since the following relations hold

$$\pi^i_{\alpha\beta} = 2\,e\,e^t_{[\alpha}\,e^i_{\beta]}, \tag{7.123}$$

while D^{ij} is the secondary constraint:

$$D^{ij} \approx [H\,,\,C^{ij}]. \tag{7.124}$$

It can be shown that $[C^{ij}\,,\,D^{kl}]$ does not vanish on the constraint hypersurface (7.121).

A covariant formulation [8, 9] (see also [202]) (with no splitting in Lorentz space-like and time-like vierbein indexes) can be given in terms of Lorentz connections and the constraints (7.121) can be promoted on a quantum level. However, the development of a consistent representation for quantum operators is much more complicated than in standard LQG (one must use Dirac brackets instead of Poisson ones [167]).

What is going on here is that the whole system of constraints is second-class. Hence in the reduced phase-space the symplectic structure is nontrivial and the constraint $G_{\alpha\beta} \approx 0$ does not induce via Poisson action a Lorentz transformation. Hence, *the phase-space structure of the Holst action is not that of a Yang-Mills theory for the Lorentz group.*

Let us quit covariance and solve explicitly the constraints $C_{ij} \approx 0$, $D^{ij} \approx 0$, in order to reduce the system of constraints (7.121) to a first-class one.

A solution for the conditions $C^{ij} \approx 0$ and $D^{ij} \approx 0$ is given by [94]

$$\begin{aligned} \pi^i_{ab} &= 2\,\chi_{[a}\,\pi^i_{b]} \\ \omega_a{}^b{}_i &= {}^{\pi}\omega_a{}^c{}_i\,T_c^{-1b}(\chi) + \chi_a\,\omega^{0b}{}_i + \chi^b\,(\omega_a{}^0{}_i - \partial_i\chi_a), \end{aligned} \tag{7.125}$$

in which $\pi^i_a = \pi^i_{0a}$, $T_c^{-1b}(\chi) = \delta^b_c - \chi^b\,\chi_c$, while ${}^{\pi}\omega_a{}^b{}_i$ reads

$${}^{\pi}\omega_a{}^b{}_i = \frac{1}{\pi^{1/2}}\,\pi^b_l\,D_i(\pi^{1/2}\pi^l_a). \tag{7.126}$$

This complicated expression can be understood as the spin connection associated with some vectors, whose densitized inverse expression is given by π^i_a. In fact, π^a_i denotes the inverse of π^i_a and π is the determinant. Once we have fixed the hypersuface (7.125) in phase-space, the formulation is not manifestly covariant, because a splitting of the internal Lorentz indexes in time-like and space-like ones is performed. However, χ_a can be promoted to be configuration variables, such that the local Lorentz frame is still generic. In order to discuss the physical interpretation for the hypersurface (7.125), let us put $\chi_a = 0$, so finding from (7.123) $\pi^i_{0a} = E^i_a$, while the conditions (7.125) read

$$\pi^i_{ab} = 0, \qquad \omega_a{}^b{}_i = \bar{\omega}_a{}^b{}_i(e). \tag{7.127}$$

In view of the relation (7.122), the conjugate variables to E^i_a are Ashtekar-Barbero variables ${}^{(\gamma)}A^a_i = {}^{(\gamma)}p^{0a}{}_{\delta\varepsilon}\omega^{\delta\varepsilon}_i$. The constraints $G_{\alpha\beta} = 0$ become

$$G_{0a} = \partial_i E^i_a - \bar{\omega}_a{}^b{}_i E^i_b \approx 0, \qquad G_{ab} = -2\omega^{\,0}_{[a\,i}E^i_{b]} \approx 0, \tag{7.128}$$

and the first condition is identically solved. The $SU(2)$ Gauss constraint is obtained as follows

$$G_a = G_{0a} + \gamma\varepsilon^{bc}_a G_{bc} = \partial_i E^i_a - \gamma\varepsilon_{ab}{}^{c\,(\gamma)}A^b_i E^i_c \approx 0\,. \tag{7.129}$$

Similarly, one can show how on the hypersurface (7.127) the constraints $\mathcal{S}$ and $\mathcal{V}_i$ in the system (7.121) reduce to the ones in (7.106).

For generic $\chi_a \neq 0$ one can write E^i_a in terms of π^i_a as follows

$$E^i_a = S^b_a \pi^i_b, \qquad S^b_a = \sqrt{1+\chi^2}\,\delta^a_b + \frac{1-\sqrt{1+\chi^2}}{\chi^2}\chi_a\chi_b. \tag{7.130}$$

The symplectic structure on the hypersurface (7.125) is nontrivial and it explicitly depends on χ_a. This fact complicated the Hamiltonian analysis, but at the end it can be shown that the conjugate variable to E^i_a reads

$$\begin{aligned}{}^{(\gamma)}A^a_i = S^{-1a}_b \Bigg((1+\chi^2)\,T^{bc}\left(\omega_{0ci} + {}^{\pi}D_i\chi_c\right) - \frac{1}{2\gamma}\varepsilon^b{}_{cd}\,{}^{\pi}\omega^{cf}{}_i\,T^{-1d}_{f} \\ + \frac{2+\chi^2-2\sqrt{1+\chi^2}}{2\gamma\chi^2}\,\varepsilon^{bcd}\,\partial_i\chi_c\,\chi_d\Bigg).\end{aligned} \tag{7.131}$$

The expression above is the generalization of Ashtekar-Barbero variables to the case where the local Lorentz frame is not fixed. Let us denote by π^a the conjugate momenta to χ_a. The constraints enforcing local Lorentz invariance are equivalent to the following conditions

$$G_a = \partial_i E^i_a - \gamma\,\varepsilon_{ab}{}^{c\,(\gamma)}A^b_i\,E^i_c \approx 0, \qquad \pi^a \approx 0\,. \tag{7.132}$$

The first condition is the Gauss constraints of the $SU(2)$ group. Therefore, the $SU(2)$ gauge symmetry arises also when the Lorentz frame is not fixed. The second condition comes from the requirement that the change of phase-space coordinates $\{\omega_i^{\alpha\beta}\,, {}^{(\gamma)}\pi^i_{\alpha\beta}\} \to \{{}^{(\gamma)}A_i^a\,, E_a^i, \chi_a\,, \pi^a\}$ is canonical. It outlines that χ_a, just like N and N^i, do not play any dynamic role.

The remaining constraints $\mathcal{S}$ and $\mathcal{V}_i$ can be expressed in terms of ${}^{(\gamma)}A_i^a$ and E_a^i only and they take the same expression as when the time gauge holds.

This analysis confirms that fixing the time-gauge condition is well-grounded and that no peculiar symmetry of the Holst action is affected by this choice. In particular since the full set of constraints is second class, there is no room for a Lorentz Yang-Mills gauge symmetry in the Holst formulation for gravity, while the actual Yang-Mills gauge group is the $SU(2)$ one.

On a quantum level, the formulation in a generic Lorentz frame just requires adding χ_a to the coordinates to be quantized and $\pi^a = 0$ to the constraints to be imposed (in a Dirac scheme). In this respect, it is possible to give a quantum formulation in a generic Lorentz frame which is equivalent to LOG [89].

We are now going to extend the analysis described in this section in the presence of matter fields. In particular here we will consider the cases with a scalar [93] and a spinor [95] fields.

7.4.4.1 *The non-minimally coupled scalar field*

A scalar field non-minimally coupled with the geometry is the prototype for modified gravity theories in a scalar-tensor representation (Brans-Dicke theories [79], $f(R)$ models [269], ...). The associated action reads (in units $\hbar = c = 8\pi G = 1$)

$$S = \int d^4x\,\sqrt{-g}\left[F(\phi)e^\mu_\delta e^\nu_\varepsilon\, {}^{(\gamma)}p^{\delta\varepsilon}{}_{\alpha\beta}R^{\alpha\beta}_{\mu\nu} + \frac{1}{2}g^{\mu\nu}K(\phi)\partial_\mu\phi\partial_\nu\phi - V(\phi)\right], \tag{7.133}$$

in which V denotes the potential of the scalar field ϕ. $F(\phi)$ and $K(\phi)$ are generic functions and they give a non-minimal coupling with geometry and a non-standard kinetic term for $F \neq 1$ and $K \neq 1$, respectively.

The scalar field enters via the function F into the definition of the conjugate momenta $\pi^i_{\alpha\beta}$, such that for the three-metric the following relation holds

$$h\,h^{ij} = \frac{1}{2F^2}\,\eta^{\alpha\gamma}\,\eta^{\beta\delta}\,\pi^i_{\alpha\beta}\,\pi^j_{\gamma\delta}. \tag{7.134}$$

The Hamiltonian analysis outlines that only the vector and the scalar constraints are modified with respect to the vacuum case, the other constraints being as in (7.121), and they now read

$$
\begin{aligned}
{}^{\phi}\mathcal{V}_i &= \mathcal{V}_i + \pi\partial_i\phi \approx 0, \\
{}^{\phi}\mathcal{S} &= \frac{1}{F}\mathcal{S} + \frac{1}{2K}{}^{\phi}\pi^2 + \frac{K}{4F^2}\pi^i_{\alpha\beta}\pi^{j\alpha\beta}\partial_i\phi\partial_j\phi + hV(\phi) \approx 0,
\end{aligned}
\tag{7.135}
$$

where ${}^{\phi}\pi$ denotes the conjugate momentum to ϕ. Hence, we can take the conditions (7.125) to solve $C^{ij} = D^{ij} = 0$, while from $G_{\alpha\beta} = 0$ the $SU(2)$ Gauss constraint and the vanishing of conjugate momenta to χ_a are obtained, just like in vacuum.

The expressions (7.135) can be simplified by the following transformation

$$
\phi \to \varphi = \int^{\varphi} F^{-1/2}(\phi)d\phi, \qquad {}^{\phi}\pi \to {}^{\varphi}\pi = F^{1/2}(\phi){}^{\phi}\pi,
\tag{7.136}
$$

such that they read

$$
\begin{aligned}
{}^{\varphi}\mathcal{S} &= \mathcal{S} + \frac{1}{2K}{}^{\varphi}\pi^2 + \frac{K}{4}{}^{\phi}h{}^{\phi}h^{ij}\partial_i\varphi\partial_j\varphi + \frac{{}^{\phi}h}{F(\phi(\varphi))^2}V(\phi(\varphi)) \approx 0, \\
{}^{\varphi}\mathcal{V}_i &= \mathcal{V}_i + {}^{\varphi}\pi\partial_i\varphi \approx 0.
\end{aligned}
\tag{7.137}
$$

The final set of constraints coincide with the one of gravity in the presence of a minimally-coupled ($F = 1$) scalar field φ (this is not surprising because the map (7.136) implements the transformation from the Jordan to the Einstein frame). However, because of the relations (7.134) the momenta ${}^{\phi}E^i_a = S^b_a\pi^i_b$ are not densitized vectors of the three-metric, but of the quantity ${}^{\phi}h_{ij} = F(\phi)h_{ij}$.

Therefore, it is possible to describe the dynamics of gravity nonminimally coupled to a scalar field by that of a "fake" geometry, whose three-metric is ${}^{\phi}h_{ij}$, and a minimally coupled scalar field. Furthermore, in LQG one can choose ${}^{\phi}E^i_a$ as basic variables, such that the operators associated with h_{ij} is not fundamental and the field ϕ enters into the quantum geometrical structure [25].

7.4.4.2 *Spinor fields*

The case with a spinor field is the most interesting one with respect to the Hamiltonian analysis in a generic local Lorentz frame, since spinors realize a nontrivial representation of the Lorentz group.

The nonminimal action for a massless spinor field reads

$$
S_{\psi} = \frac{i}{2}\int d^4x\,\sqrt{-g}(\bar{\psi}\gamma^{\mu}AD^{\psi}_{\mu}\psi - D^{\psi}_{\mu}\bar{\psi}A\gamma^{\mu}\psi),
\tag{7.138}
$$

where the matrix A contains the real parameter α as follows

$$A = 1 + i\alpha\gamma_5. \tag{7.139}$$

The Hamiltonian analysis provides the following system of constraints [95]

$$\begin{aligned}
{}^{\psi}\mathcal{S} &= \mathcal{S} + \frac{i}{\sqrt{h}}\pi^i_{\alpha\beta}\,(\bar{\Pi}\,\Sigma^{\alpha\beta}\,D^{\psi}_i\psi - D^{\psi}_i\bar{\psi}\,\Sigma^{\alpha\beta}\,\Pi) \approx 0 \\
{}^{\psi}\mathcal{V}_i &= \mathcal{V}_i + \bar{\Pi}\,D^{\psi}_i\psi + D^{\psi}_i\bar{\psi}\,\Pi \approx 0 \\
{}^{\psi}G_{\alpha\beta} &= G_{\alpha\beta} - i(\bar{\Pi}\,\Sigma_{\alpha\beta}\,\psi - \bar{\psi}\,\Sigma_{\alpha\beta}\,\Pi) \approx 0 \qquad , \\
C^{ij} &= \varepsilon^{\alpha\beta\gamma\delta}\pi^{(i}_{\alpha\beta}\,\pi^{j)}_{\gamma\delta} \approx 0 \\
{}^{\psi}D^{ij} &= D^{ij} + \frac{\gamma^2}{2(\gamma^2+1)}\left(\varepsilon^{\alpha\beta}{}_{\gamma\delta} - \frac{2}{\gamma}\delta^{\alpha\beta}_{\gamma\delta}\right)\pi^{(i}_{\alpha\beta}\pi^{j)}_{\varepsilon\zeta}(\bar{\Pi}\,\Sigma^{\varepsilon\zeta}\Sigma^{\gamma\delta}\psi + \bar{\psi}\Sigma^{\gamma\delta}\Sigma^{\varepsilon\zeta}\Pi) \approx 0
\end{aligned} \tag{7.140}$$

in which $\bar{\Pi}$ and Π denotes the conjugate momenta to ψ and $\bar{\psi}$, respectively, whose explicit expressions read

$$\bar{\Pi} = \frac{i}{2}\sqrt{-g}\,\bar{\psi}\gamma^{\alpha}\,A\,e^0_{\alpha}, \qquad \Pi = -\frac{i}{2}\sqrt{-g}\,e^0_{\alpha}\,A\,\gamma^{\alpha}\,\psi. \tag{7.141}$$

It is worth noting how the spinor enters into the Gauss constraints as a source term. This is due to the fact that spinor fields transform non-trivially under local Lorentz transformations. Moreover, ${}^{\psi}D^{ij}$ changes with respect to the vacuum case, since the spinor part of the scalar constraint contains the spin connections $\omega^{\alpha\beta}_{\mu}$ and it does not commute with C^{ij}.

A solution for $C^{ij} \approx 0$ and ${}^{\psi}D^{ij} \approx 0$ can now be written as

$$\begin{aligned}
\pi^i_{ab} &= 2\chi_{[a}\pi^i_{b]}, \\
\omega_a{}^b{}_i &= {}^{\pi}\omega_a{}^c{}_i T^{-1b}_c + \chi_a\omega^{0b}{}_i + \chi^b(\omega_a{}^0{}_i - \partial_i\chi_a) + {}^{\psi}\omega_a{}^b{}_i,
\end{aligned} \tag{7.142}$$

which differ from the ones in vacuum (7.125) because of the additional term

$$\begin{aligned}
{}^{\psi}\omega^{ab}_i &= +\frac{1}{4}\frac{\gamma(\gamma-\alpha)}{(\gamma^2+1)\sqrt{1+\chi^2}}\varepsilon^{ab}{}_c\pi^c_i(J^0 + \chi_d J^d) \\
&\quad -\frac{1}{2}\frac{\gamma(1+\alpha\gamma)}{\gamma^2+1}\pi^c_i T^{-1[a}_c\eta^{b]d}(J_d - \chi_d J^0),
\end{aligned} \tag{7.143}$$

J^α being proportional to the axial current $J^\alpha = \sqrt{h}\bar{\psi}\gamma_5\gamma^\alpha\psi$. The presence of ${}^\psi\omega^{ab}_{\ \ i}$ can be understood from the analysis made in section 7.2.1: spinors provide a non-vanishing torsion contribution which modifies the expression of the spin connections.

The analysis of constraints can now be carried on as in the previous cases. The conjugate variables to E^i_a are given by

$$
{}^{(\gamma)}\widetilde{A}^a_i = S^{-1a}_b \bigg[(1+\chi^2)T^{bc}\left(\omega_{0ci} + {}^\pi D_i\chi_c\right) - \frac{1}{2\gamma}\varepsilon^b_{\ cd}({}^\pi\omega^{cf}_{\ \ i}T^{-1d}_{\ \ f}
$$
$$
+{}^\psi\omega^{cd}_{\ \ i}) + \frac{2+\chi^2-2\sqrt{1+\chi^2}}{2\gamma\chi^2}\varepsilon^{bcd}\partial_i\chi_c\chi_d\bigg], \qquad (7.144)
$$

where ${}^\pi D_i\chi_c = \partial_i\chi_c - {}^\pi\omega_c^{\ b}{}_i\chi_b - \frac{1}{1+\chi^2}{}^\psi\omega_c^{\ b}{}_i\chi_b$.

A convenient set of spinor variables is provided by the "boosted" ones ψ^* [8]

$$
\psi^* = e^{-i\chi^a\Sigma_{0a}}\psi, \qquad (7.145)
$$

and by using these variables the constraints ${}^\psi G_{\alpha\beta} \approx 0$ can now be written as the $SU(2)$ Gauss constraints with a source given by the current $J^*_a = \sqrt{h}\bar{\psi}^*\gamma_a\gamma_5\psi^*$ plus the vanishing of π^a [95], *i.e.*

$$
\partial_i E^i_a - \gamma\varepsilon_{ab}^{\ \ c}\,{}^{(\gamma)}\widetilde{A}^b_i\,E^i_c = -\frac{\gamma}{2}J^*_a, \qquad \pi^a = 0. \qquad (7.146)
$$

The case $\alpha = \gamma$ can be regarded as outstanding because the full kinematical sector coincides with that of a background independent $SU(2)$ gauge theory. In fact the vector constraint reads

$$
{}^\psi\mathcal{V}_i = \mathcal{V}_i + \frac{i}{2}\sqrt{h}(\bar{\psi}^*\,\gamma^0\,A\,D^A_i\psi^* - D^A_i\bar{\psi}^*\,A\,\gamma^0\,\psi^*), \qquad (7.147)
$$

where the derivative $D^A_i\psi$ takes the same expression as the covariant derivative of a $SU(2)$ Yang-Mills gauge theory, *i.e.*

$$
D^A_i\psi = \partial_i\psi - \frac{i}{2}\gamma^{(\gamma)}\widetilde{A}^a_i T_a\psi \qquad (7.148)
$$

T_a being the rotation generators:

$$
T_a = \varepsilon_a^{\ bc}\Sigma_{bc}. \qquad (7.149)
$$

The scalar constraint takes the following expression

$$
{}^\psi\mathcal{S} = \mathcal{S} - \frac{i}{2}\widetilde{\pi}^i_a\left(\bar{\psi}^*\,\gamma^a\,A\,D^A_i\psi^* - D^A_i\bar{\psi}^*\,A\,\gamma^a\,\psi^*\right) - \frac{1+\gamma^2}{16\sqrt{h}}\,J^*_a\,J^{*a}, \qquad (7.150)
$$

and one sees how it contains the (perturbative non-renormalizable) four-fermion terms. Therefore, the interaction between the spinor field and the geometry resembles the one of a Yang-Mills $SU(2)$ gauge theory for $\alpha = \gamma$.

[8]The star here does not denote complex conjugation.

7.4.5 *The Kodama functional as a classical solution of the constraints*

One of the main issue in the determination of the gravitational field dynamics is the search for observables, *i.e.* of phase space functionals which Poisson commute with the full set of constraints (7.106). However, in the presence of a cosmological constant, a solution is known, the so-called *Kodama state* [186, 187]. Let us consider the Ashtekar case $\gamma = i$: in the presence of a cosmological constant Λ the scalar constraint reads

$$\mathcal{S} = -\varepsilon_{abc}\frac{E^{ai}E^{bj}}{\bar{e}}(F^c_{ij} + \frac{\Lambda}{3}\varepsilon_{ijk}E^{ck}), \tag{7.151}$$

while other constraints are not modified. A solution to $[\mathcal{S}, \Psi[A]] = 0$ is given by the Kodama functional [268]

$$\Psi[A] = e^{\frac{2}{3\Lambda}\int_\Sigma Y_{CS}[A]\sqrt{h}d^3x}, \tag{7.152}$$

Y_{CS} being the Chern-Simons form of the connections, *i.e.*

$$Y_{CS}[A] = \frac{1}{2}\varepsilon^{ijk}\delta_{ab}A^a_i\nabla_j A^b_k + \frac{2}{3}\varepsilon_{abc}\varepsilon^{ijk}A^a_i A^b_j A^c_k. \tag{7.153}$$

The functional is also manifestly invariant under diffeomorphisms and (small) $SU(2)$ transformations, such that the Poisson actions of other constraints also vanish. Hence, $\Psi[A]$ is a observable.

The Kodama functional can be seen as a quantum state, by addressing the WKB approximation (see section 5.1.6.2) in a canonical (Wheeler-deWitt) quantization in terms of Ashtekar variables. In this respect, it describes a semiclassical de-Sitter spacetime.

The main applications of the Kodama functional are devoted to cosmological models and to the study of the quantum behavior of homogeneous spacetimes [121, 241]. A convincing link with the LQG framework is still missing (it cannot be written in terms of holonomies), even though the extension to arbitrary values of the Immirzi parameter has been given in [249, 250].

Although the Kodama functional is the only known observable for gravity in the Ashtekar-Barbero formulation, nevertheless its role in canonical Quantum Gravity has not been completely developed [135].

Chapter 8

Loop Quantum Gravity

In section 7.4 we presented a reformulation of General Relativity, which provides a phase-space structure similar to the one of Yang-Mills gauge theories. Such a reformulation can be obtained from geometrodynamics by a canonical transformation in phase-space. The Stone-von Neumann theorem insures that a canonical transformation provides a unitary equivalent quantum description with respect to the original one. Hence, the canonical quantization of gravity phase-space with coordinates A_i^a and E_a^i, instead of h_{ij} and Π^{ij}, is not expected to provide any new real insight into the Quantum Gravity problem.

Loop Quantum Gravity (LQG) is based on an additional technical tool: the quantization of the holonomy-flux algebra. The choice of holonomies and fluxes as phase space coordinates has a deep impact on a quantum level. Using the GNS construction it is possible to define the so-called kinematical Hilbert space, in which the elements of the holonomy-flux algebra are associated with (bounded) operators. In this space, one can verify a posteriori how the holonomy operators violate one of the hypotheses of the Stone-von Neumann theorem, namely weak continuity. The description of the kinematical Hilbert space can be conveniently given in terms of spin-network states, which form a basis, and the two kinematical constraints, the vector and the Gauss ones, can be solved. The geometric operators have discrete spectra, such that the quantum space is endowed with quantum geometry as expected.

The definition of the kinematical Hilbert spaces places Loop Quantum Gravity closer to a complete Quantum Gravity model than the Wheeler-DeWitt formulation. However, the final theory has not been achieved yet. Although the scalar constraint can be properly defined in the gauge invariant Hilbert space, nevertheless an analytic expression cannot be given.

This fact prevents the solution of the dynamics for the quantum geometry. Other issues concern the "off-shell" constraint algebra, the definition of the scalar product, the construction of semi-classical states and the physical meaning of the Immirzi parameter.

8.1 Smeared variables

The connections are objects to be integrated on curves (they are one forms). Let us denote by $A_i = A_i^a \tau_a$ the connection taking values in the algebra of the $(SU(2))$ gauge group, τ_a being the (Hermitian) generators of such an algebra. A curve is described by a function α from a closed set of the real line (which can be taken as $[0, 1]$) taking values on the spatial manifold Σ_{x^0} as follows

$$x^i = \alpha^i(t), \qquad t \in [0, 1]. \tag{8.1}$$

The integration is usually performed on paths, which are equivalence classes of piece-wise analytic oriented curves under re-parametrizations[1]. A loop is a closed path, while a graph is a collection of paths. A graph can also be seen as a collection of edges, *i.e.* of analytic paths, and of vertexes, which are the points separating different edges. It is possible to show that the space of loops with a common point can be endowed with a group structure (such that the whole space of loops can be seen as a groupoid [283]). Therefore, one introduces the parallel transport of A_i along a curve $x^i = \alpha^i(t)$ (*holonomies*), which is defined as follows

$$U_\alpha(A) = P\left\{\exp i \int_\alpha A_i(\alpha^k(t)) \frac{d\alpha^i}{dt} dt\right\}$$

$$= \sum_{n=0}^{+\infty} \frac{i^n}{n!} \int_0^1 A_{i_1}^{a_1}(t_1) \frac{d\alpha^{i_1}}{dt_1} dt_1 \, \tau_{a_1} \int_0^{t_1} A_{i_2}^{a_2}(t_2) \frac{d\alpha^{i_2}}{dt_2} dt_2 \, \tau_{a_2} \ldots$$

$$\ldots \times \int_0^{t_{n-1}} A_{i_n}^{a_n}(t_n) \frac{d\alpha^{i_n}}{dt_n} dt_n \, \tau_{a_n}. \tag{8.2}$$

Holonomies are elements of the $SU(2)$ group, just like gauge transformations Λ. The nice feature of $U_\alpha(A)$ is their behavior under $SU(2)$ gauge transformations $\Lambda(x)$, *i.e.*

$$U'_\alpha(A') = \Lambda(f) U_\alpha(A) \Lambda^{-1}(i), \tag{8.3}$$

[1] Two curves $\alpha(t)$, $\alpha'(t')$ are identified if there exists an invertible function $s : [0, 1] \to [0, 1]$ such that $\alpha'(s(t)) = \alpha(t)$.

where i and f are the initial and final points of α, respectively. For instance, a gauge-invariant quantity is obtained by taking the trace of $U_\alpha(A)$ along a loop α (Wilson loop)

$$W_\alpha(A) = Tr(U_\alpha(A)), \tag{8.4}$$

and this can be seen from

$$\begin{aligned} W_\alpha(A') &= Tr(U_\alpha(A')) = Tr(\Lambda(f)U_\alpha(A)\Lambda^{-1}(i)) \\ &= Tr(\Lambda^{-1}(i)\Lambda(f)U_\alpha(A)) = Tr(U_\alpha(A)) = W_\alpha(A), \end{aligned} \tag{8.5}$$

where the cyclic property of the trace and the coincidence of the initial and final point of α have been used.

8.1.1 *Why a reformulation in terms of holonomies?*

The framework in which Wilson loops were introduced for the first time is the path-integral formulation of QCD on a lattice, where they provide some tools to study the confinement of quarks. The main feature of this approach is that the potential between two static quarks can be obtained from the expectation value of the Wilson loops connecting these two particles [253, 297]. The result is that a linear potential is obtained, which produces the confinement.

Moreover, quantum states can be labeled by closed and open loops, with quarks at ending points. Such states represent lines of non-Abelian electric fluxes and they turn out to be eigenstates of the Hamiltonian in the strong coupling limit [188].

Giles [151] stressed how starting from the Wilson loops of a gauge connection on the full (flat) spacetime, the connections can be entirely reconstructed, up to a gauge transformation. Therefore the knowledge of $h_\alpha(A)$ on any loop α gives all gauge-invariant information about a Yang-Mills theory on a flat spacetime. A formulation in terms of loop variables has been used for the quantization of gauge theories only on the lattice, since no substantial simplification occurs by passing from the space of connections to the loop space in a continuous spacetime. Almost 15 years after their introduction, it was recognized how Wilson loops and their generalizations can be useful for a background independent quantization via the definition of holonomies on knot-states, *i.e.* on equivalence classes of loops under diffeomorphisms (as we will discuss in section 8.3.2).

The first technical indication that a proper quantization of gravity should be based on something else than the standard canonical quantization procedure came from the fact that canonical commutation relations cannot be implemented for non-trivial phase-space topologies. Skipping some technicalities (which can be found in [174]), we rather prefer to give an example [177] which mimics some features of the gravity phase-space (in particular the requirement that the spatial metric is positive defined).

Let us consider a system with one positive coordinate q, $q > 0$, and conjugate momentum p. After the quantization, the states are defined in the Hilbert space of square integrable functions on q, with support on the positive real axis. By imposing the canonical commutation relations, q and p are promoted to hermitian operators for which the following commutation relation holds

$$[\hat{q}\,,\,\hat{p}] = i\hbar, \tag{8.6}$$

which implies $\hat{p}$ to be the generator of q-translation, *i.e.* $\phi(q+\varepsilon) = U_\varepsilon\phi(q) = e^{i\varepsilon p}\phi(q)$. However, in this scheme, U_ε is no longer a unitary operator, since the scalar product is not invariant under translations, *i.e.*

$$\int_0^\infty (U_\varepsilon\phi_1(q))^\dagger U_\varepsilon\phi_2(q)dq = \int_0^\infty \phi_1^\dagger(q+\varepsilon)\phi_2(q+\varepsilon)dq \neq \int_0^\infty \phi_1^\dagger(q)\phi_2(q)dq. \tag{8.7}$$

This feature outlines that the canonical commutation relation must be replaced. In particular, a quantization based on the following commutation relation

$$[\hat{q}, \hat{p}] = i\hbar\hat{q}\,, \tag{8.8}$$

is well grounded.

An analogous, but much more complicated, analysis performed by Isham [178] leads to similar conclusions concerning the use of canonical commutation relations for the quantization of General Relativity. A key-point of this analysis is the assumption that the configuration variables belong to a vector space. Inspired by this result, Rovelli and Smolin [258] introduced the idea of quantizing the holonomy-flux algebra, instead of canonical commutation relations. However, a direct link between their work and Isham's analysis has not been established, mainly due to the fact that the space of holonomies is not a vector space.

The use of holonomies has also spread out over Quantum Gravity approaches not directly related with LQG. For instance, holonomies have been used as basic variables to quantize topological defects in three-dimensional gravity [126, 221, 222], in which no local degrees of freedom are present. These models in three spacetime dimensions predict a deformed particle kinematics and they could sustain some scenarios with non-commutative geometries. They can also provide some hints on the four-dimensional case [192] via the Mac-Dowell and Mansouri formulation [205].

8.2 Hilbert space representation of the holonomy-flux algebra

In this section, we demonstrate how spin-network states arise in the loop quantization of the gravitational field. We will define the Hilbert space by applying the GNS construction to the holonomy-flux algebra and we will get spin-network states as basis elements.

8.2.1 *Holonomy-flux algebra*

The basic idea is to use the holonomies (8.2) along all the possible edges e of the hypersurface Σ_{x^0} as the coordinates of the configuration space. Other phase space variables are introduced such that the Poisson algebra is well-defined, *i.e.* non-distributional. This is achieved by smearing E^i_a in two dimensions: given a surface σ defined parametrically by $x^i = X^i(u,v)$, one defines $E(\sigma)$ as

$$E_a(\sigma) = \int_\sigma \varepsilon_{ijk} E^i_a d\sigma^{jk} = \int_\sigma n_i E^i_a \, dudv, \tag{8.9}$$

being u and v coordinates on the surface σ and n_i the normal vector in the adopted parametrization.

Starting from the symplectic structure (7.109), the Poisson brackets between $U_\alpha(A)$ and $E_a(\sigma)$ can be evaluated. The variational derivatives of $U_\alpha(A)$ and $E(\sigma)$ reads as follows

$$\frac{\delta U_\alpha(A)}{\delta A^a_i(x^j)} = i\int_0^1 \delta^{(3)}(x-\alpha(t))\frac{d\alpha^i}{dt}(U_\alpha)^t_0(A)\tau_a(U_\alpha)^1_t(A)\,dt \tag{8.10}$$

$$\frac{\delta E(\sigma)_a}{\delta E^i_b(x^j)} = \int_\sigma n_i \delta^{(3)}(x^j - X^j(u;v))\delta^a_b \, dudv, \tag{8.11}$$

$(U_\alpha)^b_a(A)$ being the holonomy of A along $\alpha = \alpha(s)$ from the point $s=a$ to $s=b$, and the Poisson brackets between holonomies and fluxes are

$$[E_a(\sigma)\,,\, U_\alpha(A)] = i\frac{8\pi G}{c^3}\gamma\int d^3x \frac{\delta E_a(\sigma)}{\delta E^i_b(x)}\frac{U_\alpha(A)}{\delta A^b_i(x)}$$

$$= i\frac{8\pi G}{c^3}\gamma\int\left[\int_0^1 \delta^{(3)}(x-\alpha(t))\frac{d\alpha^i}{dt}(U_\alpha)^t_0(A)\tau_a(U_\alpha)^1_t(A)dt\right] d^3x$$

$$\left[\int_\sigma n_i\delta^{(3)}(x^j - X^J(u;v))\delta^a_b\, dudv\right]$$

$$= i\frac{8\pi G}{\gamma}c^3 \int_0^1 dt \int_\sigma \delta^{(3)}(\alpha^j(t) - X^j(u,v))n_i \frac{d\alpha^i}{dt}(U_\alpha)_0^t(A)\tau_a(U_\alpha)_t^1(A)\, dudv.$$

If σ and α have no common points, one ends up with expressions containing δ with different support, thus the Poisson brackets vanish. This is the case also if α is tangent to σ in the intersection point, because of the scalar product $n_i \frac{d\alpha^i}{dt}$. Hence given a graph α with a finite number of intersections with the surface σ, we can always split α such that the intersections occur only at initial or final points of some edges. Let us denote by e' and e'' the edges intersecting σ at initial and final points, respectively, the Poisson brackets between the fluxes and the holonomies along these edges read[2]

$$[E_a(\sigma)\,,\, U_{e'}(A)] = i\frac{4\pi G}{c^3}\gamma o(e',\sigma)\tau_a U_{e'}(A) \tag{8.13}$$

$$[E_a(\sigma)\,,\, U_{e''}(A)] = i\frac{4\pi G}{c^3}\gamma o(e'',\sigma) U_{e''}(A)\tau_a, \tag{8.14}$$

$o(\alpha, \sigma)$ being the sign of $n_i \frac{d\alpha^i}{dt}$ at the intersection point, with $sign(0) = 0$. The Poisson brackets of the fluxes with the whole $U_\alpha(A)$ follow by applying the Leibniz rule after having written the holonomy in terms of the (matrix) product of holonomies along e', e'' and those edges which have no intersection with σ.

8.2.2 *Kinematical Hilbert space*

In what follows we will discuss the quantization in the connection representation, but the action of operators can be seen in a more intuitive way as an action on loops [110, 258]. This fact is based on the basic equivalence between the connection and the loop representation [109].

A generic phase-space function is now a function of holonomies and fluxes. In particular, let us consider the so-called *cylindrical functions*, *i.e.* the continuous functions of holonomies

$$f_\alpha(A) = f(U_{e_1}, U_{e_2} \ldots, U_{e_n}), \tag{8.15}$$

f being a continuous function in the n group-valued entries $\{U_{e_i}\}$, while the finite collection of edges $e_1, e_2 \ldots, e_n$ forms the graph α. The space of

[2] We defined the δ-function centered in $t = 0$ such that

$$\int_0^1 dt\delta(t)f(t) = \frac{1}{2}f(0), \tag{8.12}$$

and similarly for $t = 1$.

all possible cylindrical functions over all possible graphs is denoted by Cyl and it is possible to endow it with a norm. Such a norm is induced from the $SU(2)$ Haar measure and a Cauchy completition $\overline{Cyl}$ can be defined. The main technical difficulty in $\overline{Cyl}$ is cylindrical consistency, *i.e.* to define the norm such that the operations of removing/adding some edges and intersection/union of graphs are consistently implemented. This can be done via a projective limit. At the end, one deals with the space $\overline{Cyl}$ as the classical arena to be quantized.

A cyclic representation of the algebra of cylindrical functions on a Hilbert space can be obtained by the GNS construction (see section 5.3), taking the following state ω

$$\omega(f_\alpha(A), E_1, .., E_M) = \begin{cases} \int d\mu(g_1)..d\mu(g_N) f_\alpha(g_1, .., g_N) & M = 0 \\ 0 & M \neq 0 \end{cases}, \tag{8.16}$$

μ being the Haar measure of the $SU(2)$ group. We denote such a space by

$$\mathrm{H}^{Kin} = L^2(\bar{A}, d\mu). \tag{8.17}$$

The space $\bar{A}$ is the Gelfand spectrum of the C^* algebra[3] of cylindrical functions [28] and it can be constructed by replacing cylindrical functions (8.15) with generic homomorphisms from the set of piecewise analytic paths to $SU(2)$ elements. Actually, this is an enlargement of the classical configuration space (as usual in quantum theories).

The measure $d\mu$, known as *Ashtekar-Lewandowski measure*, is induced from the $SU(2)$ Haar one [29] and it is automatically invariant under $SU(2)$ gauge transformations and background independent (no reference to any background structure). According to the LOST theorem [200], such a representation of the holonomy-flux algebra is the only cyclic one with $SU(2)$ invariant and background independent measure, having also an essentially self-adjoint flux operator (other representations have been realized by violating some of the hypothesis of the LOST theorem, see [117, 190, 289]).

In the end, the whole Hilbert space H^{Kin} can be realized as a direct sum

$$\mathrm{H}^{Kin} = \oplus_\alpha \mathrm{H}^{Kin}_\alpha, \tag{8.19}$$

in which H^{Kin}_α is based on a fixed graph α and it is the space of square integrable functions of E-copies of the $SU(2)$ group, E being the total

[3] A C^*-algebra is a * algebra for which

$$\|A\,A^*\| = \|A\|\,\|A^*\|\,. \tag{8.18}$$

number of edges of α:

$$\mathrm{H}_{\alpha}^{Kin} = L^2((SU(2))^E, d\mu_E), \tag{8.20}$$

with $d\mu_E$ the direct product of E-$SU(2)$ Haar measures. Since one deals with functions of $SU(2)$ group elements, one can expand in terms of irreducible representations. The terms of this expansion are $SU(2)$ spin-network states. A $SU(2)$ spin-network state reads (7.32)

$$\psi_S(A) = \psi_{\alpha,\{\rho\}}(A) = \prod_{e\in\alpha} \sqrt{D_{\rho_e}}\rho_e(U_e(A)), \tag{8.21}$$

ρ_e being D_{ρ_e}-dimensional irreducible $SU(2)$ representations. Each spin network is labeled by the graph α and the $SU(2)$ quantum numbers associated with the edges $e \in \alpha$. These quantum numbers are the spins j_e of each irreducible representation ρ_e, which determine the dimension $D_{\rho_e} = 2j_e + 1$, and the row and column indexes (m_e, n_e) of the matrix $\rho_e(U_e(A))$. The latter are magnetic numbers, *i.e.*

$$m_e, n_e = -j_e, -j_e + 1, .., j_e - 1, j_e\,.$$

Spin-network states form an orthonormal basis for $\mathrm{H}_{\alpha}^{Kin}$ *and, via the direct sum over all graphs, for the whole* H^{Kin}, *i.e.*

$$\langle\psi_S(A)|\psi_{S'}(A)\rangle = \delta_{S,S'} = \delta_{\alpha,\alpha'}\,\delta_{\{\rho\},\{\rho'\}}\,, \tag{8.22}$$

in which $\delta_{\{\rho\},\{\rho'\}}$ reads explicitly

$$\delta_{\{\rho\},\{\rho'\}} = \prod_e \delta_{j_e,j'_e}\,\delta_{m_e,m'_e}\,\delta_{n_e,n'_e}\,. \tag{8.23}$$

8.2.2.1 *Holonomy operators*

Phase-space functions are cylindrical functions, thus continuous functions of holonomies $U_{\alpha'}(A)$. Each holonomy acts on $\psi_{\alpha,\{\rho\}}(A)$ via the multiplication of the corresponding $SU(2)$ group elements. The expansion of the new state in terms of spin-network states is obtained by expanding $U_{\alpha'}(A)$ in irreducible representation, *i.e.*

$$U_{\alpha'}(A) = \prod_{e'\in\alpha'} \sum_{\{\rho\}} c_{e',\{\rho\}}\rho_{e'}(U_{e'}(A))\,. \tag{8.24}$$

For a given edge $e' \in \alpha'$, one finds that

- if e' is not contained in α, the holonomy operator adds the edge e' to the original graph,

- if $e' = e \in \alpha$ the action of the holonomy operator can be obtained by evaluating (via $SU(2)$ recoupling theory[4]) the product of the representation $\rho_{e'}(U_{e'}(A))$ with $\rho_e(U_e(A))$ inside (8.21). This will give generically a new expansion in terms of irreducible representations.

Given the operators associated with the holonomies, one can try to construct the representation for the connections operators as follows

$$\hat{A}_i(x) = \lim_{e_i \to x} \frac{U_{e_i}(A) - I}{\mu(e_i)}, \tag{8.25}$$

e_i being the edge starting in x and parallel to the direction i, while $\mu(e_i)$ denotes its length. This definition is ambiguous unless one specifies the spin j_{e_i} of U_{e_i}[5] (a natural choice is to take the fundamental representation, *i.e.* $j_{e_i} = 1/2$). However, even by fixing the spin number of $U_{e_i}(A)$, which means that $U_{e_i}(A)$ equals to a certain $\rho_{e_i}(U_{e_i}(A))$, the limit (8.25) is not defined in H^{Kin}. The reason is that the operator U_{e_i} changes the $SU(2)$ quantum numbers (both spin and magnetic ones) associated with e_i or eventually the graph if e_i is not contained in α. In the latter case, the scalar product $\langle\psi_S(A)|U_{e_i}(A)|\psi_S(A)\rangle$ simply vanishes for $e_i \neq x$ and is 1 for $e_i = x$. In the former, let us consider a spin network $\psi_{e,\{j,m,n\}}$ based at the edge e containing e_i, $e_i \subset e$, and let us write it as the product of two spin-networks $\psi_{e',\{j,m,p\}}$ and $\psi_{e_i,\{j,p,n\}}$ based at e_i and e', with $e = e_i \cup e'$. The operator $\rho(U_{e_i}(A))$ acts on $\psi_{e_i,\{j,p,n\}}$ via multiplication and the result provides a new expansion in irreducible $SU(2)$ representations via recoupling theory. The scalar product $\langle\psi_S(A)|U_{e_i}(A)|\psi_S(A)\rangle$ gives the coefficients of such an expansion for which the quantum numbers of the final state are the same as those of the original spin-network $\psi_{e_i,\{j,p,n\}}$. This coefficient generically differs from 1. Moreover, the result is independent of the length of the edges e_i. Therefore the limit (8.25) is not defined, since its argument vanishes for $e_i = x$ while it generically differs from 0 (and goes as $1/\mu(e_i)$) for $e_i \neq x$. This implies that *no operator associated with the connection $A_i(x)$ can be defined.* We are in the presence of the same kind of violation of the Stone-von Neumann theorem (weak-continuity) as in polymer quantization (see section 5.4). This is the reason why we expect inequivalent results with respect to quantum geometrodynamics.

[4]Recoupling theory provides the tools to construct the sum of the $SU(2)$ representations. In the case of two representations, the sum is given in terms of Clebsch-Gordan coefficients.

[5]The magnetic numbers are determined by the matrix indexes of $A_i(x)$.

8.2.2.2 *Fluxes operators*

The action of the operator $E(\sigma)$ in H^{Kin} is defined as $-i\hbar$ times the classical Poisson brackets between fluxes and cylindrical functions, which can be evaluated from the relations (8.13), (8.14). Given a generic state f_α based at a graph α, let us denote by e'/e'' those edges of α whose initial/final points belong to σ, such that the action of flux operators reads

$$\hat{E}_a(\sigma)[f_\alpha] = 4\pi l_P^2\gamma\left(\sum_{e'} o(e',\sigma)L_a^{e'} f_\alpha - \sum_{e''} o(e'',\sigma)R_a^{e''} f_\alpha\right), \tag{8.26}$$

L_a^e and R_a^e being the left- and right-invariant vector fields associated with the $SU(2)$ group element U_e, *i.e.*

$$L_a^e f(\ldots U_e \ldots) = \frac{d}{dt} f(\ldots e^{t\tau_a} U_e \ldots)|_{t=0}, \tag{8.27}$$

$$R_a^e f(\ldots U_e \ldots) = \frac{d}{dt} f(\ldots U_e e^{-t\tau_a} \ldots)|_{t=0}. \tag{8.28}$$

The operators (8.26) are essentially self-adjoint and their action on spin-network states (8.21) read

$$\begin{aligned}\hat{E}_a(\sigma)\psi_{\alpha,\{\rho\}} = 4\pi l_P^2\gamma\Bigg[&\sum_{e'\in\alpha}\left(\prod_{e\neq e'}\sqrt{D_{\rho_e}}\rho_e\right)o(e',\sigma)\sqrt{D_{\rho_{e'}}}\tau_a^{(\rho_{e'})}\rho_{e'}(U_{e''})\\ &+\sum_{e''\in\alpha}\left(\prod_{e\neq e''}\sqrt{D_{\rho_e}}\rho_e\right)o(e'',\sigma)\sqrt{D_{\rho_{e''}}}\rho_{e''}(U_{e''})\tau_a^{(\rho_{e''})}\Bigg]\psi_{\alpha,\{\rho\}},\end{aligned} \tag{8.29}$$

$\tau_a^{(\rho)}$ being $SU(2)$ generators in the representation ρ. The relation above can be obtained from (8.26) by using the fact that ρ realizes a faithful representation of the associated group: $\rho(e^{t\tau_a}h) = \rho(e^{t\tau_a})\rho(h)$.

8.3 Kinematical constraints

In H^{Kin} it is possible to define all (continuous) functions of the phase space as quantum operators. The next step is the imposition of the constraints.

8.3.1 *Solution of the Gauss constraint: invariant spin-networks*

By the Dirac prescription, one looks for the solutions of the condition

$$\hat{G}_a f_\alpha(A) = 0. \tag{8.30}$$

Since a generic state can be expanded in spin-network states as follows

$$f_\alpha(A) = \sum_S c_S \psi_S(A), \tag{8.31}$$

let us impose the condition (8.30) on a generic spin-network state $\psi_S(A)$. This way, we will find the basis elements in the gauge-invariant Hilbert space.

The constraint $\hat{G}_a$ generates $SU(2)$ gauge transformations as in (8.3). Hence, $\hat{G}_a$ acts at vertexes only. Therefore, the solutions of (8.30) can be constructed from invariant spin-network states, which can be realized by inserting at vertexes invariant intertwiners I_v. In the expression (8.21) at each vertex v one has some free indices $\{m_1, \ldots, m_r\}$ associated with the spin-$\{j_1, \ldots, j_r\}$ representations of the outgoing edges and some other free indices $\{m'_1, \ldots, m'_s\}$ associated with the spin-$\{j'_1, \ldots, j'_r\}$ representations of the incoming edges. The invariant intertwiners map the representations of outgoing edges into those ones of incoming edges, thus they can be written as $I_v = I_{v\{m_1,\ldots,m_r,m'_1,\ldots,m'_s\}}(\{j_1, \ldots, j_r, j'_1, \ldots, j'_s\})$, and they are invariant under arbitrary $SU(2)$ transformations U acting on all indices, *i.e.*

$$I_{v\{l_1,\ldots,l_r,l'_1,\ldots,l'_s\}} U^{j_1}_{l_1 m_1} \ldots U^{j_r}_{l_r m_r} U^{\dagger j'_1}_{m'_1 l'_1} \ldots U^{\dagger j'_s}_{m'_s l'_s} = I_{v\{m_1,\ldots,m_r,m'_1,\ldots,m'_s\}},$$

U^j being the transformation U in the spin-j representation. Hence, $SU(2)$-*invariant states are those states which can be written as linear combinations of invariant spin-network states* (see figure 8.1)

$$f_\alpha(A) = \sum_S c_S \psi_S^{g-inv}, \qquad \psi_S^{g-inv} = \left(\prod_v I_v\right) \cdot \left(\prod_e D_{\rho_e} \rho_e(U_e(A))\right), \tag{8.32}$$

where $\cdot$ denotes index contraction.

The invariant intertwiners map the sum of the representations of the incoming vertexes into the sum of the representations of the outgoing ones. They can be constructed by repeated sums of couples of $SU(2)$ representations, in the end projecting the resulting representation into the fundamental one. The invariant intertwiners are labeled not only by the set of

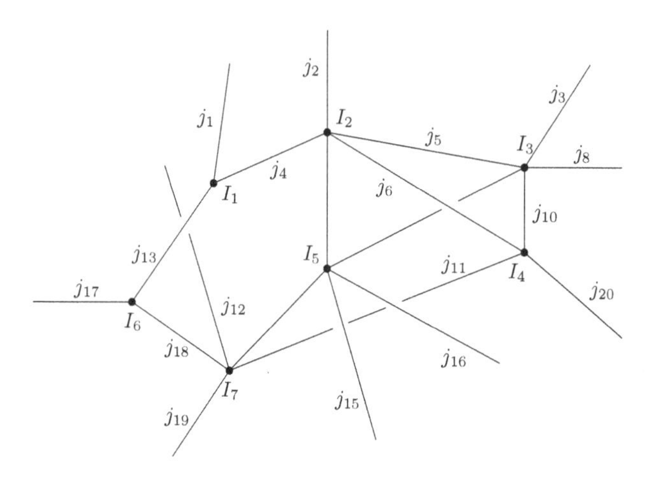

Figure 8.1 A $SU(2)$ invariant spin network, whose edges are labeled by the spin number j of the irreducible $SU(2)$ representations and with invariant intertwiners I at vertices (for clarity we drew straight edges, but generically this is not the case).

incoming and outgoing spins but also by the intermediate spins coming out after each sum. Henceforth, it is very complicated to handle such intertwiners of big valence. It has been outlined in [130] the existence of a $U(N)$ symmetry associated with an N-valence intertwiner, which can give some hints on the geometrical structures associated with these objects (see the next section).

8.3.2 *Three-diffeomorphisms invariance and s-knots*

The vector constraint $\mathcal{V}_i$ is associated with three-diffeomorphisms invariance. This can be seen by considering the following smeared combination of G_a and $\mathcal{V}_i$

$$\hat{V}[\vec{n}] = \int d^3x\, n^i(\hat{\mathcal{V}}_i - A^a_i\hat{G}_a) = \int d^3x\, n^i[E^j_a\partial_i A^a_j - \partial_j(E^j_a A^a_i)], \quad (8.33)$$

n^i being test functions, and by explicitly computing its action on a generic holonomy $U_\alpha(A)$ one gets

$$\begin{aligned} \hat{V}[\vec{n}]U_\alpha(A) &= -8\pi\gamma l_P^2 \int d^3x \left(n^i\partial_i A^a_j \frac{\delta}{\delta A^a_j} + \partial_j n^i A^a_i \frac{\delta}{\delta A^a_j} \right) U_\alpha(A) \\ &= -8\pi\gamma l_P^2 \int \frac{d\alpha^j}{dt} U^t_0 \left(n^i\partial_i A^a_j + \partial_j n^i A^a_i \right) |_{\alpha(t)}\tau_a U^1_t dt, \quad (8.34) \end{aligned}$$

where in the second line we used (8.10). The expression above coincides with the transformation of U_α under the action of an infinitesimal diffeomorphisms $\varphi(\vec{n}) : x'^i = x^i + n^i$:

$$U_{\varphi(\vec{n})(\alpha)} - U_\alpha = P\left\{\exp\left[\int A_i^a(x'^i)\frac{dx'^i(\alpha)}{dt}\tau_a\right]\right\} - P\left\{\exp\left[\int A_i^a(x^i)\frac{d\alpha^i}{dt}\tau_a\right]\right\}$$
$$= \int U_0^t\left(n^j\partial_j A_i^a\frac{d\alpha^i}{dt} + A_i^a(x^i)\partial_j n^i\frac{d\alpha^j}{dt}\right)\tau_a U_t^1 dt. \qquad (8.35)$$

Hence in order to solve the vector constraint in the gauge-invariant Hilbert space, one must find those functions of holonomies which are invariant under diffeomorphisms. It is worth noting that the expression (8.34) is not a holonomy. This means that the action of $\hat{V}[\vec{n}]$ takes out of the space of functions of holonomies. Hence, $\hat{V}[\vec{n}]$ is not defined in Cyl. For this reason, one imposes the invariance under the action $\hat{U}_{\varphi(\vec{n})} = e^{i\hat{V}[\vec{n}]}$ of finite diffeomorphisms $\varphi(\vec{n})$, such that one solves

$$\hat{U}_{\varphi(\vec{n})} f_\alpha = f_\alpha\,, \qquad \forall\vec{n}. \qquad (8.36)$$

A formal solution to this condition can be found by applying the group averaging procedure discussed in section 5.5.4 to the vector constraint, *i.e.* by taking a state f_α and summing over all the graph obtained from α by finite diffeomorphisms. This summation is infinite and a well-defined meaning can be given only in the dual Hilbert space [32]. Let us consider for simplicity a spin-network state ψ_S and dualize it as follows

$$\psi_S^* = \langle\psi_S|\ldots\rangle, \qquad (8.37)$$

which means that ψ_S^* takes an element of the Hilbert space and it evaluates the scalar product with ψ_S. If one now considers the sum over all graph $\{\alpha\}$ related to α by a diffeomorphism (*s-knots*), one gets

$$\psi_{\{S\}}^* = \sum_{\beta\in\{\alpha\}}\psi_{\beta,\{\rho\}}^* = \sum_{\beta\in\{\alpha\}}\langle\psi_{\beta,\{\rho\}}|\ldots\rangle. \qquad (8.38)$$

The action of this operator in H^{Kin} is well-defined, no matter the sum extends on an infinite (non numerable) space. In fact, given a generic element f_α, the expression $\psi_{\beta,\{\rho\}}^* f_\alpha$ is non-vanishing only for $\beta = \alpha$, thus a finite number of terms contributes to $\psi_{\{S\}}^* f_\alpha$.

Therefore, *a solution to both the Gauss and the vector constraints can be given in the dual Hilbert space to* H^{Kin} and it can be written as a linear combination of the dual of invariant spin-network states defined over s-knots. The space spanned by these objects is denoted by H^{Diff} and it is the natural arena in which the dynamic issue has to be discussed.

8.4 Geometrical operators: discrete spectra

Since the Hilbert space H^{Diff} is invariant under three-diffeomorphisms and not under general coordinate transformations, one can define the area and the volume operators on it. One expects that the introduction of matter can give an invariant character to these quantities under all four-dimensional diffeomorphisms, such that the area and the volume becomes actual physical observables.

8.4.1 *Area operator*

Let us consider a surface σ, characterized by a normal vector n_i and a parametrization given by some coordinates u, v on it. The area of σ in terms of inverse densitized triads reads

$$A(\sigma) = \int_\sigma \sqrt{\delta^{ab} E^i_a n_i E^i_b n_j} dudv. \tag{8.39}$$

Let us compute the action of this operator on a spin-network state $\psi_S(A)$. It is easy to recognize that the only non-vanishing contributions come from spin networks based at edges having isolated intersections with σ. Hence, let us take a partition of σ in sub-surfaces σ_A such that $\bigcup_A \sigma_A = \sigma$ and each σ_A contains at most one intersection with the graph α. The whole area can be written as the sum of the areas $A(\sigma_A)$ of each σ_A,

$$A(\sigma) = \sum_A A(\sigma_A). \tag{8.40}$$

The surface σ_A has only one intersection x_A with α. The regularization of $A(\sigma_A)$ is based on the idea [30] to approximate each $E^i_a(x)$ in the expression (8.39) (for σ_A) with some smeared quantities $E_a(S_x)$ realized via a test-function $f_\varepsilon(x, y)$, which goes to $\delta^{(2)}(x-y)$ as $\varepsilon \to 0$, as follows

$$E_a(S_x) = \int_{\sigma_A} E^i_a n_i f_\varepsilon(x, y(u, v))\, dudv. \tag{8.41}$$

Hence, one makes the following approximation

$$\sqrt{\delta^{ab} E^i_a n_i E^i_b n_j}|_{x=x(u,v)} \sim \sqrt{\delta^{ab} E_a(S_x) E_b(S_x)}. \tag{8.42}$$

Let us now assume that no vertex of α lies on σ and that no edge of α belongs to σ, such that at each intersection x_A one has an incoming and an outgoing edge. Hence, let us call e'_A and e''_A the edges whose initial and final points coincide with $x_A = \alpha \cap \sigma_A$, respectively. The point x_A can be seen as a two-valence vertex between e'_A and e''_A, whose associated invariant intertwiner is $\delta_{mm'}\delta_{jj'}$ such that $\rho_{e'_A}$ and $\rho_{e''_A}$ have the same spin numbers j_A. The action of the operator above on spin-network states can be easily computed from (8.29), thus finding

$$\sqrt{\delta^{ab}\hat{E}_a(S_x)\hat{E}_b(S_x)}\psi_{\alpha,\{\rho\}}$$
$$=8\pi\gamma l_P^2 f_\varepsilon(x,x_A)\left(\prod_{e\neq e'_A,e''_A} D_{\rho_e\rho_e}\right)D_{\rho_{e'_A}}D_{\rho_{e''_A}}\left(\rho_{e'_A}\sqrt{\delta^{ab}\tau_a\tau_b}\,\rho_{e''_A}\right), \tag{8.43}$$

where the generators τ_a are in the spin j_A representation. The area of $A(\sigma_A)$ is obtained by integrating over σ_A thus finding

$$\hat{A}_\varepsilon(\sigma_A)\psi_{\alpha,\{\rho\}}$$
$$= 8\pi\gamma l_P^2\int_{\sigma_A} f_\varepsilon(x,x_A)\left(\prod_{e\neq e'_A,e''_A} D_{\rho_e\rho_e}\right)D_{\rho_{e'_A}}D_{\rho_{e''_A}}\left(\rho_{e'_A}\sqrt{\delta^{ab}\tau_a\tau_b}\,\rho_{e''_A}\right)dudv, \tag{8.44}$$

and by taking the limit $\varepsilon \to 0$, $f_\varepsilon(x,x_i)$ tends to $\delta^{(2)}(x-x_i)$ so the integration disappears to give the following regularized expression

$$\hat{A}(\sigma_A)\psi_{\alpha,\{\rho\}} = 8\pi\gamma l_P^2\left(\prod_{e\neq e'_A,e''_A} D_{\rho_e\rho_e}\right)D_{\rho_{e'_A}}D_{\rho_{e''_A}}\left(\rho_{e'_A}\sqrt{\delta^{ab}\tau_a\tau_b}\,\rho_{e''_A}\right). \tag{8.45}$$

Since $\delta^{ab}\tau_a\tau_b = j_A(j_A+1)\mathbb{I}$ is the Casimir of the $SU(2)$ group, the area operator is diagonal in the spin network basis and by summing over A the final expression reads

$$\hat{A}(\sigma)\psi_{\alpha,\{\rho\}} = 8\pi\gamma l_P^2\sum_A\sqrt{j_A(j_A+1)}\psi_{\alpha,\{\rho\}}. \tag{8.46}$$

In the presence of edges belonging to σ, one also has some isolated intersections x_B (coinciding with beginning and ending points of such edges) with only incoming or outgoing edges, e'_B, e''_B. As a consequence, the action of the area operator on spin networks based on $e_{B'}$ and $e_{B''}$ provides half of the contribution with respect to spin networks based on $e_A = e'_A \cup e''_A$, so finding

$$\hat{A}(\sigma)\psi_{\alpha,\{\rho\}} = 8\pi\gamma l_P^2$$
$$\cdot\left(\sum_A\sqrt{j_A(j_A+1)} + \frac{1}{2}\sum_{B'}\sqrt{j_{B'}(j_{B'}+1)} + \frac{1}{2}\sum_{B''}\sqrt{j_{B''}(j_{B''}+1)}\right)\psi_{\alpha,\{\rho\}}. \tag{8.47}$$

The extension to the $SU(2)$- and diffeo-invariant space is straightforward (just insert invariant intertwiners, take the dual and sum over all the graphs in the same knot). In the general case in which the intersections between α and σ can coincide also with some vertexes of the graph α, one can use the index A for all kind of intersections and label the state via the spin $j_A^{(u)}$ ($j_A^{(d)}$) of the sum of the representations of the edges lying "above" ("below") S and the spin $j_A^{(u+d)}$ of the sum $j_A^u \oplus j_A^d$. Hence, the following expression for the action of the area operator can be written

$$\hat{A}(\sigma)\psi_{\alpha,\{\rho\}} = 4\pi\gamma l_P^2 \cdot \left(\sum_A \sqrt{2j_A^{(u)}(j_A^{(u)}+1) + 2j_A^{(d)}(j_A^{(d)}+1) - j_A^{(u+d)}(j_A^{(u+d)}+1)}\right)\psi_{\alpha,\{\rho\}}. \tag{8.48}$$

$A(\sigma)$ is self-adjoint in H^{Diff} and

- *spin-network states are eigenstates of the area operator,*
- *edges carry quanta of area,*
- *the area spectrum is discrete.*

The last point represents one of the most outstanding results of the whole Loop Quantum Gravity framework. In fact, starting from a continuous formulation, the discretization of the area operator has been derived entirely from the quantization procedure. This achievement confirms the expectation that a consistent Quantum Gravity scenario predicts the existence of a fundamental geometric discreteness.

8.4.2 *Volume operator*

While edges carry quanta of area determined by the spin numbers, *the vertexes carry quanta of volume which depend on the intertwiner.* The expression of the volume of a region Ω in terms of E_a^i reads

$$V(\Omega) = \int_\Omega \sqrt{h}d^3x = \int_\Omega \sqrt{|E|}d^3x, \tag{8.49}$$

with

$$E = \frac{1}{3!}\varepsilon_{ijk}\varepsilon^{abc}E_a^iE_b^jE_c^k. \tag{8.50}$$

Let us consider the action of V on a spin-network state $\psi_{\alpha,\{\rho\}}(A)$ based on a graph α which contains only one vertex v inside Ω and with n outgoing edges e_i for $i = 1,\ldots,n$. Hence, one finds [31]

$$\hat{V}(\Omega)\,\psi_{\alpha,\{\rho\}} = (8\pi\gamma)^{\frac{3}{2}} l_P^3 \sqrt{|\hat{q}|}\,\psi_{\alpha,\{\rho\}}$$

$$\hat{q}\,\psi_{\alpha,\{\rho\}} = \varepsilon^{abc} \sum_{e,e',e''} o(v,e,e',e'')\, L_a^e\, L_b^{e'}\, L_c^{e''}\, \psi_{\alpha,\{\rho\}}, \qquad (8.51)$$

where $o(v,e,e',e'')$ equals $+1$ or -1 if the tangent vectors to e, e' and e'' in v have a positive or negative orientation with respect to $V(\Omega)$ and it vanishes if the tangent vectors are not independent. The sum extends over all the edges of α emanating from v. For incoming edges the left invariant vector field L_a^e has to be replaced by the opposite of the right invariant one $-R_a^e$.

Generically, one must sum the expression above over all the vertexes contained in the region Ω.

Indeed, the regularization procedure of the volume operator is much more subtle than the one used for the area operator. This is due to the graph-dependence of the discretized pattern adopted and to the factor $o(v,e,e',e'')$. In fact, the regularization is based on subdividing Ω in elementary graph-dependent cells, whose volume can be written in terms of inverse (densitized) triads smeared over three two-surfaces passing through the interior of the cells. The regulator is removed by letting the cell be smaller and smaller. Finally, the regularized volume operator gives finite results, but the factor $o(v,e,e',e'')$ depends on the introduced background structure (it is not even diffeomorphism-covariant). In [31] the regulator is removed after performing a proper average over the background structure and the resulting expression transforms covariantly under diffeomorphisms. However, this procedure leaves an undetermined constant k_0 in front of the volume operator (8.51), which has been fixed in [145] via a consistency check on the equivalence between a quantization based on triads and one based on fluxes.

The extension of the operator $\hat{V}$ in H^{Diff} can be done as for the area inserting invariant intertwiners and constructing the s-knot. As a consequence, the left- (eventually right-) invariant vector fields in the expression (8.51) act on intertwiner indices. Hence, *the action of the volume on invariant spin-network states turns into an action on invariant intertwiners.* This is the reason why the volume is determined by the intertwiners.

The issue with the volume operator is that *no-closed analytical expression exists for its matrix elements in the basis of invariant spin-network states.* This is due to the square root inside the expression (8.51) (which becomes a double square root because of the modulus, $\sqrt{|q|} = (q^2)^{1/4}$). As a consequence a set of eigenvectors has not been found yet, even though it can be demonstrated that the spectrum is discrete. For instance, some eigenvalues have been evaluated via numerical analysis [84].

8.5 The scalar constraint operator

The implementation of the scalar constraint (7.117) is the most challenging issue of any approach towards a quantum theory for the gravitational field. *In LQG the canonical quantization program leads to a well-defined scalar constraint as a quantum operator* [138, 276].

One of the first issues on a quantum level is the factor $\frac{1}{\bar{e}}$, which is not polynomial in phase-space variables. However, one can use Thiemann's trick, which is based on the following classical identity

$$\varepsilon^{abc}\frac{E^i_a\,E^j_b}{\bar{e}}(x) = \frac{c^3}{4\pi G\gamma}\varepsilon^{ijk}[V(\Omega_x), A^c_k(x)], \tag{8.52}$$

where $V(\Omega_x)$ is an arbitrary volume containing the point x. Let us also introduce the quantity K,

$$K = \int_\Sigma K^a_i\,E^i_a\,d^3x, \tag{8.53}$$

for which one has[6]

$$K^a_i(x) = \frac{c^3}{8\pi G\gamma}[K, A^a_i(x)]. \tag{8.54}$$

The full scalar constraint can be written as

$$\mathcal{S}(x) = \mathcal{S}_E(x) + \mathcal{S}_L(x), \tag{8.55}$$

[6]The relation (8.54) is due to the fact that

$$\int_\Sigma \delta\Gamma^a_i\,E^i_a\,d^3x = -\frac{1}{2}\int_\Sigma \partial_i(\varepsilon^{ijk}\delta e^a_j e^a_k) = 0,$$

where we set the variation to vanish at the spatial boundary. In fact one has

$$[K, A^a_i(y)] = \left[-\frac{1}{\gamma}\int_\Sigma A^b_j\,E^j_b\,d^3x, A^a_i(y)\right] + \left[\frac{1}{\gamma}\int_\Sigma \Gamma^b_j\,E^j_b\,d^3x, A^a_i(y)\right]$$
$$= -\frac{1}{\gamma}\int_\Sigma A^b_j(x)\,[E^j_b(x), A^a_i(y)]\,d^3x + \frac{1}{\gamma}\int_\Sigma \Gamma^b_j(x)[E^j_b(x), A^a_i(y)]\,d^3x = \frac{8\pi G\gamma}{c^3}K^a_i(y).$$

The relation (6) can be used to demonstrate that one can change γ via a canonical transformation [274].

with

$$\mathcal{S}_E(x) = \frac{c^3}{2\pi G\gamma}\varepsilon^{ijk}Tr(F_{ij}(x)[V(\Omega_x), A_k(x)])$$

$$\mathcal{S}_L(x) = 8i(\gamma^2+1)\left(\frac{c^3}{8\pi G\gamma}\right)^3 \varepsilon^{ijk}\, Tr([A_i(x), K][A_j(x), K][V, A_k(x)]), \tag{8.56}$$

A_k and F_{ij} being $A^a_k \tau_a$ and $F^a_{ij}\tau_a$, where we used

$$Tr(\tau_a\,\tau_b) = \frac{1}{2}\,\delta_{ab} \quad , \qquad Tr(\tau_a\,\tau_b\,\tau_c) = \frac{i}{4}\,\varepsilon_{abc} + d_{abc}\,, \tag{8.57}$$

d_{abc} being symmetric under the exchange of all indexes. The two terms S_E and S_L are known as the Euclidean and the Lorentzian parts of the scalar constraint, respectively.

In order to quantize the expression (8.55), it must be rewritten in terms of holonomies and fluxes. In particular, the connections A_i and its curvature can be obtained by a limit procedure on holonomies along an edge or a loop with decreasing length. Hence, the idea is to consider the smeared version of the scalar constraint, $\mathcal{S}[N]$, N being the lapse function, and to define a discrete version $\mathcal{S}_\Delta[N]$ (defined in terms of some holonomies) whose limit is $\mathcal{S}[N]$ as the triangulation Δ gets finer and finer. This way an actual regularization of $\mathcal{S}[N]$ is realized, such that we can quantize the resulting expression simply by promoting phase-space variables to operators and by replacing Poisson brackets with commutators.

Indeed this can be done only in a graph-dependent way, where the graph is the one of the quantum states on which the operator $\hat{\mathcal{S}}[N]$ acts. Hence, the regularization is of purely quantum character. Let us take a graph α and let us consider a triangulation Δ, such that each vertex v in α is also a vertex for some tetrahedra T_v. The tetrahedra T_v have some edges s_a which belong to the edges emanating from v and some edges s_{ab} which connect s_a with s_b (see figure 8.2).

The scalar constraint operator $\hat{\mathcal{S}}_\Delta[N]$ acts only at vertexes, since it contains the volume operator, and it can be constructed as the sum of single operators over the vertexes v, which in turn are written as a sum of single operators on each tetrahedra T_v. For each T_v the idea is to approximate the curvature F_{ij} via the holonomy of the connection along the closed curve $r_{ab} = s_a \circ s_{ab} \circ s_b^{-1}$ and the connection with the holonomy along an edge s_c. Finally, the following quantum expression replaces the smeared Euclidean

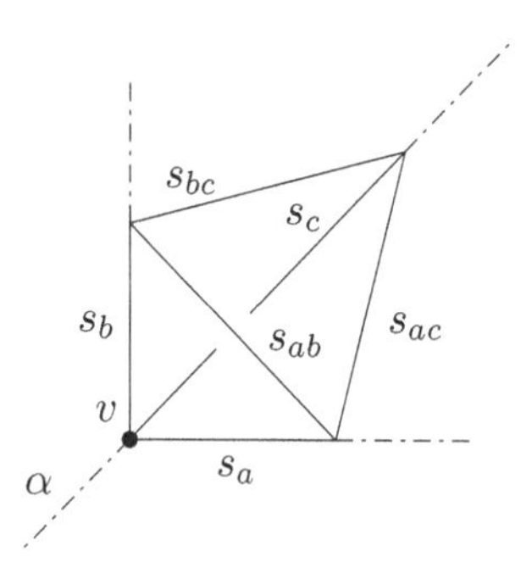

Figure 8.2 A tetrahedron (solid lines) based at the vertex v of a graph α (dashed lines).

scalar constraint

$$\hat{\mathcal{S}}_{E,\Delta}[N]\psi_\alpha = \sum_{v\in\alpha}\sum_{T_v\in\Delta} N(v)\frac{-i}{12\pi l_P^2\gamma}\varepsilon^{abc}\,Tr\left(U_{r_{ab}}U_{s_c}[\hat{V}(\Omega_v),U_{s_c}^{-1}]\right)\psi_\alpha, \tag{8.58}$$

where ψ_α denotes a generic element of H_{Kin} based at the graph α, while the sum extends over all vertexes, over all the tetrahedra sharing the same vertex v and over all the triple of segments s_a, s_b, s_c for a given T_v.

It can be shown that the classical quantity $\mathcal{S}_{E,\Delta}[N]$ goes to the first term in the expression (8.55) as the length ε of the segments s_a vanishes, *i.e.* as the triangulation becomes finer and finer, while the tetrahedra T_v shrink to v. A similar expression can be written for the Lorentzian part $\mathcal{S}_L \to \mathcal{S}_{L,\Delta}$ by quantizing the following classical formula

$$K = -\frac{c^3}{8\pi G\gamma}[V, S_E[1]], \tag{8.59}$$

thus in the end one can define

$$\hat{\mathcal{S}}_\Delta[N] = \hat{\mathcal{S}}_{E,\Delta}[N] + \hat{\mathcal{S}}_{L,\Delta}[N]. \tag{8.60}$$

The resulting expression for the scalar constraint operator can be regularized in H^{Diff} [276]. The reason is that the length of the segments s_a acts as a regulator and it can be changed by a diffeomorphsim. Thus, diffeomorphism-invariant states do not depend on the regulator, which can be removed. In fact, to find a solution of the scalar constraint in H^{Diff} one must look for those distributions over s-knots $\psi^*_{\{S\}}$ for which the following condition holds

$$\psi^*_{\{S\}}(\hat{\mathcal{S}}_\Delta[N]f_\alpha) = 0, \tag{8.61}$$

f_α being a generic element of the $SU(2)$-invariant Hilbert space. Since the distribution ψ^* is defined over s-knots, its value on a given function cannot be changed by a diffeomorphism. It is precisely the action of diffeomorphisms what makes the triangulation finer and finer, thus the expression (8.61) provides a well-defined limit as the tetrahedra T_v shrink to v.

In the same way, the behavior of the commutator $[\hat{\mathcal{S}}[N_1], \hat{\mathcal{S}}[N_2]]$ can be analyzed, finding that it vanishes in H^{Diff}. This is consistent with the classical algebra of constraints restricted to diffeo-invariant functionals and it implies the absence of anomalies.

An alternative prescription for the regularization of the scalar constraint has been given by Alesci and Rovelli [6] for four-valence vertexes. In their scheme the action of $\hat{\mathcal{S}}$ changes also the number of vertexes of the graph. The problem with this proposal is the algebra (the commutator of two scalar constraints is not correctly reproduced on a quantum level).

Therefore, the investigation on the dynamics in LQG reduces to study the action of a finite sum of operators on cylindrical functions. This involves a rather technical analysis, but, in a theoretical point of view, just a finite number of operations has to be performed.

In particular, one finds that the operator $\hat{\mathcal{S}}$ changes graphs by removing or adding edges and quantum numbers by raising or lowering quanta of spin at edges, just like in quantum field theories the operators raise or lower the number of particles. A different and complementary approach to solve the quantum dynamics of the gravitational field is given by covariant formulations, like that of spin-foam models.

8.5.1 *Spin foams and the Hamiltonian constraint*

In the canonical formulation, from a quantum-mechanical point of view, the Hamiltonian operator $H_{N,\vec{N}}(x^0)$, composed of the smeared scalar and vector constraints, $H_{N,\vec{N}}(x^0) = \mathcal{S}[N(x^0)] + \vec{\mathcal{V}}[\vec{N}(x^0)]$, can be interpreted as the generator of quantum evolution from the initial hypersurface $\Sigma_i(x^0 = 0)$ to the final hypersurface $\Sigma_f(x^0 = 1)$, parametrized by the proper time evolution $U(T)$,

$$U(T) = \int_T dN d\vec{N} U_{N,\vec{N}} = \int_T dN d\vec{N} e^{-i\int_0^1 dx^0\, H_{N,\vec{N}}(x^0)}. \tag{8.62}$$

The evolution operator $U(T)$ encodes the dynamics of the gravitational field, and its expansion in powers of T can be shown to be finite order by order. The calculation of the matrix element of such an operator between two states of the gravitational field is strictly analogous to that followed in the familiar calculation of the S-matrix elements in a gauge theory, and reads explicitly

$$\langle s_f|U(T)|s_i\rangle = \langle s_f|s_i\rangle$$
$$+ (-iT)\left[\sum_{\alpha\in s_i} A_\alpha(s_i)\langle s_f|D_\alpha|s_i\rangle + \sum_{\alpha\in s_f} A_\alpha(s_f)\langle s_f|D_\alpha^{+}|s_i\rangle\right]$$
$$+ \frac{(-iT)^2}{2!}\sum_{\alpha\in s_i}\sum_{\alpha'\in s'} A_\alpha(s_i)A_{\alpha'}(s')\langle s_f|D_{\alpha'}|s'\rangle\langle s'|D_\alpha|s_i\rangle + \dots . \tag{8.63}$$

Such a result can be obtained by splitting the calculation in several steps, *i.e.*,

(1) the evolution from the initial hypersurface to the final one is expressed as a sum over intermediate hypersurfaces, as sketched in (7.53). The intermediate hypersurfaces differ by a small coordinate time, and the time evolution between two hypersurfaces can be written in terms of the diffeomorphism that describes the shift between them, so that $U_{N,\vec{N}} = D(g)U_{0,\vec{N}}$;
(2) the expansion of $U_{0,\vec{N}}$ and the insertion of the identical projector where needed leads to a sum, where, at each order n, the operator D acts n times. Its action on the states is given by the coefficients A, which can be evaluated in terms of the explicit form of the scalar constraint;
(3) $U(T)$ can be worked out of $U_{0,\vec{N}}$ after integrating over the lapse and the shift. The first integration follows directly, as the integrand does not depend on N, and the second one corresponds to the implementation of the diffeomorphism constraint.

As a result, the matrix elements of the operator U read as (8.63), where generic spin-network states have been substituted with the corresponding s-knot states. Analyzing the geometrical meaning of the intermediate states, in which the sum has been split up, allows one to recognize (8.63) as a sum over spin foams. In fact, the time evolution of a generic surface s_i describes a "cylinder", whose time slicing are spin-network states belonging to the same s-knot, unless any interaction occurs. When operating on such a state, the Hamiltonian constraint generates a new state with one new edge and two new vertexes, *i.e.*, this structure is the elementary interaction vertex of the theory, as suggested by the comparison with ordinary gauge theories. As a generalization, the Hamiltonian constraint acts adding one dimension to the spin-network state, and such a new direction can be interpreted as time, because of the geometrical construction of (8.63), thus opening the way for the interpretation of these new states as spin foams.

In fact, at the n-th order of sum, n new dimensions are added, and the sum can be written as the sum of topologically inequivalent term, where the weight of each vertex is determined by the coefficient of the Hamiltonian constraint. Furthermore, if the irreducible representation of the gauge group carried by the edge of each spin network is taken into account during the addition of the new vertexes, the resulting geometrical objects fit the definition of spin foams given in section 7.1.3.2, *i.e.*, the implementation of the Hamiltonian constraint leads naturally to the sum over spin foams. Another approach to four-dimensional spin foams can be found in [44] and its developments, where a four-dimensional state is constructed by the spin covering of the group $SO(4)$. The description of spin foam models and of its recent achievements is out of the scope of this book. We just want to mention that there are currently two main models, the Freidel-Krasnov one [129] and the Engle-Pereira-Rovelli-Livine one [119] (see [7] for the correspondence between physical states in LQG and in the one-vertex expansion of spin foam models for four-valence graphs).

8.6 Open issues in Loop Quantum Gravity

The main technical difficulty of the LQG program is the absence of a closed analytic expression for the matrix elements of the scalar constraint operator in the spin network basis. This is due to the presence of the volume operator which involves two square roots in its definition.

Indeed, there are some ambiguities in the definition of the operator $\hat{S}$ (for instance see [138, 247]). Some of them are linked to the regularization procedure itself (as for instance the choice of the representation for the holonomies in (8.58)). The question is highly technical and the debate is open on the naturalness of some choices and on which ambiguities may have a physical significance (for a technical discussions about these topics see [236, 282]). We will not discuss these issues, but we focus our attention on other points such as the implementation of the constraint algebra, the semiclassical limit, the definition of the scalar product in the physical Hilbert space and the physical meaning of the Immirzi parameter.

8.6.1 *Algebra of the constraints*

The quantum implementation of the constraint algebra (6.17) can probe the absence of anomalies[7], thus the consistency of the adopted quantization procedure. The first problem one faces is the impossibility to implement infinitesimal three-diffeomorphisms in the space of $\bar{A}$ representations. Hence, one refers to finite diffeomorphisms only and the commutation relations can be rewritten as follows

$$\begin{aligned} \hat{U}_\varphi \hat{U}_{\varphi'} \hat{U}_\varphi^{-1} &= \hat{U}_{\varphi \circ \varphi' \circ \varphi^{-1}} \\ \hat{U}_\varphi \hat{\mathcal{S}}(f) \hat{U}_\varphi^{-1} &= \hat{\mathcal{S}}(\varphi \circ f) \end{aligned} \tag{8.64}$$

φ being an arbitrary three-diffeomorphism and $\hat{U}_\varphi$ its representation. The first relation holds exactly on H_{Kin}, while the second is reproduced modulo three-diffeomorphisms [283].

However, the main question is the implementation of the last commutation relation, which in the $SU(2)$ invariant subspace coincides with the one for the super-Hamiltonian (6.17) in geometrodynamics, *i.e.*

$$[\mathcal{S}(f), \mathcal{S}(f')] = \mathcal{V}[\vec{g}], \qquad g^i = h^{ij} \left(f \partial_j f' - f' \partial_j f \right). \tag{8.65}$$

Although the operator corresponding to the right-hand side of the expression above can be defined on H_{Kin} [275], nevertheless the equivalence with the left-hand side holds only on H^{Diff} [199], where both the left and right hand sides vanish. Whether this is a trivial way to reproduce the algebra of constraints is a question under debate.

Recently, a new formulation has been proposed, based on choosing different operator topology and density weight for the constraints with respect to the standard LQG formulation. This way, the generator of diffeomorphisms can be implemented in the new kinematical Hilbert space [290], allowing the algebra of the constraints to be reproduced off-shell in a simplified model [287].

A different idea is to quantize gravity after having identified proper matter fields as clock-like fields, chosen in order to deal with a "deparametrised" systems. The well-known example in geometrodynamics is the Brown-Kuchař mechanism [83] (see section 6.4.3.2), which provides a physical Hamiltonian strongly commuting with itself and with the diffeomorphisms generator. The use of deparametrised systems in LQG is a quite recent subject of investigation which can give important simplifications (see for instance [172], while a complete review is given in [150]), but, just like

[7] We refer to the notion of anomaly given in [167], *i.e.* that the quantum constraint algebra coincides with the classical one.

in geometrodynamics, it is affected by the ambiguity in the choice of the clock-like fields (different choices provide different dynamic scenarios and there is no compelling physical reason to choose a certain field).

8.6.2 *Semiclassical limit*

The main difficulty in the development of the low-energy sector of LQG consists in combining semiclassical tools with General Covariance. In [22], Ashtekar, Bombelli and Corichi pointed out that the "group averaging" procedure could be a powerful tool in this direction, at least in the linearized case.

In fact, in some relevant cases starting from kinematical states which behave semiclassically (coherent states), by a "group averaging" one reduces fluctuations around expectation values [22]. Hence, in such cases "the group averaging" turns out to be a well-defined procedure to develop semiclassical physical states. It is worth noting the application of this framework to the Bianchi I cosmological model [76]. However, these results have been obtained with constraints much simpler than those of General Relativity (linear and a particular class of quadratic functions of phase-space variables).

Earlier attempts to develop semiclassical states in LQG were based on defining functionals with a well-defined geometric structure [156] or which diagonalize holonomy operators [18]. The issue of combining both these properties leads to apply the complexifier technique [278, 279, 284–286] (see section 5.1.7), where the complexifier is a function of fluxes or area operators and semiclassical states are developed from spin-network states.

The resulting quantum states are well defined in H^{Kin} and the extension in H^{Diff} is not obvious. It has been shown in [284] that if semiclassical states are defined in H^{Kin} and then $SU(2)$ gauge invariance is implemented, the resulting state is sufficiently peaked to be a proper semiclassical state. Then, in [56] the projection to the $SU(2)$-invariant space has been realized via the restriction to an overcomplete basis of intertwiners, the Livine-Speziale ones [203].

Indeed, a consistency check on the viability of the semiclassical states has been proposed by testing the expectation value of the volume operator and comparing it with the classical expression. In particular, when the complexifier is a function of the area operators, there is no chance to find out the classical values [124]. On the contrary, in the case of fluxes the classical expression for the volume is reproduced only for 6-valence graphs [125]. In [42] this result is taken as an indication that cubic lattices only provide an appropriate discretization for the spatial manifold and it is suggested that path-integral approaches have to be based on cubulations (while in spin-foam models simplicial triangulations are taken).

A further issue concerns the action of the scalar constraint, which adds extra-edges and it generically takes a semiclassical state into a different one. This can be avoided only if the extra-edges belong to the graph the semiclassical state is based on. Since the definition of the scalar constraint operator involves a limiting procedure, a good semiclassical state must be defined before taking the limit and such that it contains all the edges entering into the scalar constraint (non graph-changing scalar constraint). Such semiclassical states find a natural implementation in Algebraic Quantum Gravity (see section 8.7).

8.6.3 *Physical scalar product*

The introduction of a scalar product is an intriguing point in any quantum system with constraints. Here, because of the complexity of the scalar constraint operator, it is a very difficult task to analyze its kernel [275, 277] (the physical Hilbert space H^{phys}) and to equip such a space with a scalar product (physical scalar product). In fact, since the zero eigenvalue belongs to the continuous part of the spectrum, the corresponding eigenvector is not normalizable in the kinematical Hilbert space. Hence, the kinematical scalar product cannot be extended to the physical Hilbert space.

A possible definition for the physical scalar product can be found from the kinematical one thanks to the fact that the scalar constraint is a well-defined operator in H^{Diff} [199, 257]. Let us consider the projector $P_{\mathcal{S}}$ to the physical Hilbert space, *i.e.*[8]

$$P_{\mathcal{S}} = \delta(\hat{\mathcal{S}}) = \int DN\, e^{i\hat{\mathcal{S}}[N]}, \qquad P_{\mathcal{S}} : \mathrm{H}^{Diff} \to \mathrm{H}^{phys}. \tag{8.66}$$

Given two elements $|\psi_1\rangle, |\psi_2\rangle \in \mathrm{H}^{Diff}$ one can define the physical scalar product as

$$\langle P_{\mathcal{S}}\psi_1|P_{\mathcal{S}}\psi_2\rangle_{phys} = \langle P_{\mathcal{S}}\psi_1|P_{\mathcal{S}}\psi_2\rangle = \langle \psi_1|P_{\mathcal{S}}\psi_2\rangle. \tag{8.67}$$

Unfortunately, the technical obstructions which prevent the analysis of the scalar constraint also forbid the explicit construction of the operator $P_{\mathcal{S}}$.

[8] A proper definition can be given to this formal expression.

It is worth noting that the definition of the physical scalar product is not an issue for deparametrised systems, since in that case the scalar constraint defines a physical Hamiltonian and the kinematical scalar product can be retained.

8.6.4 *On the physical meaning of the Immirzi parameter*

We have seen in section 7.4 how the Immirzi parameter [173], which can be introduced as a factor in front of the Holst modification (7.87), plays no role classically. Nevertheless, after the quantization, it enters the spectra of observables, so it modifies physical predictions. This proves that quantizations in different γ-sectors are inequivalent.

The interpretation of this ambiguity is far from being understood. Rovelli and Thiemann [260] analyzed the way this parameter comes out. They conclude that such an ambiguity is a consequence of two basic features:

- the affine structure of the configuration space, in particular the presence of two variables (the extrinsic curvature and spin connections) playing the role of connections,
- the quantization procedure, based on taking holonomies as fundamental variables.

In fact, as pointed out by Corichi and Krasnov [103], the quantization of a $U(1)$ gauge theory in the holonomy-flux representation provides an ambiguity (due to the possibility to rescale connections) which fixes the quanta of the electric charge. This is not the case for non-Abelian gauge theories, since the connections cannot be rescaled without violating gauge invariance. However, if another connection is present, it can be combined with the old one and the resulting theory presents a residual ambiguity.

The Immirzi ambiguity is often associated with the so-called θ-sector in QCD. It consists in the CP-violating term $\theta\varepsilon^{\mu\nu\rho\sigma}F^a_{\mu\nu}F^a_{\rho\sigma}$, which can be added to the Lagrangian density. Since such a term is a boundary contribution, it does not modify the classical dynamics; nevertheless, after quantization, the parameter θ can be fixed in order to account for the observed CP violations in K-decays. However, the Immirzi parameter multiplies the Holst modification, which is not a boundary contribution since it merely vanishes on-shell.

Indeed, an actual topological interpretation has been proposed for γ by noting that the Holst modification is contained into the Nieh-Yan topological density I_{NY}:

$$I_{NY} = \partial_\mu(\varepsilon^{\mu\nu\rho\sigma} e^\alpha_\nu D_\rho e_{\alpha\sigma}) = \varepsilon^{\mu\nu\rho\sigma} \left(D_\mu e^\alpha_\nu D_\rho e_{\alpha\sigma} - \frac{1}{2} e_{\alpha\mu} e_{\beta\nu} R^{\alpha\beta}_{\rho\sigma} \right). \tag{8.68}$$

If one starts from the Einstein-Hilbert action plus the integral of the density above, then one finds a theory which in vacuum is equivalent to the Holst formulation for gravity, but the Immirzi parameter is actually a constant in front of a topological term [107].

If spinors are present, they enter the definition of the topological Nieh-Yan term. As a consequence, the dynamics is different whether one advocates or not a topological interpretation for γ. In the former case the dynamic system coincides with the Einstein-Dirac one [216, 217], while in the latter case the value of the Immirzi parameter determines the amount of the four fermion interaction term [248].

The prevailing idea in the LQG community is that the Immirzi parameter is a new fundamental constant, which has to be fixed from the macroscopic limit. The early attempts in this direction were based on the comparison between the *black hole entropy* given by Bekenstein equation [48] and the analogous expression in LQG. The idea is that if the fundamental description of the spatial manifold is given by spin networks, then the collection of edges crossing the horizon are the microscopic degrees of freedom determining the local properties of the horizon itself [256]. These edges can be seen as punctures of the horizon and have been described in terms of a Chern-Simons theory [21]. The statistical mechanical treatment of such a dynamic system provides an expression for the horizon entropy proportional to the area at the leading order in the macroscopic limit [120], with a coefficient depending on γ, which can be tuned to reproduce the 1/4 factor proper of the Bekenstein formula [3, 141].

However, from two different points of view, one aiming to regard the black hole entropy as an entanglement entropy [55], the other based on implementing the statistical mechanics analysis from the definition of a local energy close to the horizon [142], it has been found that the Bekenstein equation is reproduced without any assumption on the value of γ. These results outline that one cannot fix the Immirzi parameter from the study of the black hole entropy in a quantum space.

Finally, there have also been some attempts to treat γ as a dynamic field (see [86, 219, 273]).

8.7 Master Constraint and Algebraic Quantum Gravity

The *master constraint* program proposed by Thiemann [281] is based on quantizing instead of $\mathcal{S}$ the master constraint operator $\mathbf{M}$:

$$\mathbf{M} = \int \frac{\mathcal{S}^2}{\sqrt{h}} d^3x\,. \tag{8.69}$$

The condition $\mathbf{M} = 0$ is equivalent to $\int d^3x f(x)\mathcal{S} = 0$ for any test function f, thus it defines the same hypersurface in the phase space as the scalar constraint. Moreover, the following two conditions are equivalent on the constraint hypersurface

$$\left[\int f(x)\mathcal{S}d^3x,\, F\right] = 0 \Leftrightarrow [F,\, [F,\, \mathbf{M}]] = 0 \tag{8.70}$$

F being a generic function of phase-space variables. Hence, the set of weak Dirac observables for the scalar constraint can be inferred by "double-commutators" with the master constraint.

Furthermore, since $\mathbf{M}$ is three-diffeomorphism invariant, its commutator with spatial diffeomorphisms vanishes and the whole algebra of constraints does not contain structure functions (Lie algebra). This in turn implies that the group averaging technique can be used to implement $\mathbf{M} = 0$ on a quantum level.

As soon as $\mathbf{M}$ is quantized, the full Hilbert space can be written as the direct sum of the spaces spanned by the eigenvectors of the corresponding operator. This way the issue of finding the physical space reduces to determine the kernel of $\hat{\mathbf{M}}$. It is also possible to define the master constraint operator such that its action is not graph-changing (it implies to start with an anomalous $\mathcal{S}$).

The quantization of the master constraint can be tested by proper semiclassical states in the context of *Algebraic Quantum Gravity* [146–148]. This theory is based on transporting the machinery of LQG on a fundamental algebraic graph, from which all the graphs can be derived via an embedding. In Algebraic Quantum Gravity one deals with one graph only, on which normalizable coherent states can be properly defined. In this respect, the expectation value of $\hat{\mathbf{M}}$ has been evaluated, finding at the leading order in the semiclassical expansion the proper classical expression (see [149] for an application to reduced phase-space quantization). The main issue with such an approach is how to move from algebric graphs to embedded one, thus in establishing a relation with the states in LQG.

8.8 The picture of quantum spacetime

The LQG quantization program gives a description of the quantum spacetime quite different from that coming from Wheeler-DeWitt quantization and much closer to a Quantum Field Theory for the geometry.

In Wheeler-DeWitt approach, the functional space one deals with is that of three-geometries. No discrete structure arises in this context and quantum states are defined only formally. The link with General Relativity is easily established, but the quantization program is still lacking.

On the contrary, LQG starts from a phase space similar to that of a gauge theory. The configuration variables are non-local objects, holonomies, whose physical meaning is the same as Faraday lines for the electro-magnetic field. Quantum states are linear combinations of spin-network states, which realize an orthonormal basis of the non-separable kinematical Hilbert space. Geometric quantities have discrete spectra, thus quantum geometry is discrete (it is a twisted geometry [131, 259]) as expected in a realistic Quantum Gravity scenario (see figure 8.3).

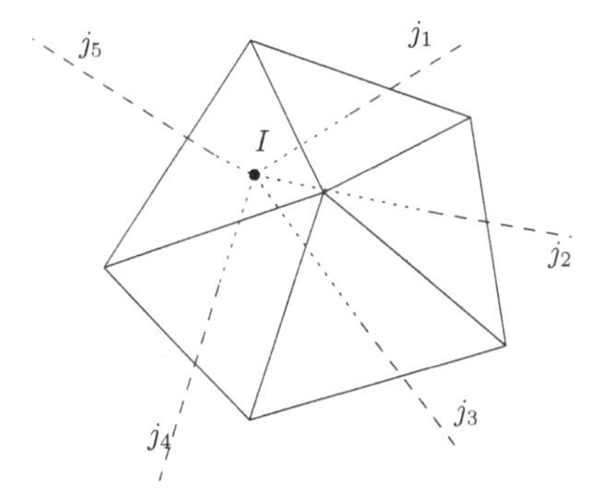

Figure 8.3 The (dual) quantum space (solid lines) associated with a spin network (dashed lines): the areas of quantum surfaces are determined by the spin numbers j's carried by the dual edges and the volume enclosed is given by the intertwiner I of the dual vertex.

However, the quantum dynamics has not been solved and the connection with low-energy physics is still an open issue. The problems are not only technical. A satisfying measurement theory for a background-independent model is still lacking and this fact could prevent a real connection with experiments. For instance, there are hints that a non-commutative structure for space-coordinates arises [26], but there are no indications whether this can provide testable predictions, such as violations of the Lorentz symmetry (for the link between Lorentz violations and measure theory in Quantum Gravity see [208]). This sort of violation, and generically all the phenomenology related with non-commutative scenarios [11], would be extremely interesting in view of the comparison with experiments [101].

Chapter 9

Quantum Cosmology

The implementation of Canonical Quantum Gravity on a cosmological setting offers a valuable arena to test the consistency and the predictability of the basic approaches we have previously discussed.

Before presenting some of the most important successes of canonical quantum cosmology and, at the same time, outlining its significant shortcomings, it is worth noting some non-trivial open questions about the legitimation of implementing quantum physics on a cosmological system.

In what follows we put these fundamental puzzles of quantum cosmology in a somewhat coherent order.

- The detectability of quantum mechanics is based on the simple statement that, for a large enough number of repeated experiments, the frequency of a given phenomenon approaches its probability to happen. Now, it is clear that, on a cosmological level, this basic statement loses its full validity. We live in a specific realization of our Universe and any perspective to be able to measure a frequency of its appearance, as well as of other observables, must be abandoned. Therefore, the predictability of a quantum theory of the Universe must be intended essentially as a mathematical notion, having physical sense *a posteriori* only. For instance, if we are able to predict that the probability amplitude for the transition between a given Universe configuration and a homogeneous and isotropic Universe is particularly high (~ 1), we can only claim that our Universe could be a realization of that scenario, but there is no chance to attribute a privileged physical channel to this perspective.
- It is not possible to apply in quantum cosmology the ordinary notion of "measurement" adopted in quantum mechanics. In fact, as discussed

in [195], a measurement on a given quantum system must be performed by a classical observer, *i.e.* by a detection apparatus whose dynamics is governed by classical laws. Such a measurement induces the wave function to collapse into a given eigenstate, allowing for a deterministic (although probabilistic) output.
In a primordial quantum Universe, it appears unrealistic to apply this notion of measurement, simply because no system in such a Universe can be described by a classical evolution. This difficulty can apparently be overcome by observing that today we live in a classical Universe and that, nonetheless, we can, in principle, get information from the Big-Bang singularity, *i.e.* from the Planck era. However, such an information, apart from being poor, since we can at best observe the spectrum of primordial fluctuation around an isotropic Universe, suffers two very important limitations. First of all, the quantum Universe we are in contact with is only one of the possible realizations of the quantum cosmological picture, as discussed above. Then, we have to stress that the Planck era is in our past light-cone, so we can receive signals from it, but we cannot send signals to it and this point appears very hard to be reconciled with the idea of an interaction between a classical observer and a quantum system, responsible for the Universe wave function collapse.

- The wave function of the Universe is referred to the Universe as a whole, *de facto* to its physical degrees of freedom, coinciding in the minisuperspace to the independent cosmic scale factors. Therefore, this formalism does not provide any information on the notion of causality across the Universe and some of the fundamental features characterizing the classical primordial Universe have no precise counterpart in the quantum regime. This limit does not affect quantum cosmology only, but it is a basic one in quantum gravity approaches, since it is connected to the non-trivial question of which meaning can have a free test particle moving on a fluctuating background. However, on a cosmological level, the absence of a precise quantum notion of causality is very relevant when we try to make predictions on how the Universe was born and when we address the semiclassical limit of the Universe from a physical point of view (which is the perturbation spectrum to the isotropic Universe, which regions had time to be in thermal contact among each other, which are the initial conditions for the inflation process, etc.). These simple considerations make clear that the wave function of the Universe can acquire a precise physical meaning only if we are able to properly define a quantum operator, whose eigenvalues fix the possible observable causal size in the Planckian Universe. In the absence of this formalism, we cannot describe the quantum evolution of the horizon scale and this fact must be considered a serious limitation of the minisuperspace approach.

- The canonical approach reduces in the minisuperspace model to the implementation of the super-Hamiltonian constraint to the Universe wave function. In fact, as an effect of the homogeneity request, underlying the minisuperspace model, the supermomentum constraint, associated with the three-diffeomorphisms invariance, is automatically satisfied. Despite the great simplification of the quantum cosmological setting with respect to the general quantum dynamics of the gravitational field, a series of technical and conceptual problems (calling attention for their solution) remain. Among these, the basic one is probably the validity itself of the minisuperspace representation, *i.e.* the possibility to commute, at least on a physical ground, the quantization procedure and the request of space homogeneity. This question is rather puzzling and still stands as unsolved, especially because of the difficulty to recover a precise quantum notion of homogeneity to be imposed on a generic cosmological field in order to freeze out the spatial degrees of freedom, reducing the quantum dynamics of the Universe to that of a finite dimensional mechanical system. Another important open question is the problem of defining a time evolution of the quantum Universe, which results from the vanishing behavior of the super-Hamiltonian.
- The most important expectation of the canonical quantization of the cosmological dynamics is the perspective of removing the initial singularity, replacing it with a regular regime, possibly associated with a primordial Big-Bounce of the Universe. A related question concerns the possibility to construct a satisfactory classical limit for the Universe dynamics, as long as the space volume (here a good internal time variable) takes enough large values.
- The Wheeler-DeWitt equation is essentially unable to give a proper solution of the initial singularity puzzle and also the origin of a classical Universe appears a rather ambiguous question. On the other hand, Loop Quantum cosmology seems to be able to clearly provide an initial Big-Bounce and a suitable classical limit. However two main points must be taken into account: i) the minisuperspace version of LQG contains a very severe simplification of the fundamental $SU(2)$ symmetry of the theory, whose impact on the predictability of the model could be

relevant (see section 9.6.2); ii) the Wheeler-DeWitt equation in minisuperspace provides a coherent and well-grounded quantum picture for the early Universe.

In what follows, we develop the minisuperspace models of both these two canonical approaches (the Wheeler-DeWitt and the LQG ones), focusing our attention on their main successes and shortcomings, reviewed on a general ground more than in the details of the specific issues.

9.1 The minisuperspace model

The basic idea of *minisuperspace* arises from the possibility to restrict the general problem of the quantum gravitational field to simpler highly symmetric spacetimes, reducing the dynamics to a finite dimensional scheme and the quantization to a natural Dirac prescription for the wave function of the system.

The most relevant minisuperspace application corresponds to the case of a homogeneous Universe, which is described by a class of different cosmologies, known as the nine *Bianchi models*. They describe Universes in which the space points are all equivalent, but the independent spatial directions scale in time with a different law, inducing a given degree of anisotropy. It is worth noting that only three models in the Bianchi classification allow the isotropic limit (Bianchi I, V, IX, mapped into the flat, open and closed Friedmann-Leimetré-Robertson-Walker models, respectively). Therefore, the anisotropy of Bianchi Universes is not only of a dynamic nature, due to different scaling of different directions, but it can also have an intrinsic geometrical origin.

The relevance of the classical and quantum dynamics of Bianchi models consists in the role these geometries could have played in the very primordial Universe (pre-inflationary scenario) and, more specifically, in the dynamic features of Bianchi type VIII and IX models near the singularity, which are the prototype for the behavior of a generic inhomogeneous Universe (having no specific symmetries).

In what follows, we will focus our attention on the Bianchi type IX model because it admits an isotropic limit (also naturally reached in its classical dynamics [155]). However, we will keep in mind the idea that every consideration we will trace for such a cosmology, can reliably resemble the local behavior of a generic Universe, *i.e.* as limited to its homogeneous patches, at least as long as we are close enough to the initial singularity.

The line element of a homogeneous Bianchi Universe takes the form

$$ds^2 = -N^2(t)dt^2 + a^2(t)(\omega^l)^2 + b^2(t)(\omega^m)^2 + c^2(t)(\omega^n)^2\,, \tag{9.1}$$

where the cosmic scale factors a, b, c determine the Universe evolution and are classically obtained as solutions of the Einstein's equations, which are reduced by the homogeneity constraint to an ordinary quasilinear second order system in time.

The 1-forms $\omega^a = \omega^a_i\, dx^i$ $(a = l, m, n)$ fix the particular geometric morphology of the considered Bianchi model. This information is contained in the structure constants associated to the specific isometry group, which coincides with a Lie group of motion. In fact, the components of the 1-forms obey the conditions

$$\partial_{[i}\omega^a_{j]} = -\frac{1}{2}C^a_{bc}\,\omega^b_i\,\omega^c_j\,. \tag{9.2}$$

The Bianchi classification comes out from the Jacobi identity associated to the structure constants. The most general case (which mimics the generic behavior) corresponds to have all the possible constants with three different indices all different from zero, and this is the case for the Bianchi VIII and IX types only. Thus, the geometry of the Universe enters the dynamics only through some constant terms (*i.e.* the spatial curvature depends on time only) and the spatial gradients are frozen during the classical and quantum dynamics.

On a quantum level, the Universe wave function must be taken over the three-geometries, which here are equivalent to the three scale factors, since the 1-forms are invariant under three-diffeomorphisms. As a consequence the minisuperspace wave function must be written as $\psi = \psi(a, b, c)$ and it is immediate to recognize that the general functional case is highly simplified to a three-dimensional ordinary mechanic system, thus allowing us to overcome some of the puzzling features of the full canonical approach. In the case of minisuperspace models, *the evolution of the cosmological systems resembles that of a relativistic particle*, whose generalized coordinates are just the three cosmic scale factors. In this respect, the Klein-Gordon-like nature of the Wheeler-DeWitt equation leads to search for a convenient representation, which outlines how *the Universe volume has the peculiar character of a good internal time* (being associated with a negative kinetic term). Furthermore, in the case of Bianchi models, a good quantum dynamic picture can be traced by means of the coincidence of the so-called multi-time approach (see section 6.4.3.3) with the set up of a natural separation of the physical degrees of freedom of the Universe (its anisotropies) from the embedding variable, here the time-like Universe volume (see the analysis in terms of the Misner variables [224]).

We conclude this section, observing that the generic inhomogeneous case corresponds, in the limit of the Belinski-Lifshitz-Khalatnikov conjecture validity [50] (see also [229, 230]) (*i.e.* the parametric nature of the space coordinates near the singularity), to reduce the superspace to the product of the minisuperspaces in the different quasi-homogeneous patches. On the classical level, such a decoupling of space points in the asymptotic regime to the singularity has been demonstrated [183], while on a canonical quantum stage, it stands as a mere, though well-grounded, hypothesis.

9.2 General behavior of Bianchi models

We now analyze some general features of the classical and quantum dynamics of Bianchi models, without entering the details of each type of cosmology.

Inserting the line element (9.1) into the general form of the Einstein-Hilbert action (4.2) and taking the integral over the spatial coordinates (the space dependence is fixed for each type), the dynamics of Bianchi models is reduced to a three-dimensional Lagrangian system, whose generalized coordinates are the scale factors a, b, c. Hence, performing the Legendre transformation, it is possible to set up the Hamiltonian formulation of such a mechanical system, which, according to the general General Relativity features, will be characterized by a vanishing Hamiltonian function and a non-positive definite kinetic term.

However, the construction of the Hamiltonian representation of Bianchi models dynamics is performed by using a different set of configuration variables with respect to the scale factors. Let us consider the new set of variables, defined via the relations

$$\ln a = \alpha + \beta_+ + \sqrt{3}\beta_- \,, \quad \ln b = \alpha + \beta_+ - \sqrt{3}\beta_- \,, \quad \ln c = \alpha - 2\beta_+ \,. \tag{9.3}$$

In this new framework, the line element (9.1) can be rewritten as

$$ds^2 = -N^2 dt^2 + e^{2\alpha}\left(e^{2\beta}\right)_{ab}\omega^a\omega^b \,, \tag{9.4}$$

where the indices $a, b, c = l, m, n$ and the β-matrix is diagonal: $\beta = \mathrm{diag}\{\beta + \sqrt{3}\beta_-, \beta_+ - \sqrt{3}\beta_-, -2\beta_+\}$.

This form of the metric clarifies how the choice of these variables, known as *Misner variables* [224], has a precise physical meaning, separating the isotropic (volume-like) contribution (related to α) and the anisotropic degrees of freedom (related to $\beta_\pm$), *i.e.* the two gravitational degrees of freedom.

However, Misner variables also have a convenient dynamic implication, corresponding to the diagonal form acquired by the kinetic term of the super-Hamiltonian. In fact, adopting the variables $\alpha, \beta_\pm$, the Hamiltonian formulation of the Bianchi model is summarized by the following action

$$S_B = \int dt \left\{ p\frac{d\alpha}{dt} + p_+\frac{d\beta_+}{dt} + p_-\frac{d\beta_-}{dt} - cNe^{-3\alpha}H \right\}, \tag{9.5}$$

where c is a constant depending on fundamental constants and on the particular space integral for the considered type, p, $p_\pm$ are the conjugate momenta to α, $\beta_\pm$, respectively, and the super-Hamiltonian $\mathcal{H}$ takes the form

$$\mathcal{H} = -p^2 + p_+^2 + p_-^2 + e^{4\alpha}V_B(\beta_\pm). \tag{9.6}$$

Here $V_B(\beta_\pm)$ denotes a potential term, different for each Bianchi type and due to the spatial curvature.

Variating this action with respect to the lapse function, we get the Hamiltonian constraint $\mathcal{H} = 0$, which plays the role of a first integral for the Hamilton equations, *i.e.* it is a constraint for the initial data problem. Among the remaining equations, we focus our attention on the following one (obtained by the variation of p)

$$\dot{\alpha} = -2cN\,e^{-3\alpha}\,p. \tag{9.7}$$

It is straightforward to recognize that, if we deal with an expanding Universe ($\dot{\alpha} > 0$), we must require that the momentum p be negative (N being positive by definition). As a consequence, for an expanding Universe, the Hamiltonian constraint (9.6) can be solved as follows

$$p = -H_{ADM} = -\sqrt{p_+^2 + p_-^2 + e^{4\alpha}V_B(\beta_\pm)}. \tag{9.8}$$

Now, we can also set the gauge condition $\dot{\alpha} = 1$, *i.e.* by the last two relations above, we fix the form of the lapse function as

$$N = N_{ADM} = \frac{e^{3\alpha}}{2cH_{ADM}}. \tag{9.9}$$

Hence, the action for the Bianchi models can be rewritten in the reduced form

$$S_{ADM} = \int d\alpha \left\{ p_+\beta'_+ + p_-\beta'_- - H_{ADM} \right\}, \tag{9.10}$$

where $\beta'_\pm \equiv \frac{d\beta_\pm}{d\alpha}$.

The Hamiltonian equations, associated to this action, take the form

$$\begin{aligned} \beta'_{\pm} &= \frac{p_{\pm}}{H_{ADM}} \\ p'_{\pm} &= -\frac{e^{4\alpha}}{2H_{ADM}}\frac{\partial V_B}{\partial \beta_{\pm}} \\ \frac{dH_{ADM}}{d\alpha} &= \frac{\partial H_{ADM}}{\partial \alpha}\,. \end{aligned} \tag{9.11}$$

Performing this ADM-reduction in the Bianchi model dynamics is equivalent to characterize the volume-like variable α as the time variable labeling the system evolution and the anisotropies $\beta_{\pm}$ like the physical degrees of freedom of the cosmological gravitational field. This way the Hamiltonian constraint is solved and the morphology of Bianchi models resembles that of an ordinary two-dimensional mechanical system. Actually, the Hamiltonian contains a square root and it is a non-local quantity, making hard both the classical and, overall, the quantum treatment. We observe that in the present scheme, the cosmological singularity is placed in the limit $\alpha \to -\infty$.

9.2.1 *Quantum Picture*

In order to construct the Wheeler-DeWitt equation corresponding to the Bianchi models, we have to promote all the canonical variables to operators, acting on the Universe wave function $\psi = \psi(\alpha, \beta_{\pm})$, which is annihilated by the Hamiltonian constraint, according to the Dirac prescription. As a result, we deal with the Wheeler-DeWitt equation of the form

$$\left\{\hbar^2\left[\frac{\partial^2}{\partial\alpha^2} - \frac{\partial^2}{\partial\beta_+^2} - \frac{\partial^2}{\partial\beta_-^2}\right] + e^{4\alpha}V_B(\beta_{\pm})\right\}\psi(\alpha,\beta_{\pm}) = 0\,. \tag{9.12}$$

This equation has the structure of a Klein-Gordon one, in the presence of an external potential term. Here, according to the pseudo-Riemannian nature of the minisupermetric, *i.e.*

$$G^{msm}_{ab} = \frac{1}{c}e^{3\alpha}\eta_{ab} \quad , \quad \eta_{ab} = \mathrm{diag}\{-1,1,1\}\,, \tag{9.13}$$

the variable α (associated to the Universe volume) is a time-like coordinate, while $\beta_{\pm}$ are the spatial coordinates in this quantum picture of the system.

The probability density ρ_p, associated to this quantum equation, takes the natural form

$$\rho_p \propto \psi^*\frac{\partial\psi}{\partial\alpha} - \frac{\partial\psi^*}{\partial\alpha}\psi\,. \tag{9.14}$$

This quantity is clearly non-positive, unless a frequency separation procedure is performed, and this, in general, is forbidden by the presence of the time dependent potential term.

The best analogy to interpret this quantum equation is the scattering of a relativistic particle against a non-perturbative potential, below the threshold of couple creation (the possibility to deal with many-particle problem is equivalent to the perspective of multi-Universe dynamic quantum theory).

Nonetheless, the idea of a second quantization of the equation (9.12) has been pursued (see [88, 122, 144, 244]), leading to the suggestive scenario of Universe creation, *i.e.* configurations with different values of the conjugate momentum associated to the anisotropy variables $\beta_\pm$. This line of investigation opens very intriguing scenarios on the quantum Universe birth and evolution, but it appears ill-grounded on a physical point of view, especially because, the causal structure, emerging in the minisuperspace, has no clear meaning and the whole second quantization procedure stands as a rather formal paradigm.

A different quantum picture of the Bianchi models emerges if we consider the ADM reduction of the dynamics, which does not contain any Hamiltonian constraint and therefore it is associated to a Schrödinger dynamics (corresponding to impose, on a quantum level, the relation between the momentum p and the Hamiltonian H_{ADM}), *i.e.*

$$i\hbar\frac{\partial\psi}{\partial\alpha} = \hat{H}_{ADM}\psi = \left[\sqrt{-\hbar^2\left(\frac{\partial^2}{\partial\beta_+^2} + \frac{\partial^2}{\partial\beta_-^2}\right) + e^{4\alpha}V(\beta_\pm)}\right]\psi\,. \tag{9.15}$$

This equation has the unpleasant feature to contain a non-local Hamiltonian operator, as a consequence of the square root. Furthermore, the time dependence of the potential prevents, in general, a simple reduction of the problem to a time independent Schrödinger equation, which could be easily addressed by a squaring procedure. Despite the equation (9.15) formally represents the natural solution of the frequency separation problem of the Wheeler-DeWitt equation (9.12), nonetheless, in the absence of simplifying approximations, it remains intractable in practice. Finally, we observe how the equivalence between the full covariant and the ADM-reduced approaches is hard to be stated in the exact case. Actually, they correspond to deal with profound different quantum point of view, the quantization of the Bianchi model geometry (in the Wheeler-DeWitt formulation) and the quantum description of the physical gravitational degrees of freedom $\beta_\pm$ versus the embedding coordinate α (in the Schrödinger equation).

9.2.2 *Matter contribution*

Let us now briefly discuss the role played by matter fields in this quantum picture, especially because it is crucial in ensuring the existence of a late isotropic dynamics of the Universe (on a classical level, a solution of the Einstein's equations, describing an isotropic Universe, cannot exist in vacuum).

Despite the synchronous system is a geodesic one, the homogeneity of the Bianchi model (*i.e.* the absence of a spatial gradient of the pressure) allows to deal with a comoving reference frame, even when the cosmological fluid is no longer a dust.

For a given equation of state for the Universe matter content $p = w\rho$, the contribution $\mathcal{H}^m$ of this matter to the total super-Hamiltonian function takes the form

$$\mathcal{H}^m_w = \frac{\mu^2_w}{e^{3(w-1)\alpha}}\,, \tag{9.16}$$

where μ^2_w is a (positive) constant, depending on the initial conditions. During its evolution, the Universe is characterized by different phases, in which the dominant matter contribution is associated to different values of the parameter w (which therefore must be regarded as a piecewise constant time function). Among the most important of such phases, we recall in time order (see section 2.4):

- i) the vacuum energy phase ($w = -1$), which corresponds to the de Sitter evolution of the Universe during the inflation;
- ii) the radiation dominated era ($w = 1/3$), when the Universe is sufficiently hot to neglect the rest mass of all the species with respect to their energy;
- iii) the matter dominated era ($w = 0$), describing a later Universe in which the non-relativistic matter dominates;
- iv) the late acceleration phase ($w < -1/3$), characterizing the Universe since about seven billions of years and associated with an accelerated expansion.

The de Sitter phase is placed in the Universe history near the Grand Unification temperature $T_{GUT} \sim 10^{15} GeV$ ($t_{GUT} \sim 10^{-34} s$), while the equivalence time between the radiation and matter contribution is much later ($t_{eq} \sim 10^{11} s$). However, some of these contributions (for instance, first the vacuum energy and radiation and then the radiation and matter) co-exist and they must additively be included in the super-Hamiltonian.

Unless mass Planck particles are postulated, the matter contribution appears rather inadequate to the quantum era (the Planck era) of the Universe, while the radiation and vacuum energy terms are expected to be included, though the latter cannot dominate in such an era (the de Sitter phase, associated to the inflation scenario, is essentially a classical process). An additional contribution that could be very important during the Planckian evolution of the Universe is given by a non-interactive massless scalar field ($w = 1$), which corresponds to the very early behavior of the inflaton field, when its potential (self-interaction) term is negligible. The kinetic energy of the inflaton has a behavior isomorphic to the one of the anisotropic variables $\beta_\pm$ and it is often accompanied by the vacuum energy contribution in order to mimic an inflationary behavior of the Universe, simple enough to be properly addressed on a quantum and classical level simultaneously. As we shall briefly discuss later (see section 9.4), the non-interactive scalar field has an important dynamic role for the Bianchi model evolution, since it is able to remove the chaotic behavior of Bianchi type VIII and IX models. This fact has relevant implications also on the quantum dynamics of these models. Therefore it is useful to explicitly describe the real self-interactive scalar field ϕ with the interaction energy density $V(\phi)$, instead of mimic its limiting behavior by the w parameter. Rescaling properly the scalar field, the super-Hamiltonian can be written as follows

$$\mathcal{H} = \mathcal{H}^g + \mathcal{H}^\phi = ce^{-3\alpha}\left(-p^2 + p_+^2 + p_-^2 + p_\phi^2 + e^{4\alpha}V_B(\beta_\pm) + e^{6\alpha}V(\phi)\right), \tag{9.17}$$

where $\mathcal{H}^g$ and $\mathcal{H}^\phi$ are the gravitational and scalar field contributions, respectively.

Then, if we re-define $\phi \equiv \beta_0$, the ADM-reduced action can be stated in a symmetric form with respect to the β_r ($r = 0, \pm$) variables, *i.e.*

$$S_{ADM} = \int d\alpha \left\{p_r\beta_r' - \left[p_r p_r + W(\alpha, \beta_r)\right]^{1/2}\right\}, \tag{9.18}$$

where the repeated index r is intended to be summed, p_r denote the conjugate momenta with respect to the β_r variables and the total potential W has be defined in the form

$$W(\alpha, \beta_r) \equiv e^{4\alpha}V_B(\beta_\pm) + e^{6\alpha}V(\beta_0). \tag{9.19}$$

Clearly, when $V(\phi) \simeq const. = \mu_{-1}^2$, we get a cosmological constant into the dynamic problem. Finally, we observe that, near the cosmological singularity ($\alpha \to -\infty$), the potential term of the scalar field (hence, the cosmological constant too) decreases more rapidly than the anisotropy potential (the Bianchi model spatial curvature). As a consequence, in this limit, it is natural to approximate the Universe evolution in terms of a Bianchi model in the presence of a non-interactive massless scalar field. This approximation classically has been verified for the most common inflationary potentials (see for instance the review on inflation [78]), but it is worth noting that for very steep potential terms (for instance exponential ones), it could happen that this scheme fails. On a quantum level, we will keep the simplification above as the most natural one and therefore we will disregard the effect of $V(\phi)$ in the Planck era.

9.3 Bianchi I model

In the Bianchi classification, the type I corresponds to $V_B \equiv 0$ and therefore the reduced ADM-Hamiltonian (9.8) takes the simplified form

$$H_{ADM} = \sqrt{p_r p_r} = \sqrt{p_+^2 + p_-^2 + p_\phi^2}\,. \tag{9.20}$$

Thus, the Hamilton equations, associated to this dynamics (known as *Kasner solution* [182]), acquire the form

$$\frac{d\beta_r}{d\alpha} = \frac{p_r}{\sqrt{p_s p_s}} \tag{9.21}$$

$$\frac{dp_r}{d\alpha} = 0 \Rightarrow p_r = const.\,, \tag{9.22}$$

from which the following explicit linear solution is obtained

$$\beta_r = \Pi_r \alpha + \bar{\beta}_r\,, \tag{9.23}$$

where the $\bar{\beta}_r$ are integration constants and we defined the new quantities

$$\Pi_r \equiv \frac{p_r}{\sqrt{p_s p_s}} = const. \Rightarrow \Pi_r \Pi_r = 1\,. \tag{9.24}$$

This solution describes the exact behavior of the Bianchi I model in the presence of a scalar field and its vacuum limit is directly obtained when $\Pi_\phi \equiv 0$. Each time interval, during which the dynamics of a generic Bianchi model is described by the solution above, is called *Kasner epoch* (or Kasner regime).

From a quantum point of view, the quantum dynamics of a Bianchi I model is described by the Wheeler-DeWitt equation (9.12), setting again $V_B \equiv 0$, *i.e.*

$$\hbar^2 \left[\frac{\partial^2}{\partial \alpha^2} - \frac{\partial^2}{\partial \beta_+^2} - \frac{\partial^2}{\partial \beta_-^2}\right] \psi(\alpha, \beta_\pm) = 0\,. \tag{9.25}$$

This is a Klein-Fock equation, describing the evolution of a massless scalar field, and its plane wave solution can be written as

$$\psi_k = \phi_k \exp\left[i\left(k_r\beta_r - k\alpha\right)\right] , \tag{9.26}$$

where k_r are constant wave-numbers, ϕ_k is a constant complex amplitude and eventually $k = \pm\sqrt{k_r k_r}$ (according to what was discussed in the ADM reduction procedure, the case of an expanding Universe corresponds to the choice of the sign $(-)$).

Since the Wheeler-DeWitt equation (9.25) is linear, the superposition principle holds and the general solution, describing the quantum dynamics of the Bianchi type I Universe, admits the following Fourier representation

$$\psi(\alpha, \beta_r) = \int d^3k \left\{\phi(k_r)\exp\left[i\left(k_r\beta_r + \sqrt{k_r k_r}\alpha\right)\right]\right\} , \tag{9.27}$$

where the integral is extended over all the admissible values of the wave-numbers k_r ($k_r \in (-\infty, \infty)$), and the function ϕ must be fixed by an initial condition in correspondence to a given initial Universe volume α_0, *i.e.* $\psi(\alpha_0, \beta_r) = \psi_0(\beta_r)$ (we note that $\phi(k_r)$ is obtained as the Fourier transform of $\psi_0(\beta_r)$).

It is important to stress that, since this wave-packet is equivalent to that of a relativistic massless particle, it is well-known that it unavoidably spreads as α increases. In other words, even if we start in α_0 with a narrow peaked (for instance Gaussian) profile, after a sufficiently long time (as the Universe expands enough), the wave-packet becomes more and more flattened. This fact does not erase at all the possibilities for a classical trajectory, since the spreading could define a sufficiently thin funnel [92]. Nonetheless, this situation opens the serious question about the possibility of a real classical limit of this quantum Universe. Actually, this picture clarifies how for the exact Bianchi I quantum Universe, such a possibility is forbidden, in the sense that the mean values continuously spread around the classical trajectory and a classical Bianchi I dynamics of the form (9.23) is never approached. We will address again this question in section 9.4.2, devoted to the Bianchi IX quantum dynamics and we will propose a solution for the classical limit puzzle in that more general case, when the potential term is present and it can play a significant role in this respect.

The choice of the minus sign in the determination of the quantity k is clearly a frequency separation procedure, here possible because the potential term vanishes identically. We now compare such a quantization scheme with the Schrödinger equation (9.15), which for the Bianchi I model reads as

$$i\frac{\partial\psi}{\partial\alpha} = \sqrt{-\frac{\partial}{\partial\beta_r}\frac{\partial}{\partial\beta_r}}\psi\,. \tag{9.28}$$

Hence, setting $\psi(\alpha, \beta_r) = e^{-ik\alpha}\chi(\beta_r)$, we get an eigenvalue problem, which, once squared, provides

$$\frac{\partial}{\partial\beta_r}\frac{\partial\chi}{\partial\beta_r} = -k^2\chi\,. \tag{9.29}$$

Taking $\chi(\beta_r) = \phi_k e^{ik_r\beta_r}$ (by using a coherent notation with the previous Wheeler-DeWitt case), we get $k = -\sqrt{k_r k_r}$ (now the choice of the minus sign is forced by the structure of the Schrödinger equation (9.28)). Then, constructing the corresponding wave-packet (by integration over k_r, as above), we clearly obtain the expansion (9.27), which tells us the correspondence existing, for the Bianchi I model, between the separation of frequencies in the Wheeler-DeWitt approach and the Schrödinger equation (for the latter the question of its locality is implicitly addressed). However, this correspondence between the two approaches cannot be directly extrapolated to the case $V_B \neq 0$, for which we recall that the frequencies separation cannot generally be addressed.

We conclude this section, observing that in the case $\beta_\pm \equiv 0$, the Bianchi I dynamics reduces to the flat Robertson-Walker geometry, which exists only in the presence of matter. Here matter is represented by the scalar field, but also the perfect fluid term $\mathcal{H}^m_{w<1}$, which can be included, plays an important role far enough from the initial singularity.

9.4 Bianchi IX model

The Bianchi IX model, together with the type VIII, is the most general geometry allowed by the homogeneity constraint which, in practice corresponds to deal with all the three constants C^c_{ab}, for $a \neq b \neq c$, different from zero. Near the singularity ($\alpha \to -\infty$), types VIII and IX models have an isomorphic classical dynamics, but as the Universe volume increases, only Bianchi IX admits an isotropic limit, *i.e.* the closed Robertson-Walker geometry (obtained for $\beta_\pm \equiv 0$). In fact, the Bianchi type IX is associated to a physical space which is invariant under the $SO(3)$ group of motion and therefore it is isomorphic to a three-dimensional hypersphere, but whose time evolution (expansion) is different along the three different coordinate lines.

Redefining the origin of the α variable as $\alpha \to \alpha - (1/4)\ln(6\pi)$ [226], the constant c [1] in the action (9.5) results fixed to the value $\sqrt{3\pi/8}$ and the model potential takes the explicit form

$$V_{IX} = \frac{1}{3}e^{-8\beta_+} - \frac{4}{3}e^{-2\beta_+}\cosh(2\sqrt{3}\beta_-) + \frac{2}{3}e^{4\beta_+}\left(\cosh(4\sqrt{3}\beta_-) - 1\right). \tag{9.30}$$

If we now introduce the notation

$$D \equiv e^{3\alpha} \tag{9.31}$$

$$Q_1 \equiv \frac{1}{3} + \frac{\beta_+ + \sqrt{3}\beta_-}{3\alpha} \tag{9.32}$$

$$Q_2 \equiv \frac{1}{3} + \frac{\beta_+ - \sqrt{3}\beta_-}{3\alpha} \tag{9.33}$$

$$Q_3 \equiv \frac{1}{3} - \frac{2\beta_+}{3\alpha}, \tag{9.34}$$

then, the potential term appearing in the Hamiltonian scheme, can be written as

$$\begin{aligned} e^{4\alpha}V_{IX} =& \frac{1}{3}\left(D^{4Q_1} + D^{4Q_2} + D^{4Q_3}\right) - \frac{2}{3}\Big(D^{2(Q_1+Q_2)} \\ &+ D^{2(Q_1+Q_3)} + D^{2(Q_2+Q_3)}\Big). \end{aligned} \tag{9.35}$$

Since $Q_i + Q_j < 2\,max(Q_i, Q_j)$ $(i \neq j)$, it is clear that the first parentheses of the expression above dominates and therefore, in the asymptotic limit to the singularity $\alpha \to -\infty$, the potential term reduces to an infinite well of the form

$$\lim_{\alpha \to -\infty} e^{4\alpha}V_{IX} \equiv U_\infty = \sum_{i=1}^{3}\Theta_\infty(Q_i), \tag{9.36}$$

where $\Theta_\infty(x > 0) = 0$ and $\Theta_\infty(x < 0) = \infty$. These potential walls have the form of an equilateral curvilinear triangle, with three corners at infinity (see figure 9.1).

Since such corners asymptotically correspond to the equality of two scale factors, *e.g.* one is fixed by the condition $\beta_- = 0$ and the other two are obtained for the rotational invariance of the picture (by acting with a rotation of $2\pi/3$). The equality of two scale factors leads to the Taub solution [272], which is dynamically unstable toward the general Bianchi IX behavior. Therefore, the triangular potential well can be regarded as dynamically closed, cutting in practice the three corners.

[1] Here we follow [226] in which geometric units $c = G = 1$ are adopted.

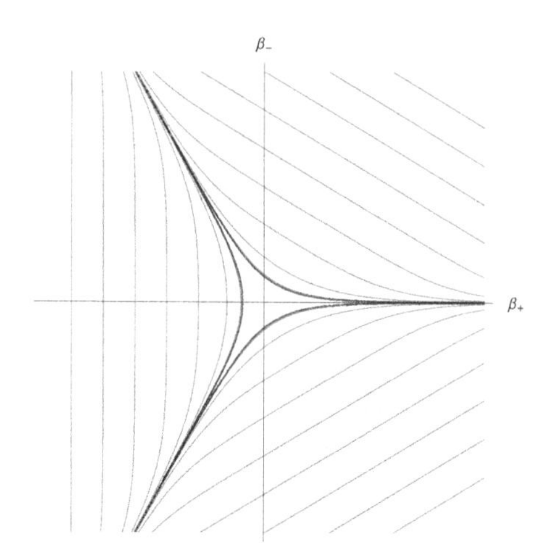

Figure 9.1 Equipotential lines of the Bianchi IX model in the plane $\beta_+\beta_-$: the bold lines are the walls of the infinite potential well in the limit $\alpha \to \infty$.

Clearly the potential walls move as α goes forward and in particular they expand, leaving a potential-free region of area proportional to α^2. To determine the velocity of the walls, we can study the simplest case, the wall $Q_3 = 0$ and then extrapolate its behavior to the other walls by the symmetries of the equilateral triangle. The condition $Q_3 = 0$ leads to the following behavior of the corresponding wall

$$-8\beta_+^{wall} + 4\alpha = 0 \Rightarrow \beta_+^{wall} = \frac{1}{2}\alpha\,,\ \beta_- = const.\,, \tag{9.37}$$

which implies that the potential well moves outward with the speed

$$\left|\frac{d\vec{\beta}^{wall}}{d\alpha}\right| = \frac{1}{2}\,. \tag{9.38}$$

Since far from the infinite walls, the point-Universe $\vec{\beta}^f = \{\beta_+^f, \beta_-^f\}$ move freely, according to the Bianchi I dynamics (9.23), we have to compare the wall velocity above with the following free-motion one

$$\left|\frac{d\vec{\beta}^f}{d\alpha}\right| = \sqrt{\left(\frac{d\beta_+}{d\alpha}\right)^2 + \left(\frac{d\beta_-}{d\alpha}\right)^2} = \sqrt{\Pi_+^2 + \Pi_-^2} = \sqrt{1 - \Pi_\phi^2}\,. \tag{9.39}$$

Thus, if we are in vacuum ($\Pi_\phi = 0$), the speed of the point-Universe, far from the potential walls, is equal to unity and therefore it exceeds the wall velocity. More specifically the speed of the point-Universe along the normal (outward oriented) to the walls is $\cos\theta$ (θ being the angle between the velocity vector and such a normal) and therefore the point-Universe reaches the potential only if $\theta < \pi/3$. However, the symmetry of the domain ensures that for $\theta > \pi/3$, the bounce is against another potential wall (corresponding to $\theta < \pi/3$).

This means that *the vacuum Bianchi IX classical evolution is constituted by an infinite series of Kasner regimes* (free motions of the point-Universe), *associated with continuous scatterings against the infinite potential walls.* The effect of the bounces against the potential can be studied with respect to a specific wall, *e.g.* the simplest $Q_3 = 0$ case. If we indicate with θ_i and θ_f the initial and final angles respectively, we obtain the following reflection law [224]

$$\sin\theta_f - \sin\theta_i = \frac{1}{2}\sin(\theta_i + \theta_f)\,. \tag{9.40}$$

This reflection rule is equivalent to the Belinski-Khalatnikov-Lifshitz map [49] and it possesses stochastic properties (Misner called this approximated representation of the Bianchi IX dynamics via infinite bounces, the *Mixmaster Universe* [224]). For instance, using a different set of variables (the so-called Misner-Chitré variables [226]), it is possible to demonstrate that the point-Universe performs ergodic trajectories in the configuration space and, as a consequence of the exponential instability of such trajectories, together with the compactness of the domain, *the dynamics results to be chaotic* [229].

If the scalar field is present, the situation is completely different, since for $\Pi_\phi \neq 0$, it is possible that the point-Universe is too slow to reach the potential walls. In that case, the existence of a stable Kasner regime toward the initial singularity comes out. Even if, in the very beginning, the parameters θ and Π_ϕ are not in the region ensuring the absence of a bounce, they will, soon or later, reach it, in view of the trajectories instability, which brings the system to invade all the available configuration space and then the parameter space too.

We conclude this brief review of the classical Bianchi IX dynamics by observing that considering a matter contribution in the form of a perfect fluid, associated to $w < 1$, does not qualitatively alter the scheme traced above nor in vacuum, nor in the presence of a scalar field.

9.4.1 *Quantum dynamics*

The exact solution of the Wheeler-DeWitt equation (9.12) in the presence of the potential term V_{IX} (9.30) is not known in analytic form and also the numerical simulations of the quantum evolution from an assigned initial condition do not offer a clear scenario of the Bianchi IX quantum behavior near the cosmological singularity (see for instance [133, 134]).

Nonetheless, following the original Misner work [225], we can extract some information about the asymptotic quantum Mixmaster Universe in vacuum, by using the ADM reduced quantum equation (9.15), approximated via the infinite potential well U_∞. Such an equation takes the explicit form

$$i\hbar\frac{\partial\psi}{\partial\alpha} = \left\{\sqrt{-\hbar^2\left(\frac{\partial^2}{\partial\beta_+^2} + \frac{\partial^2}{\partial\beta_-^2}\right) + U_\infty(\alpha, \beta_\pm)}\right\}\psi\,. \tag{9.41}$$

Taking the wave function in the form $\psi = \exp\{-\frac{i}{\hbar}\int^\alpha E(\alpha')d\alpha'\}\chi(\alpha, \beta_\pm)$ and squaring the eigenvalue problem, we eventually get

$$\left(\hbar^2\Delta_\pm - U_\infty\right)\chi = -E^2(\alpha)\chi\,, \tag{9.42}$$

where $\Delta_\pm$ denotes the Laplacian operator in the variables $\beta_\pm$. This simplification of the eigenvalue problem is possible if we make the adiabatic assumption (to be verified a posteriori) that the derivative of χ with respect to the variable α can be neglected.

This problem is isomorphic to the one of a free particle in an infinite potential well (α is a parametric variable), but it cannot analytically be solved because of the complexity of the triangular boundary conditions. Therefore, we can infer that the qualitative properties of the quantum Mixmaster Universe are preserved if we replace the triangular domain by a squared region, having the same area, *i.e.* of size $L_\alpha = l \mid \alpha \mid$, where l can be geometrically estimated, but its numerical value is irrelevant in such a qualitative analysis of the quantum dynamics.

Since the origin of the coordinates ($\beta_\pm = 0$ is placed in the center of the triangle, *i.e.* of the squared domain, the eigenfunctions χ take the simple form

$$\chi_{m,n}(\alpha, \beta_\pm) = A_{m,n}\sin\left(\frac{2\pi m\beta_+}{L_\alpha}\right)\sin\left(\frac{2\pi n\beta_-}{L_\alpha}\right)\,, \tag{9.43}$$

where $A_{m,n}$ is a constant amplitude and m and n are occupation numbers. The specific form of these modes is not so meaningful since they refer to an approximated domain of the potential term, nonetheless two main features are relevant:

i) The eigenvalue takes the explicit expression

$$E_{m,n}(\alpha) = \hbar\frac{2\pi}{L_\alpha}\sqrt{m^2 + n^2} \sim \frac{1}{\mid\alpha\mid}\,. \tag{9.44}$$

ii) The derivative of $\chi_{m,n}$ with respect to α can actually be disregarded, because it would generate a term $\sim (1/\alpha^2)\mathcal{O}(\chi_{m,n})$, which, for $\alpha \to -\infty$, is negligible in comparison to $E_{m,n}\chi_{m,n}$.

Now, a careful analysis of the classical dynamics allows to show [225] that the quantity $H_{ADM} \mid \alpha \mid$ is a constant of the motion. In fact, if we label by r the value of the dynamic quantities before a generic bounce, then we have $H^r_{ADM} \mid \alpha^r \mid = H^{r+1}_{ADM} \mid \alpha^{r+1} \mid$ and this relation can be indefinitely iterated. If we apply a basic principle of correspondence between classical and quantum mechanics, we can replace the classical constant of the motion H_{ADM} by $E_{m\gg1,n\gg1}$, *i.e.* the quantum Hamiltonian eigenvalue corresponding to very high occupation number approaches the classical value of the Hamiltonian itself. As an immediate implication, we get the following surprising result

$$E_{m\gg1,n\gg1} \mid \alpha \mid = \hbar\frac{2\pi}{l}\sqrt{m^2+n^2}\,,\; m \gg 1, n \gg 1\,, \qquad (9.45)$$

that, in other words, states the possibility to deal with high occupation numbers, *i.e.* semi-classical states, arbitrarily close to the initial singularity $\mid \alpha \mid \to \infty$ (the quantity $\sqrt{m^2+n^2}$ results in this way to be a constant of the motion).

However, we have to stress that this result does not imply that the quantum evolution remains localized around the classical trajectory (the cumbersome and chaotic matching of the infinite free motions between two bounces against the potential well), because a given wave packet spreads also during the free motion and the potential reflection is expected to enhance this effect. However the question about which is the asymptotic form of the wave function close to the singularity, when we start from a localized, for instance Gaussian, packet, is still open. It is not clear if the quantum Mixmaster model exhibits quantum chaos features or not and if the asymptotic state is really independent from the initial conditions.

Let us now analyze the quantum case of a Bianchi IX model in the presence of a scalar field. The classical no-bounce condition takes the form

$$\sqrt{1-\Pi_\phi^2}\,\cos\theta < \frac{1}{2} \qquad (9.46)$$

and, for $\Pi_\phi \neq 0$, it can be clearly satisfied in correspondence to the condition $\theta < \pi/3$. In other words, there are possible choices for the free Kasner parameters, such that the point-Universe is slower than the recession of the potential walls and the bounce can no longer occur.

From a quantum point of view, this means that it is possible to construct a quasi-classical wave packet, peaked around a classical no-bounce trajectory, which freely propagates, as long as it remains localized. In this sense, the box, in which the point-Universe is confined, no longer matters in the presence of the scalar field, and the evolution toward the singularity is sensitive to the initial conditions. Clearly, if the initial parameters of the quasi-classical state are out of the no-bounce zone, we can expect that a certain number of packet reflections takes place before the evolution is potential-free. If these reflections spread the packet within the box, the physical meaning of the final potential-free quantum evolution is elusive.

Thus, on a quantum level, the initial picture of a quantum Universe seems to be not yet well-traceable, since there is a strong dependence on the initial conditions. However, we observe that, if the scalar field kinetic energy dominates the anisotropy dynamics, we can estimate when a quasi-classical wave packet, corresponding to a no-bounce behavior, appears. If the packet is significantly different from zero in a domain of area $\Delta\beta$ in the $\beta_{\pm}$-plane, the condition for its existence reads as

$$p_{\phi}^2 \gg e^{4\alpha}\langle V_{IX}(\beta_{\pm})\rangle \,, \tag{9.47}$$

where $\langle V_{IX}\rangle$ denotes the mean value of the potential over the region $\Delta\beta$, taken by the packet wave function with a suitable scalar product. Given the values of p_{ϕ} (this is a quasi-classical momentum value, corresponding to $\hbar\bar{k}_{\phi}$, $\bar{k}_{\phi}$ being the wave number around which the packet is peaked) and $\Delta\beta$, and assuming the simplest Gaussian form for a localized wave packet, the condition above is clearly satisfied for a sufficiently large and negative value of the variable α, *i.e.* sufficiently close to the singularity.

We see how the primordial quantum picture of the Universe is sensitive to the presence of an inflaton scalar field and how relevant is its initial kinetic energy with respect to the anisotropy evolution. As soon as the inflaton plays a role in the Planck era, the initial state of the spatial metric appears determined by specific initial conditions, more than by the intrinsic properties of the quantum dynamics, as the vacuum case suggests.

9.4.2 *Semiclassical behavior*

We now implement the Vilenkin idea (discussed in section 6.4.3.4) for the quantum dynamics of the Mixmaster model in order to investigate the implications of a decoherence behavior of a part of the Universe. We will consider as the configuration variable approaching the quasi-classical limit before all the others, the Universe volume, *i.e.* the time-like variable α. This choice cannot be strictly motivated, but it appears as a well-grounded hypothesis in view of the peculiar nature of the Universe volume: it does not correspond to a physical degree of freedom of the cosmological gravitational field, but, as we have seen in the ADM-reduced scheme, it plays the role of a label time for the evolution of the anisotropic variables $\beta_\pm$, which are the real degrees of freedom. Let us start from the Wheeler-DeWitt equation for the Bianchi IX Universe, in the presence of ultrarelativistic matter and a cosmological constant (in this phase we can neglect the role played by the kinetic term of the scalar field), *i.e.*

$$\frac{\partial^2\psi}{\partial\alpha^2} - \frac{\partial^2\psi}{\partial\beta_+^2} - \frac{\partial^2\psi}{\partial\beta_-^2} + e^{4\alpha}V_{IX}(\beta_\pm)\psi + e^{6\alpha}\mu_{-1}^2\psi + e^{-\alpha}\mu_{1/3}^2\psi = 0\,. \quad (9.48)$$

Despite the analysis developed below applies in general, we will consider the particular limit in which the anisotropy variables $\beta_\pm$ take very small values and the quantum dynamics lives nearby the isotropic Universe. In that case the potential term takes the simple quadratic form

$$V_{IX}(\beta_\pm) = -1 + 8\left(\beta_+^2 + \beta_-^2\right) \quad (9.49)$$

and the separation of the isotropic dynamics from that of the anisotropies appears in a fully consistent scheme à la Vilenkin (see section 6.4.3.4), based on a decoherence process of the Universe volume. Thus, according to this idea, we can take the wave function in the following form

$$\psi(\alpha, \beta_\pm) = e^{iS(\alpha)/\hbar}\sigma(\alpha, \beta_\pm)\,, \quad (9.50)$$

which, once substituted into (9.48) at the zeroth order in $\hbar$ (we neglect the second derivative of σ with respect to α), leads to the following conditions for the functions S and σ:

$$-\left(\frac{\partial S}{\partial\alpha}\right)^2 - e^{4\alpha} + e^{6\alpha}\mu_{-1}^2 + e^{-\alpha}\mu_{1/3}^2 = 0 \quad (9.51)$$

$$2i\hbar\frac{\partial S}{\partial\alpha}\frac{\partial\sigma}{\partial\alpha} - \hbar^2\left(\frac{\partial^2}{\partial\beta_+^2} + \frac{\partial^2}{\partial\beta_-^2}\right)\sigma + 8e^{4\alpha}\left(\beta_+^2 + \beta_-^2\right)\sigma = 0\,. \quad (9.52)$$

We can see that (9.51) is nothing more than the classical Hamilton-Jacobi equation for a closed Robertson-Walker Universe in the presence of a cosmological term and an ultrarelativistic thermal bath contribution. Instead, the relation (9.52) describes the quantum dynamics of the anisotropy degrees of freedom and it can be easily rewritten in a more expressive form. In fact, recalling that in the Hamilton-Jacobi formalism the identification $\partial_\alpha S \equiv p$ takes place and that the momentum p can be calculated by (9.7) for $c = (3\pi/2)^{1/2}$, we get the final relation

$$\partial_\alpha S = -\frac{\dot\alpha e^{3\alpha}}{\sqrt{6\pi}N}\,. \quad (9.53)$$

Now, since the variable α follows a classical dynamics, according to (9.51), then, substituting into the equation above the form of $\partial_\alpha S$, we can, in principle, calculate the function $\alpha(t)$, of course after the function $N(t)$ has been specified. Thus, it is natural to make the following identification $\dot{\alpha}\partial_\alpha\sigma \equiv \partial_t\sigma$. As a consequence, we can rewrite equation (9.52) in the following Schrödinger form

$$i\hbar\partial_t\sigma = N(t)e^{-3\alpha(t)}\sqrt{\frac{3\pi}{2}}\left[-\hbar^2\left(\frac{\partial^2}{\partial\beta_+^2}+\frac{\partial^2}{\partial\beta_-^2}\right)\sigma + 8e^{4\alpha(t)}\left(\beta_+^2+\beta_-^2\right)\sigma\right]. \tag{9.54}$$

If we now introduce the new time variable $T(t)$, defined as

$$T(t) = \int^t N(t')e^{-3\alpha(t')}\sqrt{\frac{3\pi}{2}}dt' \tag{9.55}$$

and we define $\omega(T) \equiv \sqrt{8}e^{2\alpha(T)}$, then equation (9.54) takes the form

$$i\hbar\partial_T\sigma = -\hbar^2\left(\frac{\partial^2}{\partial\beta_+^2}+\frac{\partial^2}{\partial\beta_-^2}\right)\sigma + \omega^2(T)\left(\beta_+^2+\beta_-^2\right)\sigma\,. \tag{9.56}$$

This equation is clearly that of a quantum harmonic oscillator, having a time dependent frequency.

This result is relevant for two main reasons: i) the harmonic oscillator admits non-spreading semiclassical states and offers a reliable paradigm for the semiclassical limit of the Universe dynamics; ii) in [46] it has been shown how in such a restricted quantum scheme, the anisotropy degrees of freedom are probabilistically suppressed and the classical limit corresponds to the closed Robertson-Walker geometry.

Thus, the semiclassical Vilenkin formalism provides a valuable scenario leading the quantum era, dominated by the anisotropy dynamics, to the classical isotropic dynamics.

This scheme can even be enforced in the presence of an inflationary mechanism. In fact, if the volume exponentially expands, the frequency of the time dependent quantum oscillator increases too and the mean values of the variables $\beta_\pm$ correspondingly decrease.

However, in this picture, a relevant question remains open: how is it possible to make the transition in the quantum regime between a phase with large anisotropy fluctuations to the quasi-isotropic configuration? A clear mechanism for explaining such restriction of the quantum Bianchi IX model to a quantum fluctuation nearby the classical Robertson-Walker geometry is not known. We can only note the progressive increasing of the time factor in front of the anisotropy potential as the Universe expand. However, that this effect can be selective in the $\beta_\pm$ space is not justified.

9.4.3 *The quantum behavior of the isotropic Universe*

In this section, we will trace some basic features of the isotropic Universe (in this scheme the isotropic limit of the Bianchi IX model, *i.e.* the closed Robertson-Walker geometry, see section 1.12), which, in the spirit of our present discussion, will appear as a much simplified model, a bit poor in physical content.

In the assumption $\beta\pm \equiv 0$ (and hence $p_\pm \equiv 0$), the classical Hamiltonian constraint reduces to the simple form (we retain in the quantum description a cosmological constant μ^2_{-1})

$$-e^{-3\alpha}p^2 - e^{\alpha} + \mu^2_{-1}e^{3\alpha} = 0\,. \tag{9.57}$$

Now, we can come back to the scale factor variable $a(t)$, by implementing the canonical transformation $\{a = e^{\alpha}\,,\, ap_a = p\}$, p_a being the conjugate momentum to the variable a. In this new set of variables, the classical Hamiltonian constraint can be rewritten as

$$-\frac{p_a^2}{a} - a + \mu^2_{-1}a^3 = 0\,, \tag{9.58}$$

and it can be restated in a more expressive form, as follows

$$p_a^2 + V(a) = 0\,, \quad V(a) \equiv a^2\left(1 - \frac{a^2}{a_0^2}\right)\,, \tag{9.59}$$

where we defined $a_0 \equiv \sqrt{1/\mu^2_{-1}}$. This Hamiltonian is isomorphic to that of a non-relativistic one-dimensional particle, subjected to the potential energy V and having a vanishing total energy (here the scale factor a plays the role of the generalized coordinate) [62]. The classical evolution of the system can take place only in the region where the potential $V(a)$ is negative, *i.e.* $a > a_0$, according to the positive nature of p_a^2. It is easy to check that such a classical cosmology corresponds to a Big-Bounce feature, being associated with a collapse of the Universe up to the minimal (non-vanishing) value of the scale factor $a = a_0$ granted by the presence of the cosmological constant and the successive re-expansion towards an infinite volume configuration (if ordinary matter and radiation contribution were included in the dynamics, a turning point in the future could be present too).

However, we observe that the isolated point of the Universe phase-space $\{a = 0, p_a = 0\}$ also satisfies the Hamiltonian constraint (9.59). The value $a = 0$ is separated by the classically permitted region $a > a_0$ by a potential barrier, since the profile $V(a)$ has a maximum (positive) value at $a = a_0/\sqrt{2}$ (see figure 9.2).

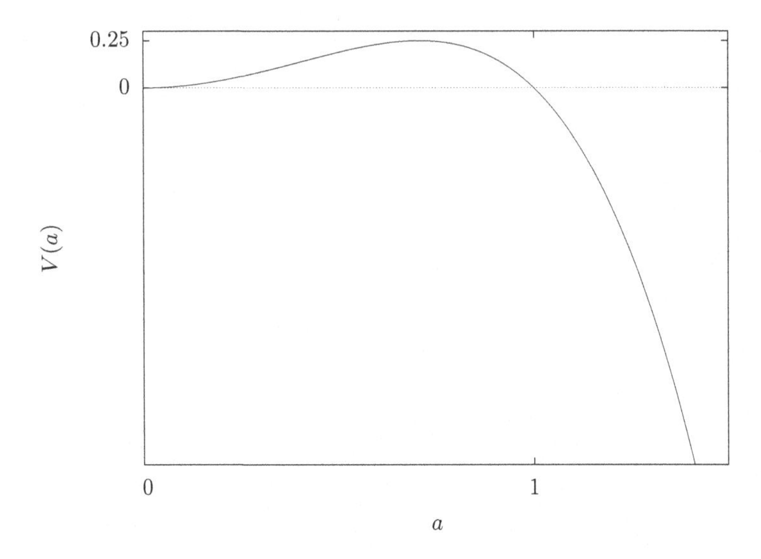

Figure 9.2 The potential $V(a)$ for $a_0 = 1$: it is worth noting the classically forbidden region $V(a) > 0$ for $0 < a < 1$.

Therefore, there is no way to classically connect such an isolated point with the classical evolution of the isotropic Universe and we also note how the values $a = 0$ and $p_a = 0$ would not be easily interpreted as a physical configuration, since they describe a zero-volume and zero-expansion Universe; for this configuration it is worth to adopt the suggestive notion of "nothing", in the sense of a phase-space point well-describing the absence of a real Universe.

The quantum dynamics of the considered model is again determined by the corresponding Wheeler-DeWitt equation, associated to the wave function $\psi = \psi(a)$, namely in the natural operator ordering (with momenta on the right)

$$-\hbar^2 \frac{\partial^2 \psi}{\partial a^2} + V(a)\psi = 0\,, \tag{9.60}$$

which can be accompanied by the Vilenkin boundary condition $\psi(a = 0) = 0$.

This equation can be analytically solved, providing a series of physical states for the quantum configuration of the present model. However, a deep question arises, as far as we recognize that such states are different from zero even in the classically forbidden region $a < a_0$. In fact, it is natural to infer that the Universe can make the transition from the "nothing" configuration to a classical permitted one by performing a tunneling process across the potential barrier. Is such a hypothesis physically well-grounded? Although such a perspective has been considered in some classical papers [162], we here address a critical point of view on the reliability of such an imaginative interpretation: i) the Universe wave function depends on the scale factor only, which we discussed in the previous section as having the proper features of a time-like (not a space-like as inferred here) variable. In this sense, the quantization of a homogeneous and isotropic model has no clear physical meaning, since no real gravitational degree of freedom is here quantized. Indeed, in the previous section, we inferred the perspective that this variable (the Universe volume) has an intrinsic classical nature, within a decoherence picture of the Bianchi IX Universe; ii) even if we accept the interpretation of the variable a as a generalized coordinate, nonetheless, the notion of a tunneling transition of the Universe appears not well-grounded. In fact, the tunneling effect is an intrinsically non-reversible process, therefore capable to see the arrow of the time. But, what is time in such a scheme? The Wheeler-DeWitt equation is affected, as discussed in section 6.4.3, by the so-called "frozen formalism" problem, *i.e.* no real evolution is present along the spacetime slicing in the quantum picture.

These difficulties could be partially addressed by introducing a relational clock, made up with a matter field, for instance a massless scalar field (the kinetic term of the inflaton field), but the possibility to deal with the Universe volume as a space-like degree of freedom of the configuration space is hard to be accepted in the spirit of the analysis we developed for the homogeneous Bianchi model (we recall that the scalar field behaves in the kinetic term of the super-Hamiltonian (9.17) like the anisotropic degrees of freedom, *i.e.* it has a space-like character). In this respect, we can infer that the quantization of a homogeneous and isotropic Universe must be mainly regarded as a toy model on which the different quantum gravity approaches can be easily tested, but whose real physical meaning is to be searched in the semi-classical limit of a more general anisotropic cosmology, where the quantum behavior of the gravitational degrees of freedom is really determined.

9.5 BKL conjecture

The classical and quantum dynamics of the Bianchi IX model is of great cosmological interest because it is commonly believed that it well mimics the local behavior of a generic inhomogeneous model near the cosmological singularity. This concept is at the core of the so-called *Belinski-Khalatnikov-Lifshitz* (BKL) conjecture [50]. On the classical level, such a conjecture has been properly argued in [184, 227], where it is shown however how the generation of increasing small spatial scales in the evolution of the system toward the singularity induces spacetime turbulence: the chaoticity of the time evolution is reflected in a stochastic behavior of the spatial dependence too, but preserving the BKL conjecture on a sufficiently small region. On a quantum level the BKL conjecture remains mainly a hypothesis to be implemented in the dynamic superspace, that is reduced to the product of local minisuperspaces, defined on sufficiently small regions of space (ideally space points). This statement is equivalent to say that the space metric evolve, near the singularity, independently in each point, according to the local value of the probability amplitude.

The physical meaning of the BKL conjecture must be recognized in the monotonic decrease that the volume element exhibits in the oscillatory regime, disregarding the complicated, but smooth in time, behavior of each space direction. In other words, the physical inhomogeneous characteristic scale of a given cosmological model decreases in time slower than the average horizon. As a result, sufficiently close to the singularity, the inhomogeneous scale becomes a super-horizon sized one and the local homogeneous behavior is recovered in each neighborhood of a space point.

The spatial line element of a generic inhomogeneous cosmological model can be written in the form

$$dl^2 = (a^2 l_i l_k + b^2 m_i m_k + c^2 n_i n_k) dx^i dx^k \,, \tag{9.61}$$

where the spatial vectors $\vec{l}$, $\vec{m}$ and $\vec{n}$ are generic ones (*i.e.* their components are generic functions of the spatial coordinates), while the scale factors a, b and c depend on space coordinates and time. Since a generic solution of the Einstein's equations has to contain four physically arbitrary functions of spatial coordinates (the gravitational field has two physical degrees of freedom and satisfies a second order partial differential problem), a *generalized Kasner solution* of the form [201]

$$a \sim t^{p_l(x)}\,, \quad b \sim t^{p_m(x)}\,, \quad c \sim t^{p_n(x)}\,, \tag{9.62}$$

$$p_l + p_m + p_n = p_l^2 + p_m^2 + p_n^2 = 1 \tag{9.63}$$

can be shown to correspond to such a requirement of generality and can be subjected to an arbitrary Cauchy initial value problem. In fact, this solution formally contains nine arbitrary functions of the space coordinates coming from the three vectors and one from the three Kasner indices (subjected to two independent conditions). These ten functions must fulfill the three constraints associated with the G_{0k} components of the vacuum Einstein's equations, the G_{00} providing the second Kasner relation on the indices. Furthermore, we can eliminate other three functions by the choice of the three space coordinates. Thus, the real number of functions available to the Cauchy problem is equal to the four, and the relevance of a generalized Kasner regime comes out in all its power towards the construction of an oscillatory regime for a generic inhomogeneous cosmological model.

Indeed, like in the Bianchi IX model, this regime is not able, in general, to reach the singularity, because the spatial curvature of a generic model always contains a disturbing term, corresponding to the negative index $p_1(x^k)$ and a transition to a new Kasner regime takes place, according to the same BKL map valid for the homogeneous case, referred to each space point (this scheme clearly holds only if the inhomogeneity scale remains super-horizon sized). From a Hamiltonian point of view, such a parametric role of the space coordinate in the dynamics of the Universe near the cosmological singularity, is translated by extending in a point-like manner the Misner representation, leading to the so-called *inhomogeneous Mixmaster model.*

In the inhomogeneous case, new features arise:

- a rotation of the coordinate vectors, associated to the BKL map, *i.e.* to the collision of the point-Universe against one of the triangular potential walls (in each point the potential profile is isomorphic to the homogeneous model),
- the time chaoticity of the model couples to the asymptotic behavior of the space points (each point follows an independent chaotic evolution from any other, even arbitrarily close), but the structure of the Mixmaster dynamics conserves locally all the properties briefly depicted in section 9.4.

To implement this generic picture on a quantum level, using Misner variable, is equivalent to decompose the wave functional $\Psi(\alpha(x), \beta_\pm(x))$ to the simplified (heuristic) form

$$\Psi\left(\alpha(x), \beta_\pm(x)\right) = \prod_x \psi_x(\alpha^x, \beta_\pm^x), \tag{9.64}$$

where α^x and $\beta_\pm^x$ are now three (local) degrees of freedom, on which the wave function ψ_x is taken. All the considerations developed in section 9.4.1 apply to this local reduced problem and we find the same issues on the Universe birth, but now referred to a roughly homogeneous causal patch of the space. The Universe emerges from the singularity in the same form at each point of space, but the quantum dynamics (*i.e.* the expectation values of the observables) are totally uncorrelated among different points.

Despite this picture of the quantum Universe is extremely suggestive and it relies on the coincidence on average between the notion of local homogeneity and causality, nonetheless, it is worth emphasizing that, differently from the classical case, here no rigorous (even a posteriori) proof exists that the spatial gradients can be on average neglected in the asymptotic limit. Therefore the Universe birth, here roughly determined in terms of Bianchi I-like solutions of the Wheeler-DeWitt equation, remains only a fascinating heuristic proposal emerging from the original BKL conjecture.

9.6 Cosmology in LQG

We have seen in the previous chapter how although LQG provides some steps further in the quantization of the gravitational field (mainly on a kinematical level) with respect to the WDW framework, nevertheless some technical issues still prevent a proper analysis of the dynamic implications. Hence, in order to extract some predictions from such a Quantum Gravity approach, researchers focused their attention to symmetry reduced models and, in particular, to those describing cosmological models. We are now going to present the main framework in which the cosmological sector of LQG has been implemented, namely *Loop Quantum Cosmology* (LQC) [38, 67]. LQC is based on mimicking the LQG quantization procedure for the relevant degrees of freedom of cosmological models. The simplification of the dynamic issue (with respect to the one of the full theory) allows to solve completely the scalar constraint in some physically relevant cases. Hence, LQC gives a complete picture for the dynamics of the early Universe, within which the replacement of the initial singularity with a bounce stands as the most outstanding result [33]. It is also worth to mention the set-up of proper initial conditions for inflation to start [39] and the phenomenological implications on the cosmic microwave background radiation spectrum [2,68] as significant achievements of LQC (even though the theoretical framework of perturbation theory in LQC is still under debate).

However, we will also outline some shortcomings of the theory, mainly concerning its foundations on the whole LQG framework. The tension between LQG and LQC is essentially due to the fact that in the latter kinematical symmetries are fixed on a classical level. On the contrary in LQG symmetries play a crucial role, since $SU(2)$ gauge invariance is responsible for the presence of invariant intertwiners at vertices, while background independence enters the regularization of the scalar constraint. In fact, a controversial point of LQC is precisely the regularization of the scalar constraint. In this respect, we will present also a novel proposal [4, 5] in which the quantum kinematics for the Bianchi I model is directly derived from the Hilbert space of the full theory, such that all the techniques adopted in LQG can also be used in its symmetric sector.

9.6.1 *Loop Quantum Cosmology*

Let us discuss the main features of LQC starting from the isotropic Robertson-Walker line element (2.1). The full geometrical structure on spatial hypersurfaces can be written as follows

$$h_{ij} = a^2(t)\gamma_{ij}, \tag{9.65}$$

in which $a = a(t)$ is the scale factor, while the fiducial metric γ_{ij} is time-independent and its form is preserved by the action of the symmetry group proper of spatial hypersurfaces (for instance in the case $k = 1$, γ_{ij} is the metric of the three-sphere with radius 2, while for $k = 0$ it is a Minkowskian one).

A convenient (but not the only one) set of dreibein vectors associated with the metric (9.65) reads

$$e^a_i = a(t)\,\omega^a_i, \tag{9.66}$$

in which ω^i_a denotes fiducial dreibein vectors of γ_{ij}, *i.e.*

$$\gamma_{ij} = \delta_{ab}\,\omega^a_i\,\omega^b_j. \tag{9.67}$$

The Ashtekar-Barbero connections and the inverse densitized triads take the following expressions

$$A^a_i = C(t)\,\omega^a_i, \qquad E^i_a = P(t)\,\sqrt{\gamma}\omega^i_a\,. \tag{9.68}$$

Hence, a useful set of phase-space coordinates is given by $\{C, P\}$, whose explicit expressions read

$$|P| = a^2, \qquad C = k + \gamma\dot{a}\,, \tag{9.69}$$

and the Poisson brackets are given by

$$[C(t), P(t)] = -\frac{8\pi G\gamma}{3c^3 V_0}, \tag{9.70}$$

V_0 being the fiducial volume of the region of space Ω_0 we are considering.

It is worth noting how both the Gauss and the vector constraints are identically solved by the expression (9.68) for the connections and the momenta. Therefore, the kinematical symmetries are lost already on a classical level.

The holonomies can be evaluated starting from those associated with straight edges parallel to ω^i_a, thus obtaining

$$U^\mu_a(A) = e^{i\mu C\tau_a}, \tag{9.71}$$

μ being the length of the edge in the fiducial metric. Form the expression above one gets that the matrix elements of holonomies are determined by *quasi-periodic functions* $N_\mu = e^{\frac{i\mu C}{2}}$. These functions are taken as basis elements of the configuration space. In particular, the algebra generated by $\{N_\mu, P\}$ plays here the same role as the holonomy-flux algebra in the full theory and by the analogous construction of the general case the Hilbert space turns out to be[2]

$$\mathrm{H} = L^2(\mathbb{R}_B, d\mu_{Bohr}), \tag{9.72}$$

$\mathbb{R}_B$ being the Bohr compactification of the real line (see section 5.4.3). An orthonormal basis is given by N_μ themselves:

$$\langle N_{\mu'}|N_\mu\rangle = \delta_{\mu',\mu}. \tag{9.73}$$

On a quantum level, the action of operators is given by

$$N_\mu\psi(C) = e^{\frac{i\mu C}{2}}\psi(C), \qquad \hat{P}\psi(C) = i\frac{8\pi\gamma l_P^2}{3V_0}\frac{d}{dC}\psi(C), \tag{9.74}$$

so that some states $|\mu\rangle$ can be defined

$$\langle C|\mu\rangle = e^{\frac{i\mu C}{2}}, \tag{9.75}$$

for which

$$N_{\mu_0}|\mu\rangle = |\mu + \mu_0\rangle, \qquad \hat{P}|\mu\rangle = -\frac{8\pi\gamma l_P^2}{6V_0}\mu|\mu\rangle. \tag{9.76}$$

It is worth noting how, because of the discrete topology (9.73), the operator corresponding to C is not defined (as in polymer quantization).

[2]This Hilbert space is the analogous of H^{Kin} in LQG, but also of H^{Diff}, since no kinematical constraint is present.

In what follows, let us restrict to the $k = 0$ case, in which the fiducial metric is flat and $\omega^i_a = \delta^i_a$ (for $k = \pm 1$ see [36, 271] and [270, 288]).

The classical scalar constraint for the flat Robertson-Walker model can be inferred from the one of the full theory (7.117), by substituting the expressions of the connections and of the momenta of the reduced model (9.68). It turns out that the Lorentzian part is proportional to the Euclidean part. This can be seen from the expression of $\mathcal{S}$ in (7.106) by noting that $\bar{\omega}^{ab}_i$, thus also $\bar{R}^{ab}_{ij}$, vanishes for $k = 0$, such that the full scalar constraint reads

$$\mathcal{S} = -\frac{E^i_a\, E^j_b}{\gamma^2\, \bar{e}}\varepsilon^{ab}_{\ \ c}\, F^c_{ij} \approx 0\,. \tag{9.77}$$

Hence, it's enough to quantize the Euclidean piece only and in this respect we have to replace E^i_a with $P\delta^i_a$, the operator $\hat{P}$ given in (9.76), and to express the curvature in terms of holonomies (9.71). Let us take the holonomies entering the scalar constraint in the fundamental representation $j = 1/2$, such that the (Hermitian) $SU(2)$ generators are $\tau_a = \frac{1}{2}\sigma_a$, σ_a being Pauli matrices. The curvature can be written as

$$F^a_{ij} = -2\,Tr\left(\lim_{\mu\to 0}\frac{U_{\Box_{bc}} - I}{\mu^2}\tau_a\right)\delta^b_i\delta^c_j, \tag{9.78}$$

$\Box_{bc}$ being the square with area μ^2 and edges parallel to the vectors δ^b_i, δ^c_i. The quantization of the term $1/e$ can be addressed as in full theory by mimicking Thiemann's trick (see section 8.5), thus finding

$$\varepsilon^{abc}\,\frac{E^i_b E^j_c}{e} = \sum_c i\,\frac{sign(P)}{2\pi\gamma G\mu V_0}\,\varepsilon^{ijk}\,\omega^c_k\,Tr\left(U^\mu_c\,[V, U^{\mu\,-1}_c]\,\tau_a\right), \tag{9.79}$$

$V = V_0|p|^{3/2}$ being the total volume. The two expressions (9.78) and (9.79) are written in terms of quasi-periodic functions and momenta, thus they can be directly quantized, getting the following quantum scalar constraint

$$\hat{\mathcal{S}} = \lim_{\mu\to 0}\hat{\mathcal{S}}^\mu, \tag{9.80}$$

$$\begin{aligned}\hat{\mathcal{S}}^\mu =& sign(P)\frac{24i}{8\pi\gamma^3\mu^3 l_P^2}\sin^2(\mu C)\left[\sin\left(\frac{\mu C}{2}\right)|\hat{P}|^{3/2}\cos\left(\frac{\mu C}{2}\right)\right.\\ &\left.-\cos\left(\frac{\mu C}{2}\right)|\hat{P}|^{3/2}\sin\left(\frac{\mu C}{2}\right)\right].\end{aligned} \tag{9.81}$$

However, this limit does not exist. In LQG, the scalar constraint is regularized over s-knot states, which belong to the dual of the kinematical Hilbert space. However, here, background independence does not hold and one cannot introduce s-knots. Hence, a different regularization must be addressed. This feature marks the main difference between the LQG and LQC quantization procedures and it is usually interpreted as reflecting the presence of an underlying discrete structure (coming from the full theory). In particular, the regularization procedure of LQC implies to evaluate $\hat{\mathcal{S}}$ at a fixed value $\mu = \bar{\mu}$ as

$$\hat{\mathcal{S}}^{reg} = \lim_{\mu \to \bar{\mu}} \hat{\mathcal{S}}^{\mu} = \hat{\mathcal{S}}^{\bar{\mu}}. \tag{9.82}$$

We will comment in the following on how to fix the actual value of $\bar{\mu}$. Let us now assume for simplicity $\bar{\mu}$ to be a constant (this choice corresponds to the one made in [34]).

The expression (9.82) can be made Hermitian by a proper factor ordering and a possible choice (see [64] for a different ordering) implies to rewrite it as

$$\begin{aligned}\hat{\mathcal{S}}^{\bar{\mu}} =& \sin(\bar{\mu}C)\left[sign(P)\frac{24i}{8\pi\gamma^3\bar{\mu}^3 l_P^2}\left(\sin\left(\frac{\bar{\mu}C}{2}\right)|\hat{P}|^{3/2}\cos\left(\frac{\bar{\mu}C}{2}\right)\right.\right.\\ &\left.\left.-\cos\left(\frac{\bar{\mu}C}{2}\right)|\hat{P}|^{3/2}\sin\left(\frac{\bar{\mu}C}{2}\right)\right)\sin(\bar{\mu}C)\right].\end{aligned} \tag{9.83}$$

If one expands a generic state in the basis $|\mu\rangle$

$$|\psi\rangle = \sum_{\mu} \psi(\mu)|\mu\rangle, \tag{9.84}$$

the scalar constraint acts as

$$\hat{\mathcal{S}}^{\bar{\mu}}\psi(\mu) = f_+(\mu)\psi(\mu + 4\bar{\mu}) + f_0(\mu)\psi(\mu) + f_-(\mu)\psi(\mu - 4\bar{\mu}), \tag{9.85}$$

with

$$f_+(\mu) = \frac{l_P}{2\bar{\mu}^3}\sqrt{\frac{8\pi}{6\gamma^3}}\left||\mu + 3\bar{\mu}|^{\frac{3}{2}} - |\mu + \bar{\mu}|^{\frac{3}{2}}\right|, \tag{9.86}$$

$$f_-(\mu) = f_+(\mu - 4\bar{\mu}) \qquad f_0(\mu) = -f_+(\mu) - f_-(\mu). \tag{9.87}$$

At this level, among all admissible graphs, one restricts the sum in (9.84) to the lattices $\mathcal{L}_\epsilon = \{\mu | \mu = \epsilon + 4n\bar{\mu}, n \in Z\}$, which are preserved under the action of $\mathcal{S}^{\bar{\mu}}$ (superselected sectors). Therefore, it is a well-defined procedure to study the dynamics within the associated Hilbert space H_ϵ (*i.e.* the one of the functionals having support on $\mathcal{L}_\epsilon$). Furthermore, H_ϵ is separable. The description in terms of a countable set of basis vectors is the main achievement with respect to Wheeler-DeWitt quantum cosmology. This also implies *the replacement of the Wheeler-DeWitt differential equation with a difference one.*

However, as we will see, there are some reasons to fix $\bar{\mu}$ as a function of μ. In this case it is convenient to label the basis $|\mu\rangle$ via the eigenfunctions $v = v(\mu)$ such that

$$\bar{\mu}\frac{dv}{d\mu} = 1. \tag{9.88}$$

This way, the operators $e^{i\frac{\bar{\mu}c}{2}}$ appearing in the scalar constraint $\mathcal{S}^{\bar{\mu}}$ drag the state $\psi(v)$ a unit affine parameter along d/dv, *i.e.*

$$e^{i\frac{\bar{\mu}c}{2}}\psi(v) = \psi(v+1). \tag{9.89}$$

The basis vectors $|v\rangle$ associated with a decomposition in terms of $\psi(v)$ are proportional to $|\mu\rangle$

$$|v\rangle = v(\mu)|\mu\rangle, \tag{9.90}$$

via a constant that can be determined by solving (9.88). As for momenta, their action is the same as in (9.76)

$$\hat{P}|v\rangle = \frac{8\pi\gamma l_P^2}{6}\mu(v)|v\rangle. \tag{9.91}$$

Let us consider the case discussed in [35]

$$\bar{\mu} = \frac{3^{3/4}}{2^{1/2}|\mu|^{1/2}}, \tag{9.92}$$

which corresponds to

$$v(\mu) = K sign(\mu)|\mu|^{3/2}, \qquad K = \frac{2^{3/2}}{3^{7/4}}, \tag{9.93}$$

and in this basis the operator $|\hat{P}|^{3/2}$ acts as follows

$$|\hat{P}|^{3/2}|v\rangle = \frac{3^{7/4}}{2^{3/2}}\left(\frac{8\pi\gamma}{6}\right)^{3/2} l_P^3|v||v\rangle. \tag{9.94}$$

The scalar constraint (9.83) now reads

$$\hat{\mathcal{S}}^{\bar{\mu}}\psi(v) = f_+(v)\psi(v+4) + f_0(v)\psi(v) + f_-(v)\psi(v-4), \tag{9.95}$$

with

$$f_+(v) = \frac{3^{5/4}}{2^{5/2}}\sqrt{\frac{8\pi}{6}}\frac{l_P}{\gamma^3/2}|v+2|\Big||v+1| - |v+3|\Big|, \tag{9.96}$$

$$f_-(v) = f_+(v-4) \qquad f_0(v) = -f_+(v) - f_-(v). \tag{9.97}$$

The superselected sectors are now the lattices $\mathcal{L}_\epsilon = \{v | v = \epsilon + 4n, n \in Z\}$.

9.6.1.1 *The Big-Bounce*

In order to build a realistic description of the Universe one has to address the issue of standard quantum cosmological paradigm, in particular the problem of time. This is usually done in LQC by introducing a massless clock-like scalar field ϕ, such that the full scalar constraint reads

$$\mathcal{S}_{tot} = \mathcal{S} + 8\pi G \frac{p_\phi^2}{|p|^{3/2}}, \tag{9.98}$$

p_ϕ being the momentum associated with ϕ. The scalar field is canonically quantized and the total wave function $\Psi = \Psi(C, \phi)$ depends both on c and ϕ. The momentum p_ϕ is replaced by $-i\hbar\frac{d}{d\phi}$.

On a lattice a well-defined meaning to the operator $1/|\hat{P}|^{3/2}$ can be given. This can be done by using the Thiemann's trick discussed in section 8.5. In particular, by means of the classical-quantum correspondence (Poisson brackets vs. commutators) one can write the following self-adjoint operator for $|\hat{P}|^{1/2}$

$$\left(\frac{1}{|\hat{P}|}\right)^{1/2} = sign(P)\frac{3}{2\pi\gamma l_P^2 \bar{\mu}}\left(e^{i\frac{\bar{\mu}C}{2}}[|\hat{P}|^{1/2}, e^{-i\frac{\bar{\mu}C}{2}}] - e^{-i\frac{\bar{\mu}C}{2}}[|\hat{P}|^{1/2}, e^{i\frac{\bar{\mu}C}{2}}]\right), \tag{9.99}$$

whose action reads

$$\left(\frac{1}{|\hat{P}|}\right)^{1/2}\psi(\mu) = \left(\frac{6}{8\pi\gamma l_P^2}\right)^{1/2}\frac{1}{\bar{\mu}}\left||\mu+\bar{\mu}|^{1/2} - |\mu-\bar{\mu}|^{1/2}\right|\psi(\mu). \tag{9.100}$$

This way, one finds

$$\left(\frac{1}{|\hat{P}|}\right)^{3/2}\psi(\mu) = B(\mu)\psi(\mu), \tag{9.101}$$

in which

$$B(\mu) = \left(\frac{6}{8\pi\gamma l_P^2}\right)^{3/2}\frac{1}{\bar{\mu}^3}\left||\mu+\bar{\mu}|^{1/2} - |\mu-\bar{\mu}|^{1/2}\right|^3, \tag{9.102}$$

while in the basis $|v\rangle$ one has

$$\left(\frac{1}{|\hat{P}|}\right)^{3/2}\psi(v) = B(v)\psi(v), \tag{9.103}$$

with

$$B(v) = \left(\frac{3}{2}\right)^3 \frac{2^{3/2}}{3^{7/4}}\left(\frac{6}{8\pi\gamma l_P^2}\right)^{3/2}|v|\left||v+1|^{1/2} - |v-1|^{1/2}\right|^3. \tag{9.104}$$

It is worth noting that $B(\mu)$ ($B(v)$) is finite even for $\mu = 0$ ($v = 0$), thus the inverse volume operator is bounded at the classical singularity. This result has been one of the first hints that LQC solves the cosmological singularity [63, 64].

Finally, one ends up with the following difference equation from the vanishing of the full scalar constraint (9.98)

$$\frac{\partial^2}{\partial\phi^2}\Psi(\mu,\phi) = \frac{c^3}{8\pi l_P^2}B^{-1}(\mu)\mathcal{S}^{\bar{\mu}}\Psi(\mu,\phi) = \Theta\Psi(\mu,\phi), \tag{9.105}$$

and the analogous one for v. From the last Klein-Gordon like equation, a Schrödinger dynamics is inferred, *i.e.*

$$i\frac{\partial}{\partial\phi}\Psi(\mu,\phi) = -\sqrt{\Theta}\Psi, \tag{9.106}$$

and the positive/negative eigenfunctions of the operator $-\sqrt{\Theta}$ can be identified with positive/negative frequency states. The whole quantum cosmological issue has been reduced to the investigation of a quantum mechanical system described by a well-defined Hamiltonian $-\sqrt{\Theta}$.

As far as the fate of the classical singularity is concerned, one starts from an initial semi-classical Universe, described by a state sharply picked around $\mu = \mu^\star \gg \bar{\mu}$, and evolves it backward in time. What happens is that the state remains semi-classical during the evolution and a bounce occurs for $\mu \sim \bar{\mu}$, followed by an expansion phase. Therefore, *the classical singularity is solved in the LQC framework.*

However within this scheme two fundamental ambiguities remain, *i.e.* $\bar{\mu}$ and ϵ. A natural requirement is to fix them according with a primitive formulation or with the phenomenological implications. For instance, the choice (9.92) corresponds to the assumption that the physical area enclosed by the minimum loop (made of edges having fiducial length $\bar{\mu}$) coincides with the minimum area gap predicted by LQG [104], *i.e.*

$$\bar{\mu}^2|P| = 2\sqrt{3}\pi\gamma l_P^2, \tag{9.107}$$

in which the right-hand side has been obtained from (8.46) for a single intersection of the kind $e_{B'}$ or $e_{B''}$ for $j = 1/2$. Generically, $\bar{\mu}$ is restricted to be

$$\bar{\mu} \propto \mu^\delta, \qquad -\frac{1}{2} \leq \delta \leq 0, \tag{9.108}$$

and $\delta \sim -1/2$ seems to be the best choice for its phenomenological implications [85, 232, 233] (see next paragraph).

Effective equations It is very impressive to outline the scenario proper of LQC in terms of effective equations. For $\mu \gg \bar{\mu}$ the coefficient $f_+(\mu)$ (9.86) and $B(\mu)$ (9.102) can be written as

$$f_+(\mu) = \frac{3l_P}{2\bar{\mu}^2}|\mu|^{1/2}\sqrt{\frac{8\pi}{6\gamma^3}}\left(1 + O\left(\frac{\bar{\mu}}{\mu}\right)\right), \tag{9.109}$$

$$B(\mu) = \left(\frac{6}{8\pi\gamma l_P^2}\right)^{3/2}\frac{1}{\mu^{3/2}}\left(1 + O\left(\frac{\bar{\mu}}{\mu}\right)\right), \tag{9.110}$$

such that at the leading order the quantum scalar constraint reads[3]

$$\hat{\mathcal{S}}_{tot} \sim -\frac{6}{\gamma^2\bar{\mu}^2}|\hat{P}|^{1/2}\sin^2(\bar{\mu}C) + 8\pi G\frac{\hat{p}_\phi^2}{|\hat{P}|^{3/2}} = 0. \tag{9.111}$$

If (and we know that it is the case, see below) a proper semiclassical state can be defined, from the Ehrenfest theorem we expect that the expectation values will follow the classical trajectory dictated by the condition above. Hence, the leading order correction in the semiclassical expansion is given by the classical expression corresponding to (9.111) and the physical Hamiltonian H_{phys} follows by multiplying times $-NV_0/(16\pi G)$ [4]

$$H_{phys} \sim \frac{3NV_0}{8\pi G\gamma^2\bar{\mu}^2}|P|^{1/2}\sin^2(\bar{\mu}C) - NV_0\frac{p_\phi^2}{2|P|^{3/2}}. \tag{9.112}$$

Let us fix $N = 1$ and evaluate $\dot{a}$ as follows

$$\dot{a} = [a, H_{phys}] = [\sqrt{|p|}, H_{phys}] = sign(P)\frac{1}{\gamma\bar{\mu}}\sin(\bar{\mu}C)\cos(\bar{\mu}C), \tag{9.113}$$

such that the Friedmann equation reads

$$H^2 = \frac{1}{|P|}\frac{1}{(\gamma\bar{\mu})^2}\sin^2(\bar{\mu}C)\cos^2(\bar{\mu}C) = \frac{1}{|P|}\frac{1}{(\gamma\bar{\mu})^2}\sin^2(\bar{\mu}C)(1 - \sin^2(\bar{\mu}C)). \tag{9.114}$$

By using (9.111), one can write $\sin^2\bar{\mu}C$ in terms of other variables, so getting the modified Friedmann equation

$$H^2 = \frac{8\pi G}{3c^3}\frac{p_\phi^2}{2|P|^3}\left(1 - \frac{8\pi G\gamma^2\bar{\mu}^2}{6P^2}p_\phi^2\right). \tag{9.115}$$

The equation can be rewritten as [266]

$$H^2 = \frac{8\pi G}{3c^3}\rho\left(1 - \frac{\rho}{\rho_{cr}}\right), \tag{9.116}$$

---

[3] It is worth noting that the same scalar constraint is obtained from the classical one (in terms of the variables C and P) by the polymer prescription (5.93) $C \to \sin(\bar{\mu}C)/\bar{\mu}$. However, in polymer the resulting expression corresponds to the exact operator on a quantum level.

[4] The factor $-N/(16\pi G)$ comes from the expression of the action in Hamiltonian variables (7.108), while V_0 is due to the spatial integration.

ρ being the matter energy density, while

$$\rho_{cr} = \frac{\sqrt{3}}{8\pi G\gamma^2 |P| \bar{\mu}^2}. \tag{9.117}$$

Hence, for $\rho \gg \rho_{cr}$ the evolution is essentially the one dictated by the Friedmann equation (2.16). On the other hand, when $\rho \approx \rho_{cr}$ the quantum corrections behave as a negative pressure term responsible for the vanishing of the Hubble parameter H. For $\rho = \rho_{cr}$ the bounce occurs. The behavior of the scale factor a vs. ϕ is presented in figure 9.3.

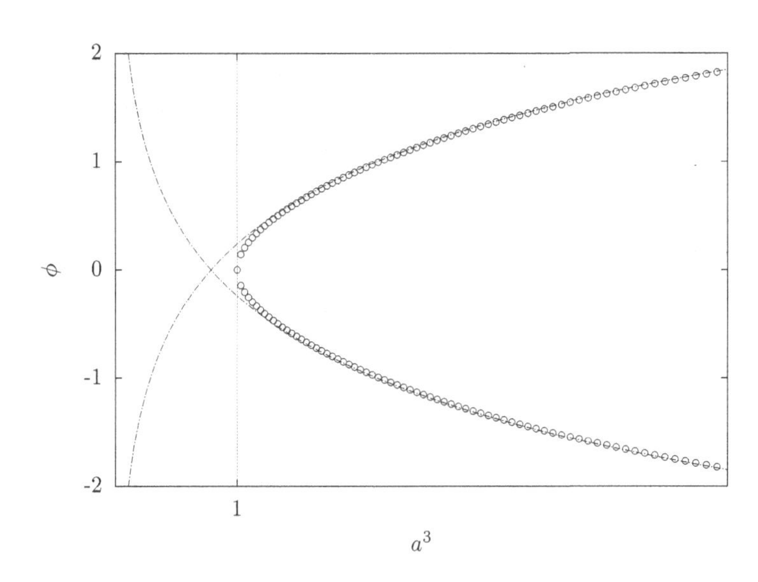

Figure 9.3 The circles denote the points of the numerical integration of $\phi(a)$ obtained from the effective Hamiltonian (9.112), while the dashed lines corresponds to the classical behavior for an expanding and contracting Universe. We see how the Universe undergoes two essentially classic phases, a contracting and an expanding ones, which are connected by the bounce (which we set at $a = 1$).

The numerical investigation on the behavior of the semiclassical states are quite well approximated by the modified Friedmann equation (9.116). In other words, the semiclassical LQC scenario with a massless and not interactive scalar field is stable under quantum corrections [65] (indeed, a strong modification has been proposed in [223]). This is not the case for a massive or interactive scalar field, for which quantum back-reaction on the expectation values cannot be neglected [66, 73].

It is worth noting how generically the critical density is a function of P, thus it depends on the scale factor. This is unwanted, because one expects that quantum effects should be relevant at the Planck scale (maybe times the Immirzi parameters), which is constant. This is the case if ρ_{cr} is constant, which coincides with the choice (9.92) for $\bar{\mu}$.

Perturbations The bounce replacing the initial singularity is the most important theoretical prediction of the LQC framework. However, we have no direct evidence of any signal coming from the early Universe. This situation may be tamed only by the eventual detection of gravitational waves. The best source of information on the early Universe we have at our disposal now is the CMBR. The high degree of isotropy and homogeneity of the CMBR is not only the greatest success (together with nucleosynthesis) of the standard cosmological model, but it also provides us with a clear map on which the behavior of perturbations can be tested. In both cases, the role of perturbations seems crucial in order to link LQC with experiments. In particular, the study of perturbations in LQC outlines that there are two main sources of corrections which could affect the spectrum predicted in the Standard Cosmological Model with a generic inflation mechanism [69, 74, 75]:

- inverse volume corrections, coming from the expansion of the operator $1/\hat{P}^{3/2}$, *i.e.* next-to-leading order terms in (9.109) and (9.110);
- holonomy corrections, due to the fact that connections have been substituted with holonomies.

The most important achievement of this analysis has been the prediction of the suppression for the power spectrum of CMBR scalar perturbations at high scales [68] coming from inverse volume corrections.

Furthermore, it has been pointed out how in LQC, essentially because quantum backreaction is under control, it is possible to naturally set initial conditions such that the Universe is sufficiently homogeneous for inflation to start [2]. However, the results presented in [2] and in [68] are based on a different treatment of perturbations and they are not equivalent. There is now no agreement within the LQC community on which scheme is the correct one.

Anisotropic models The bouncing scenario is a robust prediction of LQC, since it has been verified also in the presence of a cosmological constant [52, 180, 242] and for the most relevant anisotropic models, namely Bianchi type I, II and IX [40, 41, 211, 213, 298].

For instance, in anisotropic cases the fiducial metric is of Bianchi kind and there are three configuration variables C_a ($a = 1, 2, 3$) with conjugate momenta P_a, *i.e.*

$$E^i_a = P^a(t)\,\omega^i_a, \qquad A^a_i(t,x) = C_a(t)\,\omega^a_i, \tag{9.118}$$

ω^i_a being the components of the fiducial 1-forms of the considered Bianchi model (9.2) (the indices a are not summed). In the reduced phase-space P^a and C_a are canonically conjugate variables and the Hilbert space is constructed as the direct product of those of quasi-periodic functions $N_\mu(C_a) = e^{\frac{i\mu C_a}{2}}$, *i.e.*

$$\mathrm{H} = \bigotimes_a L^2(\mathbf{R}_{Bohr,a}, d\mu_{Bohr,a}) \tag{9.119}$$

such that a generic element can be written as

$$\psi_{\{\mu\}}(C_1, C_2, C_3) = e^{\frac{i\mu_1 C_1}{2}} \otimes e^{\frac{i\mu_2 C_2}{2}} \otimes e^{\frac{i\mu_3 C_3}{2}}. \tag{9.120}$$

The quantum scalar constraint is defined by extending the procedure adopted in the isotropic case for $k = 0$. While for Bianchi I this is straightforward (the fiducial vectors are the same as in the isotropic case for $k = 0$), for Bianchi type II and IX there are some technical complications due to the nontrivial action of the Lorentzian part of the scalar constraint and to the fact that one cannot follow the integral curves of two fiducial vectors to close loops (the fiducial metric is curved). The latter issue is present also for non-flat closed Robertson-Walker model. However, while in the isotropic case one can safely construct closed loops taking additional edges not along fiducial vectors, for anisotropic models the resulting holonomy cannot be written in terms of quasi periodic functions (see for instance the case of the Bianchi II model [41]). Hence, the classical expression of the scalar constraint is quantized by "polymerizing" the connection variables (which means writing them in terms of open loops) (see [267] for a discussion on the available inequivalent ways to carry on this procedure).

In the end, a proper scalar constraint can be defined by fixing three minimal values $\bar{\mu}_a$ for μ_a, which result to be three non-polynomial function of (P_1, P_2, P_3). This fact complicates the definition of the operator $e^{i\bar{\mu}_a C_a}$. Nevertheless, by choosing the volume of the Universe as one of the coordinates the dynamics simplifies significantly, since the scalar constraint operator simply shifts the volume and multiplies the wave functions times a function of the volume itself, while anisotropy parameters are rescaled. As a consequence, the (bouncing) scenario is qualitatively similar to the isotropic case and the cosmological singularity is still solved.

Therefore, LQC provides a well grounded formulation also for anisotropic cosmologies and the mechanism responsible for the avoidance of the initial singularity is the same as in the isotropic case. This mechanism is also expected to tame the chaotic behavior of the Bianchi IX model. The quantum gravity corrections manifest as a repulsive interaction which flattens the potential wall of Bianchi IX model at small volumes. Hence, the potential well opens and the Kasner transitions stop after a small number of Kasner epochs. Indeed, this behavior has been verified only for constant $\bar{\mu}_a$ [71, 72] and it is still a conjecture that it works also for different choices (see [298]).

Inhomogeneous case The treatment of inhomogeneous Universes in LQC has been mainly realized via a hybrid quantization scheme. The spatial metric is split into a homogeneous part, which is quantized in the LQC framework, and some inhomogeneous perturbations, for which Fock quantization can be addressed. This has been realized for a Gowdy model [81, 82, 137, 209, 210, 212], for which the qualitative and quantitative features of the bounce are not significantly modified with respect to the full homogeneous case. A different proposal is presented in [70], in which it is suggested to develop a full inhomogeneous Universe by matching several homogeneous patches. However, the resulting model is tamed by non-linearities, which prevents the solution of the full quantum dynamic system.

9.6.2 *Other approaches*

LQC is a well-settled paradigm to extract some phenomenological predictions in a cosmological setting. However, some issues related with the actual foundation from the full theory remain. The basic kinematical symmetries ($SU(2)$ gauge invariance and background independence) are fixed before quantizing. As a consequence two relevant structures proper of the LQG Hilbert space are absent: intertwiners and s-knots. In particular, the absence of the latter is responsible for the mismatch between the regularization procedures of the scalar constraint adopted in LQG and in LQC. The polymer parameter $\bar{\mu}$ has no counterpart in the full theory (where no minimum scale enters the regularization procedure), while it is crucial for the bouncing scenario. In this respect, it is worth noting the correspondence [23, 24] between LQC and the models of spin-foam cosmology [57, 77, 261], in which there is a basic scale fixed by the length of the fundamental tetrahedra filling spacetime.

The argument given in LQC for the emergence of $\bar{\mu}$ is related with the issue to reproduce the minimum non-vanishing area eigenvalue in LQG. The presence of a discrete spectrum for the area operator is due to the $SU(2)$ gauge structure, which is ignored in LQC [96]. This fact makes the argument for the emergence of $\bar{\mu}$ heuristic.

An alternative formulation for cosmology in LQG has been proposed in [4, 5] (Quantum reduced Loop Gravity, QRLG). This new framework is based on projecting the kinematical Hilbert space of the full theory down to a (reduced) space which captures the relevant degrees of freedom of a cosmological model. The definition of this reduced space is made such that a residual diffeomorphisms invariance is retained and the regularization of the scalar constraint can be performed as in LQG.

9.6.2.1 *Quantum Reduced Loop Gravity*

The restriction to a homogeneous line element breaks background independence, since no local diffeomorphism preserves the expression (2.1). Hence, in order to retain a kind of background independence, one must skip homogeneity. In this respect, the idea of QRLG is to reproduce on a quantum level the inhomogeneous extension of the Bianchi I model described by the following phase space variables

$$E^i_a = P^a(t,x)\delta^i_a, \qquad A^a_i(t,x) = C_a(t,x)\delta^a_i, \tag{9.121}$$

which differ from (9.118) in the flat case for the dependence of reduced variables P^a and C_a on spatial coordinates. Such a form for connections and momenta can be obtained exactly by considering a re-parametrized Bianchi I model, in which each scale factor contains a dependence on the associated Cartesian coordinate x^i. The same result holds approximatively under the BKL conjecture [49], in which the spatial gradients of the scale factors are negligible with respect to time-derivatives.

The conditions (9.121) imply the breaking of both $SU(2)$ gauge invariance and background independence. As for the latter, one is restricting to a diagonal metric tensor and there is a residual invariance under the transformations keeping the metric diagonal. These transformations can be seen as those diffeomorphisms (reduced diffeomorphisms) whose associated infinitesimal parameters satisfy

$$n^i = n^i(x^i). \tag{9.122}$$

Geometrically, a reduced diffeomorphism acts on an edge e_i parallel to the fiducial vector ω_i^a as the composition of an arbitrary diffeomorphism along ω_i^a and a rigid translation along other directions.

The $SU(2)$ gauge symmetry is broken and the associated gauge-fixing condition can be written as [91, 96]

$$\chi_a = \varepsilon_{ab}{}^c E^i_c \delta^b_i = 0. \tag{9.123}$$

The conditions (9.121) have been implemented in the kinematical Hilbert space of LQG by i) taking only cuboidal graphs, with edges e_i parallel to fiducial vectors ω_i^a and ii) by the restriction from $SU(2)$ to $U(1)$ group elements via the imposition of the condition (9.123).

The requirement i) can be realized via a projector $\mathcal{P}$ which acts on states ψ_e based at edges e such that $\mathcal{P}\psi_e = \psi_e$ if $e = e_i$ for some i, otherwise it vanishes. Let us call $\mathrm{H}_{\mathcal{P}}$ the resulting Hilbert space, one can project the action $U(\varphi_{\vec{n}})$ of a generic diffeomorphism $\varphi_{\vec{n}}$ from H^{Kin} down to $\mathrm{H}_{\mathcal{P}}$, thus finding that only reduced diffeomorphsism are implemented. Via reduced diffeomorphisms any closed loop in $\mathrm{H}_{\mathcal{P}}$ can be shrunk, while reduced s-knots can be defined as in the full theory (8.38) by group averaging the action of reduced diffeomorphisms.

The condition (9.123) is a gauge fixing and it does not commute with the $SU(2)$ Gauss constraint. The implementation of (9.123) and of $SU(2)$ gauge invariance is performed by mimicking the procedure adopted in spin-foam models to impose the simplicity constraints [119] (which is analogous to the Gupta-Bleuler procedure in QED [61, 157]). We will not give here the details of this procedure, but we focus on the implications. On a classical level the holonomies based at edges e_i for $i = 1, 2, 3$ that fulfill the second condition into (9.121) are elements of the $U(1)_i$ subgroups obtained by stabilizing the $SU(2)$ group around the internal directions $(1, 0, 0)$, $(0, 1, 0)$ and $(0, 0, 1)$, respectively. The quantum analogue in $\mathrm{H}_{\mathcal{P}}$ is the projection of the $SU(2)$ elements ψ_{e_i} into $\tilde{\psi}_{e_i} \in U(1)_i$. This can be done by expanding ψ_{e_i} in $SU(2)$ irreps and by projecting them on the sub-representations with maximum (or minimum) magnetic indexes along the directions i, *i.e.*

$$\tilde{\psi}_{e_i}(g) = \sum_j D_{\rho_{e_i}}{}_i\langle j, j|\rho_{e_i}(g)|j, j\rangle_i \psi^j_{e_i}, \tag{9.124}$$

$|j, m\rangle_i$ being the spin base that diagonalize the operators J^2 and J_i, while $\psi^j_{e_i}$ denotes generic coefficients of the expansion in $SU(2)$ irreps. By repeating this procedure for all edges one gets the element of the reduced Hilbert space H^R.

Fluxes are defined only across the dual surfaces S^i to ω_i^a and their action is diagonal

$$E_k(S^l)\tilde{\psi}(g)_{e_i} = 8\pi\gamma l_P^2 \delta_k^l \delta_i^k \sum_j j\, D_{\rho_{e_i}\, i}\langle j,j|\rho_{e_i}|j,j\rangle_i \psi_{e_i}^j. \qquad (9.125)$$

The action of $SU(2)$ gauge transformations results by projecting from H^{Kin} to H^R, such that some intertwiners are induced in the reduced Hilbert space (*reduced intertwiners*). These are given by projecting the usual invariant intertwiners I_v on the state $|j,j\rangle_i$ for each edge e_i emanating from v (they coincide with Livine-Speziale coherent intertwiners [204] with normals tangent to the cubical cell).

The virtue of this formulation is that the volume operator is diagonal, while the scalar constraint can be regularized as in LQG thanks to reduced diffeomorphisms invariance. These achievements allow to write *an analytic expression for the matrix elements of the scalar constraint.* Hence, in QRLG it is possible to analyze the dynamic implications for the inhomogeneous extension of the Bianchi I model and this gives for the first time the possibility to test LQG techniques in a computable framework. The comparison with LQC may sustain or reject the bouncing scenario (at least for the Bianchi I model and for the flat isotropic cases) and eventually clarify the meaning of the parameter $\bar{\mu}$.

The extension of QRLG to other Bianchi models, to a generic diagonal metric tensor, to the generic cosmological solution and in the presence of matter fields are tantalizing subjects of investigations in view of the simplifications that this scheme provides to the dynamic issue of LQG.

Bibliography

[1] Ade, P. *et al.* (2013). Planck 2013 results. I. Overview of products and scientific results, arXiv:1303.5062.
[2] Agullo, I., Ashtekar, A. and Nelson, W. (2012). A Quantum Gravity Extension of the Inflationary Scenario, *Phys.Rev.Lett.* **109**, p. 251301, doi: 10.1103/PhysRevLett.109.251301, arXiv:1209.1609.
[3] Agullo, I., Fernando Barbero, G., Borja, E. F., Diaz-Polo, J. and Villasenor, E. J. (2009). The Combinatorics of the SU(2) black hole entropy in loop quantum gravity, *Phys.Rev.* **D80**, p. 084006, doi:10.1103/PhysRevD.80.084006, arXiv:0906.4529.
[4] Alesci, E. and Cianfrani, F. (2013a). A new perspective on cosmology in Loop Quantum Gravity, *Europhys.Lett.* **104**, p. 10001, doi:10.1209/0295-5075/104/10001, arXiv:1210.4504.
[5] Alesci, E. and Cianfrani, F. (2013b). Quantum-reduced loop gravity: Cosmology, *Phys. Rev. D* **87**, p. 083521, doi:10.1103/PhysRevD.87.083521, URL `http://link.aps.org/doi/10.1103/PhysRevD.87.083521`.
[6] Alesci, E. and Rovelli, C. (2010). A Regularization of the hamiltonian constraint compatible with the spinfoam dynamics, *Phys.Rev.* **D82**, p. 044007, doi:10.1103/PhysRevD.82.044007, arXiv:1005.0817.
[7] Alesci, E., Thiemann, T. and Zipfel, A. (2012). Linking covariant and canonical LQG: New solutions to the Euclidean Scalar Constraint, *Phys.Rev.* **D86**, p. 024017, doi:10.1103/PhysRevD.86.024017, arXiv:1109.1290.
[8] Alexandrov, S. (2000). SO(4,C) covariant Ashtekar-Barbero gravity and the Immirzi parameter, *Class.Quant.Grav.* **17**, pp. 4255–4268, doi:10.1088/0264-9381/17/20/307, arXiv:gr-qc/0005085.
[9] Alexandrov, S. and Livine, E. R. (2003). SU(2) loop quantum gravity seen from covariant theory, *Phys.Rev.* **D67**, p. 044009, doi:10.1103/PhysRevD.67.044009, arXiv:gr-qc/0209105.
[10] Ambjorn, J., Jurkiewicz, J. and Loll, R. (2000). A Nonperturbative Lorentzian path integral for gravity, *Phys.Rev.Lett.* **85**, pp. 924–927, doi: 10.1103/PhysRevLett.85.924, arXiv:hep-th/0002050.
[11] Amelino-Camelia, G. (2013). Quantum Gravity Phenomenology, *Living*

Rev.Rel. **16**, p. 5, doi:10.12942/lrr-2013-5, arXiv:0806.0339.

[12] Amendola, L., Polarski, D. and Tsujikawa, S. (2007). Are f(R) dark energy models cosmologically viable ? *Phys.Rev.Lett.* **98**, p. 131302, doi:10.1103/PhysRevLett.98.131302, arXiv:astro-ph/0603703.

[13] Arnold, V. I. (1989). *Mathematical Methods of Classical Mechanics* (Springer-Verlag, Berlin).

[14] Arnowitt, R., Deser, S. and Misner, C. W. (1959). Dynamical structure and definition of energy in general relativity, *Phys. Rev.* **116**, pp. 1322–1330, doi:10.1103/PhysRev.116.1322, URL `http://link.aps.org/doi/10.1103/PhysRev.116.1322`.

[15] Arnowitt, R., Deser, S. and Misner, C. W. (1960a). *Journal of Mathematical Physics* **1**, p. 434.

[16] Arnowitt, R., Deser, S. and Misner, C. W. (1960b). Canonical variables for general relativity, *Phys. Rev.* **117**, pp. 1595–1602, doi:10.1103/PhysRev.117.1595, URL `http://link.aps.org/doi/10.1103/PhysRev.117.1595`.

[17] Arnowitt, R. L., Deser, S. and Misner, C. W. (2008). The Dynamics of general relativity, *Gen.Rel.Grav.* **40**, pp. 1997–2027, doi:10.1007/s10714-008-0661-1, arXiv:gr-qc/0405109.

[18] Arnsdorf, M. (1999). Approximating connections in loop quantum gravity, arXiv:gr-qc/9910084.

[19] Ashtekar, A. (1986). New variables for classical and quantum gravity, *Physical Review Letters* **57**, pp. 2244–2247, doi:10.1103/PhysRevLett.57.2244.

[20] Ashtekar, A. (1987). New Hamiltonian formulation of general relativity, *Phys. Rev. D* **36**, pp. 1587–1602, doi:10.1103/PhysRevD.36.1587.

[21] Ashtekar, A., Baez, J., Corichi, A. and Krasnov, K. (1998a). Quantum geometry and black hole entropy, *Phys.Rev.Lett.* **80**, pp. 904–907, doi:10.1103/PhysRevLett.80.904, arXiv:gr-qc/9710007.

[22] Ashtekar, A., Bombelli, L. and Corichi, A. (2005). Semiclassical states for constrained systems, *Phys.Rev.* **D72**, p. 025008, doi:10.1103/PhysRevD.72.025008, arXiv:gr-qc/0504052.

[23] Ashtekar, A., Campiglia, M. and Henderson, A. (2010a). Casting Loop Quantum Cosmology in the Spin Foam Paradigm, *Class.Quant.Grav.* **27**, p. 135020, doi:10.1088/0264-9381/27/13/135020, arXiv:1001.5147.

[24] Ashtekar, A., Campiglia, M. and Henderson, A. (2010b). Path Integrals and the WKB approximation in Loop Quantum Cosmology, *Phys.Rev.* **D82**, p. 124043, doi:10.1103/PhysRevD.82.124043, arXiv:1011.1024.

[25] Ashtekar, A. and Corichi, A. (2003). Nonminimal couplings, quantum geometry and black hole entropy, *Class.Quant.Grav.* **20**, pp. 4473–4484, doi:10.1088/0264-9381/20/20/310, arXiv:gr-qc/0305082.

[26] Ashtekar, A., Corichi, A. and Zapata, J. A. (1998b). Quantum theory of geometry III: Noncommutativity of Riemannian structures, *Class.Quant.Grav.* **15**, pp. 2955–2972, doi:10.1088/0264-9381/15/10/006, arXiv:gr-qc/9806041.

[27] Ashtekar, A., Fairhurst, S. and Willis, J. L. (2003). Quantum gravity, shadow states, and quantum mechanics, *Class.Quant.Grav.* **20**, pp. 1031–1062, doi:10.1088/0264-9381/20/6/302, arXiv:gr-qc/0207106.

[28] Ashtekar, A. and Lewandowski, J. (1995a). Differential geometry on the space of connections via graphs and projective limits, *J.Geom.Phys.* **17**, pp. 191–230, doi:10.1016/0393-0440(95)00028-G, arXiv:hep-th/9412073.
[29] Ashtekar, A. and Lewandowski, J. (1995b). Differential geometry on the space of connections via graphs and projective limits, *J.Geom.Phys.* **17**, pp. 191–230, doi:10.1016/0393-0440(95)00028-G, arXiv:hep-th/9412073.
[30] Ashtekar, A. and Lewandowski, J. (1997). Quantum theory of geometry. 1: Area operators, *Class.Quant.Grav.* **14**, pp. A55–A82, doi:10.1088/0264-9381/14/1A/006, arXiv:gr-qc/9602046.
[31] Ashtekar, A. and Lewandowski, J. (1998). Quantum theory of geometry. 2. Volume operators, *Adv.Theor.Math.Phys.* **1**, pp. 388–429, arXiv:gr-qc/9711031.
[32] Ashtekar, A., Lewandowski, J., Marolf, D., Mourao, J. and Thiemann, T. (1995). Quantization of diffeomorphism invariant theories of connections with local degrees of freedom, *J.Math.Phys.* **36**, pp. 6456–6493, doi:10.1063/1.531252, arXiv:gr-qc/9504018.
[33] Ashtekar, A., Pawlowski, T. and Singh, P. (2006a). Quantum nature of the big bang, *Phys.Rev.Lett.* **96**, p. 141301, doi:10.1103/PhysRevLett.96.141301, arXiv:gr-qc/0602086.
[34] Ashtekar, A., Pawlowski, T. and Singh, P. (2006b). Quantum Nature of the Big Bang: An Analytical and Numerical Investigation. I. *Phys.Rev.* **D73**, p. 124038, doi:10.1103/PhysRevD.73.124038, arXiv:gr-qc/0604013.
[35] Ashtekar, A., Pawlowski, T. and Singh, P. (2006c). Quantum Nature of the Big Bang: Improved dynamics, *Phys.Rev.* **D74**, p. 084003, doi:10.1103/PhysRevD.74.084003, arXiv:gr-qc/0607039.
[36] Ashtekar, A., Pawlowski, T., Singh, P. and Vandersloot, K. (2007). Loop quantum cosmology of k=1 FRW models, *Phys.Rev.* **D75**, p. 024035, doi:10.1103/PhysRevD.75.024035, arXiv:gr-qc/0612104.
[37] Ashtekar, A., Romano, J. D. and Tate, R. S. (1989). New Variables for Gravity: Inclusion of Matter, *Phys.Rev.* **D40**, p. 2572, doi:10.1103/PhysRevD.40.2572.
[38] Ashtekar, A. and Singh, P. (2011). Loop Quantum Cosmology: A Status Report, *Class.Quant.Grav.* **28**, p. 213001, doi:10.1088/0264-9381/28/21/213001, arXiv:1108.0893.
[39] Ashtekar, A. and Sloan, D. (2011). Probability of Inflation in Loop Quantum Cosmology, *Gen.Rel.Grav.* **43**, pp. 3619–3655, doi:10.1007/s10714-011-1246-y, arXiv:1103.2475.
[40] Ashtekar, A. and Wilson-Ewing, E. (2009a). Loop quantum cosmology of Bianchi I models, *Phys.Rev.* **D79**, p. 083535, doi:10.1103/PhysRevD.79.083535, arXiv:0903.3397.
[41] Ashtekar, A. and Wilson-Ewing, E. (2009b). Loop quantum cosmology of Bianchi type II models, *Phys.Rev.* **D80**, p. 123532, doi:10.1103/PhysRevD.80.123532, arXiv:0910.1278.
[42] Baratin, A., Flori, C. and Thiemann, T. (2012). The Holst Spin Foam Model via Cubulations, *New J.Phys.* **14**, p. 103054, doi:10.1088/1367-2630/14/10/103054, arXiv:0812.4055.

[43] Barbero G., J. F. (1996). From Euclidean to Lorentzian general relativity: The Real way, *Phys.Rev.* **D54**, pp. 1492–1499, doi:10.1103/PhysRevD.54.1492, arXiv:gr-qc/9605066.

[44] Barrett, J. W. and Crane, L. (1998). Relativistic spin networks and quantum gravity, *J.Math.Phys.* **39**, pp. 3296–3302, doi:10.1063/1.532254, arXiv:gr-qc/9709028.

[45] Barros e Sa, N. (2001). Hamiltonian analysis of general relativity with the Immirzi parameter, *Int.J.Mod.Phys.* **D10**, pp. 261–272, doi:10.1142/S0218271801000858, arXiv:gr-qc/0006013.

[46] Battisti, M. V., Belvedere, R. and Montani, G. (2009). Semiclassical suppression of the weak anisotropies of a generic Universe, *Europhys.Lett.* **86**, p. 69001, doi:10.1209/0295-5075/86/69001, arXiv:0905.3695.

[47] Bean, R., Bernat, D., Pogosian, L., Silvestri, A. and Trodden, M. (2007). Dynamics of Linear Perturbations in f(R) Gravity, *Phys.Rev.* **D75**, p. 064020, doi:10.1103/PhysRevD.75.064020, arXiv:astro-ph/0611321.

[48] Bekenstein, J. D. (1973). Black holes and entropy, *Phys.Rev.* **D7**, pp. 2333–2346, doi:10.1103/PhysRevD.7.2333.

[49] Belinskii, V. A., Khalatnikov, I. M. and Lifshitz, E. M. (1970). Oscillatory approach to a singular point in the relativistic cosmology, *Advances in Physics* **19**, pp. 525–573.

[50] Belinskii, V. A., Khalatnikov, I. M. and Lifshitz, E. M. (1982). A general solution of the Einstein equations with a time singularity, *Advances in Physics* **31**, pp. 639–667.

[51] Benini, R. and Montani, G. (2007). Inhomogeneous quantum Mixmaster: From classical toward quantum mechanics, *Classical and Quantum Gravity* **24**, pp. 387–404, arXiv.org/abs/gr-qc/0612095.

[52] Bentivegna, E. and Pawlowski, T. (2008). Anti-deSitter universe dynamics in LQC, *Phys.Rev.* **D77**, p. 124025, doi:10.1103/PhysRevD.77.124025, arXiv:0803.4446.

[53] Bertolami, O. (1992). The Concept of time in physics, URL `http://ccdb5fs.kek.jp/cgi-bin/img/allpdf?199306259`.

[54] Bertone, G., Hooper, D. and Silk, J. (2005). Particle dark matter: Evidence, candidates and constraints, *Phys.Rept.* **405**, pp. 279–390, doi:10.1016/j.physrep.2004.08.031, arXiv:hep-ph/0404175.

[55] Bianchi, E. (2012). Entropy of Non-Extremal Black Holes from Loop Gravity, arXiv:1204.5122.

[56] Bianchi, E., Magliaro, E. and Perini, C. (2010a). Coherent spin-networks, *Phys.Rev.* **D82**, p. 024012, doi:10.1103/PhysRevD.82.024012, arXiv:0912.4054.

[57] Bianchi, E., Rovelli, C. and Vidotto, F. (2010b). Towards Spinfoam Cosmology, *Phys.Rev.* **D82**, p. 084035, doi:10.1103/PhysRevD.82.084035, arXiv:1003.3483.

[58] Blagojevic, M. (2002). *Gravitation and gauge symmetries* (Institute of physics publishing).

[59] Blagojevic, M. (2003). Three lectures on Poincare gauge theory, *SFIN* **A1**, pp. 147–172, arXiv:gr-qc/0302040.

[60] Blaut, A. and Kowalski-Glikman, J. (1997). Solutions of quantum gravity coupled to the scalar field, *Phys.Lett.* **B406**, pp. 33–36, doi:10.1016/S0370-2693(97)00665-5, arXiv:gr-qc/9706076.

[61] Bleuler, K. (1950). A New method of treatment of the longitudinal and scalar photons, *Helv.Phys.Acta* **23**, pp. 567–586.

[62] Blyth, W. and Isham, C. (1975). Quantization of a Friedmann Universe Filled with a Scalar Field, *Phys.Rev.* **D11**, pp. 768–778, doi:10.1103/PhysRevD.11.768.

[63] Bojowald, M. (2001). Absence of singularity in loop quantum cosmology, *Phys.Rev.Lett.* **86**, pp. 5227–5230, doi:10.1103/PhysRevLett.86.5227, arXiv:gr-qc/0102069.

[64] Bojowald, M. (2002). Isotropic loop quantum cosmology, *Class.Quant.Grav.* **19**, pp. 2717–2742, doi:10.1088/0264-9381/19/10/313, arXiv:gr-qc/0202077.

[65] Bojowald, M. (2007). Large scale effective theory for cosmological bounces, *Phys.Rev.* **D74**, p. 081301, doi:10.1103/PhysRevD.75.081301, arXiv:gr-qc/0608100.

[66] Bojowald, M. (2008). The Dark Side of a Patchwork Universe, *Gen.Rel.Grav.* **40**, pp. 639–660, doi:10.1007/s10714-007-0558-4, arXiv:0705.4398.

[67] Bojowald, M. (2011). Quantum cosmology, *Lect.Notes Phys.* **835**, pp. pp.1–308, doi:10.1007/978-1-4419-8276-6.

[68] Bojowald, M., Calcagni, G. and Tsujikawa, S. (2011a). Observational constraints on loop quantum cosmology, *Phys.Rev.Lett.* **107**, p. 211302, doi:10.1103/PhysRevLett.107.211302, arXiv:1101.5391.

[69] Bojowald, M., Calcagni, G. and Tsujikawa, S. (2011b). Observational test of inflation in loop quantum cosmology, *JCAP* **1111**, p. 046, doi:10.1088/1475-7516/2011/11/046, arXiv:1107.1540.

[70] Bojowald, M., Chinchilli, A. L., Dantas, C. C., Jaffe, M. and Simpson, D. (2012). Non-linear (loop) quantum cosmology, *Phys.Rev.* **D86**, p. 124027, doi:10.1103/PhysRevD.86.124027, arXiv:1210.8138.

[71] Bojowald, M. and Date, G. (2004). Quantum suppression of the generic chaotic behavior close to cosmological singularities, *Phys. Rev. Lett.* **92**, p. 071302, doi:10.1103/PhysRevLett.92.071302, URL `http://link.aps.org/doi/10.1103/PhysRevLett.92.071302`.

[72] Bojowald, M., Date, G. and Hossain, G. M. (2004). The Bianchi IX model in loop quantum cosmology, *Class.Quant.Grav.* **21**, pp. 3541–3570, doi:10.1088/0264-9381/21/14/015, arXiv:gr-qc/0404039.

[73] Bojowald, M., Hernandez, H. and Skirzewski, A. (2007). Effective equations for isotropic quantum cosmology including matter, *Phys.Rev.* **D76**, p. 063511, doi:10.1103/PhysRevD.76.063511, arXiv:0706.1057.

[74] Bojowald, M. and Hossain, G. M. (2007). Cosmological vector modes and quantum gravity effects, *Class.Quant.Grav.* **24**, pp. 4801–4816, doi:10.1088/0264-9381/24/18/015, arXiv:0709.0872.

[75] Bojowald, M. and Hossain, G. M. (2008). Loop quantum gravity corrections to gravitational wave dispersion, *Phys.Rev.* **D77**, p. 023508, doi:10.1103/

PhysRevD.77.023508, arXiv:0709.2365.

[76] Bolen, B., Bombelli, L. and Corichi, A. (2004). Semiclassical states in quantum cosmology: Bianchi one coherent states, *Class.Quant.Grav.* **21**, pp. 4087–4106, doi:10.1088/0264-9381/21/17/005, arXiv:gr-qc/0404004.

[77] Borja, E. F., Diaz-Polo, J., Garay, I. and Livine, E. R. (2010). Dynamics for a 2-vertex Quantum Gravity Model, *Class.Quant.Grav.* **27**, p. 235010, doi:10.1088/0264-9381/27/23/235010, arXiv:1006.2451.

[78] Boyanovsky, D., Destri, C., De Vega, H. and Sanchez, N. (2009). The Effective Theory of Inflation in the Standard Model of the Universe and the CMB+LSS data analysis, *Int.J.Mod.Phys.* **A24**, pp. 3669–3864, doi: 10.1142/S0217751X09044553, arXiv:0901.0549.

[79] Brans, C. and Dicke, R. (1961). Mach's principle and a relativistic theory of gravitation, *Phys.Rev.* **124**, pp. 925–935, doi:10.1103/PhysRev.124.925.

[80] Brill, D. R. and Wheeler, J. A. (1957). Interaction of neutrinos and gravitational fields, *Rev.Mod.Phys.* **29**, pp. 465–479, doi:10.1103/RevModPhys. 29.465.

[81] Brizuela, D., Mena Marugan, G. and Pawlowski, T. (2010). Big Bounce and inhomogeneities, *Class.Quant.Grav.* **27**, p. 052001, doi:10.1088/0264-9381/ 27/5/052001, arXiv:0902.0697.

[82] Brizuela, D., Mena Marugan, G. A. and Pawlowski, T. (2011). Effective dynamics of the hybrid quantization of the Gowdy T^3 universe, *Phys.Rev.* **D84**, p. 124017, doi:10.1103/PhysRevD.84.124017, arXiv:1106.3793.

[83] Brown, J. D. and Kuchař, K. V. (1995). Dust as a standard of space and time in canonical quantum gravity, *Phys.Rev.* **D51**, pp. 5600–5629, doi: 10.1103/PhysRevD.51.5600, arXiv:gr-qc/9409001.

[84] Brunnemann, J. and Rideout, D. (2008). Properties of the volume operator in loop quantum gravity. I. Results, *Class.Quant.Grav.* **25**, p. 065001, doi: 10.1088/0264-9381/25/6/065001, arXiv:0706.0469.

[85] Cailleteau, T., Mielczarek, J., Barrau, A. and Grain, J. (2012). Anomaly-free scalar perturbations with holonomy corrections in loop quantum cosmology, *Class.Quant.Grav.* **29**, p. 095010, doi:10.1088/0264-9381/29/9/ 095010, arXiv:1111.3535.

[86] Calcagni, G. and Mercuri, S. (2009). The Barbero-Immirzi field in canonical formalism of pure gravity, *Phys.Rev.* **D79**, p. 084004, doi:10.1103/ PhysRevD.79.084004, arXiv:0902.0957.

[87] Castellani, L. (1982). Symmetries in Constrained Hamiltonian Systems, *Annals Phys.* **143**, p. 357, doi:10.1016/0003-4916(82)90031-8.

[88] Cavaglia, M. (1994). Can the interaction between baby universes generate a big universe? *Int.J.Mod.Phys.* **D3**, pp. 623–626, doi:10.1142/ S0218271894000757, arXiv:gr-qc/9408011.

[89] Cianfrani, F. (2011). The kinematical Hilbert space of Loop Quantum Gravity from BF theories, *Class.Quant.Grav.* **28**, p. 175014, doi:10.1088/ 0264-9381/28/17/175014, arXiv:1012.1982.

[90] Cianfrani, F., Lulli, M. and Montani, G. (2012a). Solution of the non-canonicity puzzle in General Relativity: a new Hamiltonian formulation, *Phys.Lett.* **B710**, pp. 703–709, doi:10.1016/j.physletb.2012.03.053,

arXiv:1104.0140.

[91] Cianfrani, F., Marchini, A. and Montani, G. (2012b). The picture of the Bianchi I model via gauge fixing in Loop Quantum Gravity, *Europhys.Lett.* **99**, p. 10003, doi:10.1209/0295-5075/99/10003, arXiv:1201.2588.

[92] Cianfrani, F. and Montani, G. (2008). Dirac equations in curved space-time versus Papapetrou spinning particles, *Europhys.Lett.* **84**, p. 30008, doi:10.1209/0295-5075/84/30008, arXiv:0810.0447.

[93] Cianfrani, F. and Montani, G. (2009a). Matter in Loop Quantum Gravity without time gauge: a non-minimally coupled scalar field, *Phys.Rev.* **D80**, p. 084045, doi:10.1103/PhysRevD.80.084045, arXiv:0904.4435.

[94] Cianfrani, F. and Montani, G. (2009b). Towards Loop Quantum Gravity without the time gauge, *Phys.Rev.Lett.* **102**, p. 091301, doi:10.1103/PhysRevLett.102.091301, arXiv:0811.1916.

[95] Cianfrani, F. and Montani, G. (2010). Gravity in presence of fermions as a SU(2) gauge theory, *Phys.Rev.* **D81**, p. 044015, doi:10.1103/PhysRevD.81.044015, arXiv:1001.2699.

[96] Cianfrani, F. and Montani, G. (2012). Implications of the gauge-fixing in Loop Quantum Cosmology, *Phys.Rev.* **D85**, p. 024027, doi:10.1103/PhysRevD.85.024027, arXiv:1104.4546.

[97] Cianfrani, F. and Montani, G. (2013). Dirac prescription from BRST symmetry in FRW space-time, *Phys.Rev.* **D87**, 8, p. 084025, doi:10.1103/PhysRevD.87.084025, arXiv:1301.4122.

[98] Codello, A. and Zanusso, O. (2013). On the non-local heat kernel expansion, *J.Math.Phys.* **54**, p. 013513, doi:10.1063/1.4776234, arXiv:1203.2034.

[99] Cohen-Tannoudji, C., Diu, B. and Laloe, F. (1991). *Quantum Mechanics* (Wiley-Interscience).

[100] Coleman, S. (1995). *Aspects of Symmetry* (Cambridge University Press).

[101] Collins, J., Perez, A., Sudarsky, D., Urrutia, L. and Vucetich, H. (2004). Lorentz invariance and quantum gravity: an additional fine-tuning problem? *Phys.Rev.Lett.* **93**, p. 191301, doi:10.1103/PhysRevLett.93.191301, arXiv:gr-qc/0403053.

[102] Conrady, F. (2005). Geometric spin foams, Yang-Mills theory and background-independent models, arXiv:gr-qc/0504059.

[103] Corichi, A. and Krasnov, K. V. (1998). Ambiguities in loop quantization: Area versus electric charge, *Mod.Phys.Lett.* **A13**, pp. 1339–1346, doi:10.1142/S0217732398001406, arXiv:hep-th/9703177.

[104] Corichi, A. and Singh, P. (2008). Is loop quantization in cosmology unique? *Phys.Rev.* **D78**, p. 024034, doi:10.1103/PhysRevD.78.024034, arXiv:0805.0136.

[105] Corichi, A., Vukasinac, T. and Zapata, J. A. (2007a). Hamiltonian and physical Hilbert space in polymer quantum mechanics, *Class.Quant.Grav.* **24**, pp. 1495–1512, doi:10.1088/0264-9381/24/6/008, arXiv:gr-qc/0610072.

[106] Corichi, A., Vukasinac, T. and Zapata, J. A. (2007b). Polymer Quantum Mechanics and its Continuum Limit, *Phys.Rev.* **D76**, p. 044016, doi:10.1103/PhysRevD.76.044016, arXiv:0704.0007.

[107] Date, G., Kaul, R. K. and Sengupta, S. (2009). Topological Interpretation

of Barbero-Immirzi Parameter, *Phys.Rev.* **D79**, p. 044008, doi:10.1103/PhysRevD.79.044008, arXiv:0811.4496.

[108] de la Madrid, R. (2005). The role of the rigged hilbert space in quantum mechanics, *European Journal of Physics* **26**, 2, p. 287.

[109] De Pietri, R. (1997a). On the relation between the connection and the loop representation of quantum gravity, *Class.Quant.Grav.* **14**, pp. 53–70, doi:10.1088/0264-9381/14/1/009, arXiv:gr-qc/9605064.

[110] De Pietri, R. (1997b). Spin networks and recoupling in loop quantum gravity, *Nucl.Phys.Proc.Suppl.* **57**, pp. 251–254, doi:10.1016/S0920-5632(97)00397-6, arXiv:gr-qc/9701041.

[111] DeWitt, B. S. (1967). Quantum theory of gravity I: The canonical theory, *Physical Review* **160**, pp. 1113–1148.

[112] Dirac, P. A. (1950). Generalized Hamiltonian dynamics, *Can.J.Math.* **2**, pp. 129–148, doi:10.4153/CJM-1950-012-1.

[113] Dirac, P. A. (1958). The Theory of gravitation in Hamiltonian form, *Proc.Roy.Soc.Lond.* **A246**, pp. 333–343, doi:10.1098/rspa.1958.0142.

[114] Dittrich, B. (2006). Partial and complete observables for canonical general relativity, *Class.Quant.Grav.* **23**, pp. 6155–6184, doi:10.1088/0264-9381/23/22/006, arXiv:gr-qc/0507106.

[115] Dobrev, V. (2002). Quantum mechanics with difference operators, *Reports on Mathematical Physics* **50**, pp. 409–431, doi:10.1016/S0034-4877(02)80069-6, arXiv:arXiv:quant-ph/0207077.

[116] Dolgov, A. and Kawasaki, M. (2003). Can modified gravity explain accelerated cosmic expansion? *Phys.Lett.* **B573**, pp. 1–4, doi:10.1016/j.physletb.2003.08.039, arXiv:astro-ph/0307285.

[117] Dziendzikowski, M. and Okolow, A. (2010). New diffeomorphism invariant states on a holonomy-flux algebra, *Class.Quant.Grav.* **27**, p. 225005, doi:10.1088/0264-9381/27/22/225005, arXiv:0912.1278.

[118] Einstein, A. (1916). Hamilton's Principle and the General Theory of Relativity, *Sitzungsber.Preuss.Akad.Wiss.Berlin (Math.Phys.)* **1916**, pp. 1111–1116.

[119] Engle, J., Livine, E., Pereira, R. and Rovelli, C. (2008). LQG vertex with finite Immirzi parameter, *Nucl.Phys.* **B799**, pp. 136–149, doi:10.1016/j.nuclphysb.2008.02.018, arXiv:0711.0146.

[120] Engle, J., Perez, A. and Noui, K. (2010). Black hole entropy and SU(2) Chern-Simons theory, *Phys.Rev.Lett.* **105**, p. 031302, doi:10.1103/PhysRevLett.105.031302, arXiv:0905.3168.

[121] Ezawa, K. (1996). A Semiclassical interpretation of the topological solutions for canonical quantum gravity, *Phys.Rev.* **D53**, pp. 5651–5663, doi:10.1103/PhysRevD.53.5651, arXiv:gr-qc/9512017.

[122] Fakir, R. (1990). Quantum creation of universes with nonminimal coupling, *Phys. Rev. D* **41**, pp. 3012–3023, doi:10.1103/PhysRevD.41.3012, URL `http://link.aps.org/doi/10.1103/PhysRevD.41.3012`.

[123] Fatibene, L., Francaviglia, M. and Rovelli, C. (2007). On a Covariant Formulation of the Barbero-Immirzi Connection, *Class.Quant.Grav.* **24**, pp. 3055–3066, doi:10.1088/0264-9381/24/11/017, arXiv:gr-qc/0702134.

[124] Flori, C. (2009). Semiclassical analysis of the Loop Quantum Gravity volume operator: Area Coherent States, arXiv:0904.1303.

[125] Flori, C. and Thiemann, T. (2008). Semiclassical analysis of the Loop Quantum Gravity volume operator. I. Flux Coherent States, arXiv:0812.1537.

[126] Fock, V. and Rosly, A. (1999). Poisson structure on moduli of flat connections on Riemann surfaces and r matrix, *Am.Math.Soc.Transl.* **191**, pp. 67–86, arXiv:math/9802054.

[127] Fredenhagen, K. and Reszewski, F. (2006). Polymer state approximations of Schrodinger wave functions, *Class.Quant.Grav.* **23**, pp. 6577–6584, doi: 10.1088/0264-9381/23/22/028, arXiv:gr-qc/0606090.

[128] Freidel, L. and Krasnov, K. (2000). Simple spin networks as Feynman graphs, *J.Math.Phys.* **41**, pp. 1681–1690, doi:10.1063/1.533203, arXiv:hep-th/9903192.

[129] Freidel, L. and Krasnov, K. (2008). A New Spin Foam Model for 4d Gravity, *Class.Quant.Grav.* **25**, p. 125018, doi:10.1088/0264-9381/25/12/125018, arXiv:0708.1595.

[130] Freidel, L. and Livine, E. R. (2010). The Fine Structure of SU(2) Intertwiners from U(N) Representations, *J.Math.Phys.* **51**, p. 082502, doi: 10.1063/1.3473786, arXiv:0911.3553.

[131] Freidel, L. and Speziale, S. (2010). Twisted geometries: A geometric parametrisation of SU(2) phase space, *Phys.Rev.* **D82**, p. 084040, doi: 10.1103/PhysRevD.82.084040, arXiv:1001.2748.

[132] Frieman, J., Turner, M. and Huterer, D. (2008). Dark Energy and the Accelerating Universe, *Ann.Rev.Astron.Astrophys.* **46**, pp. 385–432, doi: 10.1146/annurev.astro.46.060407.145243, arXiv:0803.0982.

[133] Furusawa, T. (1986a). Quantum chaos of Mixmaster universe, *Progress of Theoretical Physics* **75**, pp. 59–67.

[134] Furusawa, T. (1986b). Quantum chaos of Mixmaster universe. II, *Progress of Theoretical Physics* **76**, p. 67.

[135] Gambini, R., Griego, J. and Pullin, J. (1997). Chern-Simons states in spin network quantum gravity, *Phys.Lett.* **B413**, pp. 260–266, doi:10.1016/S0370-2693(97)01048-4, arXiv:gr-qc/9703042.

[136] Gamow, G. (1946). Expanding universe and the origin of elements, *Phys.Rev.* **70**, pp. 572–573, doi:10.1103/PhysRev7.0.572.

[137] Garay, L., Martin-Benito, M. and Mena Marugan, G. (2010). Inhomogeneous Loop Quantum Cosmology: Hybrid Quantization of the Gowdy Model, *Phys.Rev.* **D82**, p. 044048, doi:10.1103/PhysRevD.82.044048, arXiv:1005.5654.

[138] Gaul, M. and Rovelli, C. (2001). A Generalized Hamiltonian constraint operator in loop quantum gravity and its simplest Euclidean matrix elements, *Class.Quant.Grav.* **18**, pp. 1593–1624, doi:10.1088/0264-9381/18/9/301, arXiv:gr-qc/0011106.

[139] Gelfand, I. and Naimark, M. (1957). *Unitaere Darstellungen der klassischen Gruppen* (Akademie Verlag).

[140] Geroch, R. P. (1970). The domain of dependence, *J.Math.Phys.* **11**, pp. 437–439, doi:10.1063/1.1665157.

[141] Ghosh, A. and Mitra, P. (2005). An Improved lower bound on black hole entropy in the quantum geometry approach, *Phys.Lett.* **B616**, pp. 114–117, doi:10.1016/j.physletb.2005.05.003, arXiv:gr-qc/0411035.

[142] Ghosh, A. and Perez, A. (2011). Black hole entropy and isolated horizons thermodynamics, *Phys.Rev.Lett.* **107**, p. 241301, doi:10.1103/PhysRevLett.107.241301,10.1103/PhysRevLett.108.169901, arXiv:1107.1320.

[143] Gibbons, G. and Hawking, S. (1977). Action Integrals and Partition Functions in Quantum Gravity, *Phys.Rev.* **D15**, pp. 2752–2756, doi:10.1103/PhysRevD.15.2752.

[144] Giddings, S. B. and Strominger, A. (1989). Baby Universes, Third Quantization and the Cosmological Constant, *Nucl.Phys.* **B321**, p. 481, doi:10.1016/0550-3213(89)90353-2.

[145] Giesel, K. and Thiemann, T. (2006). Consistency check on volume and triad operator quantisation in loop quantum gravity. II. *Class.Quant.Grav.* **23**, pp. 5693–5772, doi:10.1088/0264-9381/23/18/012, arXiv:gr-qc/0507037.

[146] Giesel, K. and Thiemann, T. (2007a). Algebraic Quantum Gravity (AQG). I. Conceptual Setup, *Class.Quant.Grav.* **24**, pp. 2465–2498, doi:10.1088/0264-9381/24/10/003, arXiv:gr-qc/0607099.

[147] Giesel, K. and Thiemann, T. (2007b). Algebraic Quantum Gravity (AQG). II. Semiclassical Analysis, *Class.Quant.Grav.* **24**, pp. 2499–2564, doi:10.1088/0264-9381/24/10/004, arXiv:gr-qc/0607100.

[148] Giesel, K. and Thiemann, T. (2007c). Algebraic quantum gravity (AQG). III. Semiclassical perturbation theory, *Class.Quant.Grav.* **24**, pp. 2565–2588, doi:10.1088/0264-9381/24/10/005, arXiv:gr-qc/0607101.

[149] Giesel, K. and Thiemann, T. (2010). Algebraic quantum gravity (AQG). IV. Reduced phase space quantisation of loop quantum gravity, *Class.Quant.Grav.* **27**, p. 175009, doi:10.1088/0264-9381/27/17/175009, arXiv:0711.0119.

[150] Giesel, K. and Thiemann, T. (2012). Scalar Material Reference Systems and Loop Quantum Gravity, arXiv:1206.3807.

[151] Giles, R. (1981). Reconstruction of gauge potentials from wilson loops, *Phys. Rev. D* **24**, pp. 2160–2168, doi:10.1103/PhysRevD.24.2160, URL `http://link.aps.org/doi/10.1103/PhysRevD.24.2160`.

[152] Giulini, D. (2009). The Superspace of Geometrodynamics, *Gen.Rel.Grav.* **41**, pp. 785–815, doi:10.1007/s10714-009-0771-4, arXiv:0902.3923.

[153] Goldstein, H., Poole, C. and Safko, J. (2002). *Classical Mechanics* (Pearson Education).

[154] Greensite, J. (1991). Ehrenfest's principle in Quantum Gravity, *Nucl.Phys.* **B351**, pp. 749–766, doi:10.1016/S0550-3213(05)80043-4.

[155] Grishchuk, L., Doroshkevich, A. and Yudin, V. (1975). Long Gravitational Waves in a Closed Universe, *Zh.Eksp.Teor.Fiz.* **69**, pp. 1857–1871.

[156] Grot, N. and Rovelli, C. (1997). Weave states in loop quantum gravity, *Gen.Rel.Grav.* **29**, pp. 1039–1048, doi:10.1023/A:1018876726684.

[157] Gupta, S. N. (1950). Theory of longitudinal photons in quantum electrodynamics, *Proc.Phys.Soc.* **A63**, pp. 681–691.

[158] Haag, R. (1992). *Local quantum physics: Fields, particles, algebras*

(Springer-Verlag).

[159] Hamber, H. W., Toriumi, R. and Williams, R. M. (2012a). Wheeler-DeWitt Equation in 2 + 1 Dimensions, *Phys.Rev.* **D86**, p. 084010, doi:10.1103/PhysRevD.86.084010, arXiv:1207.3759.

[160] Hamber, H. W., Toriumi, R. and Williams, R. M. (2012b). Wheeler-DeWitt Equation in 3 + 1 Dimensions, arXiv:1212.3492.

[161] Hartle, J. B. (1991). Spacetime coarse grainings in nonrelativistic quantum mechanics, *Phys. Rev. D* **44**, pp. 3173–3196, doi:10.1103/PhysRevD.44.3173, URL `http://link.aps.org/doi/10.1103/PhysRevD.44.3173`.

[162] Hartle, J. B. and Hawking, S. W. (1983). Wave function of the universe, *Phys. Rev. D* **28**, pp. 2960–2975, doi:10.1103/PhysRevD.28.2960, URL `http://link.aps.org/doi/10.1103/PhysRevD.28.2960`.

[163] Hayashi, K. and Shirafuji, T. (1979). New general relativity, *Phys. Rev. D* **19**, pp. 3524–3553, doi:10.1103/PhysRevD.19.3524, URL `http://link.aps.org/doi/10.1103/PhysRevD.19.3524`.

[164] Hayashi, K. and Shirafuji, T. (1980). Gravity from Poincare Gauge Theory of the Fundamental Particles. 1. Linear and Quadratic Lagrangians, *Prog.Theor.Phys.* **64**, p. 866.

[165] Hehl, F., Von Der Heyde, P., Kerlick, G. and Nester, J. (1976). General Relativity with Spin and Torsion: Foundations and Prospects, *Rev.Mod.Phys.* **48**, pp. 393–416, doi:10.1103/RevModPhys.48.393.

[166] Hehl, F. W. (1979). Four lectures on Poincaré gauge field theory, .

[167] Henneaux, M. and Teitelboim, C. (1992). *Quantization of gauge systems* (Princeton University press).

[168] Hewitt, E. and Ross, K. (1963). *Abstract harmonic analysis* (Springer-Verlag).

[169] Hinshaw, G. *et al.* (2013). Nine-Year Wilkinson Microwave Anisotropy Probe (WMAP) Observations: Cosmological Parameter Results, *Astrophys.J.Suppl.* **208**, p. 19, doi:10.1088/0067-0049/208/2/19, arXiv:1212.5226.

[170] Holst, S. (1996). Barbero's Hamiltonian derived from a generalized Hilbert-Palatini action, *Phys.Rev.* **D53**, pp. 5966–5969, doi:10.1103/PhysRevD.53.5966, arXiv:gr-qc/9511026.

[171] Hubble, E. (1929). A relation between distance and radial velocity among extra-galactic nebulae, *Proc.Nat.Acad.Sci.* **15**, pp. 168–173, doi:10.1073/pnas.15.3.168.

[172] Husain, V. and Pawlowski, T. (2012). Time and a physical Hamiltonian for quantum gravity, *Phys.Rev.Lett.* **108**, p. 141301, doi:10.1103/PhysRevLett.108.141301, arXiv:1108.1145.

[173] Immirzi, G. (1997). Real and complex connections for canonical gravity, *Class.Quant.Grav.* **14**, pp. L177–L181, doi:10.1088/0264-9381/14/10/002, arXiv:gr-qc/9612030.

[174] Isham, C. (1983). Topological and global aspects of quantum theory, URL `http://ccdb5fs.kek.jp/cgi-bin/img$\_$index?8404235`.

[175] Isham, C. (1992). Canonical quantum gravity and the problem of time, arXiv:gr-qc/9210011.

[176] Isham, C. and Butterfield, J. (2000). Some possible roles for topos theory in quantum theory and quantum gravity, *Found.Phys.* **30**, pp. 1707–1735, doi:10.1023/A:1026406502316, arXiv:gr-qc/9910005.

[177] Isham, C. and Kakas, A. (1984a). A group theoretical approach to the canonical quantization of gravity. 1. Construction of the canonical group, *Class.Quant.Grav.* **1**, p. 621, doi:10.1088/0264-9381/1/6/008.

[178] Isham, C. and Kakas, A. (1984b). A group theoretical approach to the canonical quantization of gravity. 2. Unitary representations of the canonical group, *Class.Quant.Grav.* **1**, p. 633, doi:10.1088/0264-9381/1/6/009.

[179] Israel, W. (1973). *Relativity, Astrophysics and Cosmology* (Springer).

[180] Kaminski, W. and Pawlowski, T. (2010). The LQC evolution operator of FRW universe with positive cosmological constant, *Phys.Rev.* **D81**, p. 024014, doi:10.1103/PhysRevD.81.024014, arXiv:0912.0162.

[181] Kaptanoglu, S. (1981). Weakly Canonical Transformations and the Path Integral Quantization, *Phys.Lett.* **B98**, p. 77, doi:10.1016/0370-2693(81)90372-5.

[182] Kasner, E. (1921). Geometrical theorems on einstein's cosmological equations, *Advances in Physics* **43**, p. 217.

[183] Kirillov, A. and Melnikov, V. (1995). Dynamics of inhomogeneities of metric in the vicinity of a singularity in multidimensional cosmology, *Phys.Rev.* **D52**, pp. 723–729, doi:10.1103/PhysRevD.52.723, arXiv:gr-qc/9408004.

[184] Kirillov, A. A. (1993). On the question of the characteristics of the spatial distribution of metric inhomogeneities in a general solution to einstein equations in the vicinity of a cosmological singularity, *Soviet Physics JETP* **76**, p. 355.

[185] Kiriushcheva, N. and Kuzmin, S. (2011). The Hamiltonian formulation of General Relativity: Myths and reality, *Central Eur.J.Phys.* **9**, pp. 576–615, doi:10.2478/s11534-010-0072-2, arXiv:0809.0097.

[186] Kodama, H. (1988). Specialization of Ashtekar's formalism to Bianchi cosmology, *Prog.Theor.Phys.* **80**, p. 1024, doi:10.1143/PTP.80.1024.

[187] Kodama, H. (1990). Holomorphic wave function of the Universe, *Phys.Rev.* **D42**, pp. 2548–2565, doi:10.1103/PhysRevD.42.2548.

[188] Kogut, J. and Susskind, L. (1975). Hamiltonian formulation of wilson's lattice gauge theories, *Phys. Rev. D* **11**, pp. 395–408, doi:10.1103/PhysRevD.11.395, URL `http://link.aps.org/doi/10.1103/PhysRevD.11.395`.

[189] Kolb, E. W. and Turner, M. S. (1990). *The Early Universe* (Addison-Wesley).

[190] Koslowski, T. and Sahlmann, H. (2012). Loop quantum gravity vacuum with nondegenerate geometry, *SIGMA* **8**, p. 026, doi:10.3842/SIGMA.2012.026, arXiv:1109.4688.

[191] Kowalski-Glikman, J. and Meissner, K. A. (1996). A Class of exact solutions of the Wheeler-de Witt equation, *Phys.Lett.* **B376**, pp. 48–52, doi:10.1016/0370-2693(96)00268-7, arXiv:hep-th/9601062.

[192] Kowalski-Glikman, J. and Starodubtsev, A. (2008). Effective particle kinematics from Quantum Gravity, *Phys.Rev.* **D78**, p. 084039, doi:10.1103/PhysRevD.78.084039, arXiv:0808.2613.

[193] Kuchař, K. (1980). Canonical methods of quantization.
[194] Kuchař, K. (2011). Time and interpretations of quantum gravity, *Int.J. Mod.Phys.Proc.Suppl.* **D20**, pp. 3–86, doi:10.1142/S0218271811019347.
[195] Landau, L. D. and Lifshitz, E. M. (1958). *Quantum mechanics : non-relativistic theory* (Pergamon Press).
[196] Landau, L. D. and Lifshitz, E. M. (1975). *Classical Theory of Fields*, 4th edn. (Addison-Wesley, New York).
[197] Lecian, O. M., Montani, G. and Carlevaro, N. (2013). Novel Analysis of Spinor Interactions and non-Riemannian Geometry, *EPJ Plus 128,* **19**, doi:10.1140/epjp/i2013-13019-y, arXiv:1301.7708.
[198] Levi-Civita, T. (1917). Nozione di parallelismo in una varietà qualunque e consequente specificazione geometrica della curvatura Riemanniana, *Rend. Circ. Mat. Palermo* **42**, pp. 73–205.
[199] Lewandowski, J. and Marolf, D. (1998). Loop constraints: A Habitat and their algebra, *Int.J.Mod.Phys.* **D7**, pp. 299–330, doi:10.1142/S0218271898000231, arXiv:gr-qc/9710016.
[200] Lewandowski, J., Okolow, A., Sahlmann, H. and Thiemann, T. (2006). Uniqueness of diffeomorphism invariant states on holonomy-flux algebras, *Commun.Math.Phys.* **267**, pp. 703–733, doi:10.1007/s00220-006-0100-7, arXiv:gr-qc/0504147.
[201] Lifshitz, E. M. and Khalatnikov, I. M. (1963). Investigations in relativistic cosmology, *Advances in Physics* **12**, pp. 185–249, doi:10.1080/00018736300101283.
[202] Livine, E. R. (2006). Towards a Covariant Loop Quantum Gravity, arXiv:gr-qc/0608135.
[203] Livine, E. R. and Speziale, S. (2007a). A New spinfoam vertex for quantum gravity, *Phys.Rev.* **D76**, p. 084028, doi:10.1103/PhysRevD.76.084028, arXiv:0705.0674.
[204] Livine, E. R. and Speziale, S. (2007b). A New spinfoam vertex for quantum gravity, *Phys.Rev.* **D76**, p. 084028, doi:10.1103/PhysRevD.76.084028, arXiv:0705.0674.
[205] MacDowell, S. and Mansouri, F. (1977). Unified Geometric Theory of Gravity and Supergravity, *Phys.Rev.Lett.* **38**, p. 739, doi:10.1103/PhysRevLett.38.1376,10.1103/PhysRevLett.38.739.
[206] Magnano, G. (1995). Are there metric theories of gravity other than general relativity? arXiv:gr-qc/9511027.
[207] Marolf, D. (2000). Group averaging and refined algebraic quantization: Where are we now? arXiv:gr-qc/0011112.
[208] Marolf, D. and Rovelli, C. (2002). Relativistic quantum measurement, *Phys.Rev.* **D66**, p. 023510, doi:10.1103/PhysRevD.66.023510, arXiv:gr-qc/0203056.
[209] Martin-Benito, M., Garay, L. J. and Mena Marugan, G. A. (2008a). Hybrid Quantum Gowdy Cosmology: Combining Loop and Fock Quantizations, *Phys.Rev.* **D78**, p. 083516, doi:10.1103/PhysRevD.78.083516, arXiv:0804.1098.

[210] Martin-Benito, M., Martin-de Blas, D. and Mena Marugan, G. A. (2011). Matter in inhomogeneous loop quantum cosmology: the Gowdy T^3 model, *Phys.Rev.* **D83**, p. 084050, doi:10.1103/PhysRevD.83.084050, arXiv:1012.2324.

[211] Martin-Benito, M., Marugan, G. A. M. and Pawlowski, T. (2009). Physical evolution in Loop Quantum Cosmology: The Example of vacuum Bianchi I, *Phys.Rev.* **D80**, p. 084038, doi:10.1103/PhysRevD.80.084038, arXiv:0906.3751.

[212] Martin-Benito, M., Marugan, G. A. M. and Wilson-Ewing, E. (2010). Hybrid Quantization: From Bianchi I to the Gowdy Model, *Phys.Rev.* **D82**, p. 084012, doi:10.1103/PhysRevD.82.084012, arXiv:1006.2369.

[213] Martin-Benito, M., Mena Marugan, G. and Pawlowski, T. (2008b). Loop Quantization of Vacuum Bianchi I Cosmology, *Phys.Rev.* **D78**, p. 064008, doi:10.1103/PhysRevD.78.064008, arXiv:0804.3157.

[214] Melchiorri, A., Pagano, L. and Pandolfi, S. (2007). When Did Cosmic Acceleration Start ? *Phys.Rev.* **D76**, p. 041301, doi:10.1103/PhysRevD.76.041301, arXiv:0706.1314.

[215] Mena Marugan, G. A. (1995). Is the exponential of the Chern-Simons action a normalizable physical state? *Class.Quant.Grav.* **12**, pp. 435–442, doi:10.1088/0264-9381/12/2/012, arXiv:gr-qc/9402034.

[216] Mercuri, S. (2006). Fermions in Ashtekar-Barbero connections formalism for arbitrary values of the Immirzi parameter, *Phys.Rev.* **D73**, p. 084016, doi:10.1103/PhysRevD.73.084016, arXiv:gr-qc/0601013.

[217] Mercuri, S. (2008). From the Einstein-Cartan to the Ashtekar-Barbero canonical constraints, passing through the Nieh-Yan functional, *Phys.Rev.* **D77**, p. 024036, doi:10.1103/PhysRevD.77.024036, arXiv:0708.0037.

[218] Mercuri, S. and Montani, G. (2004). Dualism between physical frames and time in quantum gravity, *Mod.Phys.Lett.* **A19**, p. 1519, doi:10.1142/S0217732304014756, arXiv:gr-qc/0312077.

[219] Mercuri, S. and Taveras, V. (2009). Interaction of the Barbero-Immirzi Field with Matter and Pseudo-Scalar Perturbations, *Phys.Rev.* **D80**, p. 104007, doi:10.1103/PhysRevD.80.104007, arXiv:0903.4407.

[220] Messiah, A. (1979). *Quantum Mechanics. VOL.. 2* (North-Holland).

[221] Meusburger, C. and Schroers, B. (2003). Poisson structure and symmetry in the Chern-Simons formulation of (2+1)-dimensional gravity, *Class.Quant.Grav.* **20**, pp. 2193–2234, doi:10.1088/0264-9381/20/11/318, arXiv:gr-qc/0301108.

[222] Meusburger, C. and Schroers, B. J. (2009). Generalised Chern-Simons actions for 3d gravity and kappa-Poincare symmetry, *Nucl.Phys.* **B806**, pp. 462–488, doi:10.1016/j.nuclphysb.2008.06.023, arXiv:0805.3318.

[223] Mielczarek, J. and Szydlowski, M. (2008). Emerging singularities in the bouncing loop cosmology, *Phys.Rev.* **D77**, p. 124008, doi:10.1103/PhysRevD.77.124008, arXiv:0801.1073.

[224] Misner, C. W. (1969a). Mixmaster universe, *Phys. Rev. Lett.* **22**, pp. 1071–1074, doi:10.1103/PhysRevLett.22.1071, URL `http://link.aps.org/doi/10.1103/PhysRevLett.22.1071`.

[225] Misner, C. W. (1969b). Quantum cosmology I, *Physical Review* **186**, pp. 1319–1327.

[226] Misner, C. W., Thorne, K. S. and Wheeler, J. A. (1973). *Gravitation* (Freeman, W. H. and C., San Francisco).

[227] Montani, G. (1995). On the general behavior of the universe near the cosmological singularity, *Classical and Quantum Gravity* **12**, 10, pp. 2505–2517.

[228] Montani, G. (2002). Canonical quantization of gravity without frozen formalism, *Nucl.Phys.* **B634**, pp. 370–392, doi:10.1016/S0550-3213(02)00301-2, arXiv:gr-qc/0205032.

[229] Montani, G., Battisti, M. V., Benini, R. and Imponente, G. (2008). Classical and Quantum Features of the Mixmaster Singularity, *Int.J.Mod.Phys.* **A23**, pp. 2353–2503, doi:10.1142/S0217751X08040275, arXiv:0712.3008.

[230] Montani, G., Battisti, M. V., Benini, R. and Imponente, G. (2011). *Primordial cosmology* (World Scientific).

[231] Naimark, M. and Stern, A. (2006). *Theory of Group Representations* (Springer-Verlag).

[232] Nelson, W. and Sakellariadou, M. (2007a). Lattice Refining Loop Quantum Cosmology and Inflation, *Phys.Rev.* **D76**, p. 044015, doi:10.1103/PhysRevD.76.044015, arXiv:0706.0179.

[233] Nelson, W. and Sakellariadou, M. (2007b). Lattice refining LQC and the matter Hamiltonian, *Phys.Rev.* **D76**, p. 104003, doi:10.1103/PhysRevD.76.104003, arXiv:0707.0588.

[234] Nesti, F. and Percacci, R. (2008). Graviweak Unification, *J.Phys.* **A41**, p. 075405, doi:10.1088/1751-8113/41/7/075405, arXiv:0706.3307.

[235] Neumann, J. v. (1931). Die eindeutigkeit der schrödingerschen operatoren, *Mathematische Annalen* **104**, pp. 570–578, URL `http://eudml.org/doc/159483`.

[236] Nicolai, H., Peeters, K. and Zamaklar, M. (2005). Loop quantum gravity: An Outside view, *Class.Quant.Grav.* **22**, p. R193, doi:10.1088/0264-9381/22/19/R01, arXiv:hep-th/0501114.

[237] Nitsch, J. (1979). The macroscopic limit of the Poincaré gauge field theory of gravitation. (TALK), .

[238] Ohanian, H. C. (1976). *Gravitation and spacetime* (W. W. Norton, Incorporated).

[239] Parentani, R. (1997). Interpretation of the solutions of the Wheeler-DeWitt equation, *Phys. Rev. D* **56**, pp. 4618–4624, doi:10.1103/PhysRevD.56.4618, URL `http://link.aps.org/doi/10.1103/PhysRevD.56.4618`.

[240] Parisi, G. (1998). *Statistical Field Theory* (Westview Press).

[241] Paternoga, R. and Graham, R. (2000). Triad representation of the chern-simons state in quantum gravity, *Phys. Rev. D* **62**, p. 084005, doi:10.1103/PhysRevD.62.084005, URL `http://link.aps.org/doi/10.1103/PhysRevD.62.084005`.

[242] Pawłowski, T. and Ashtekar, A. (2012). Positive cosmological constant in loop quantum cosmology, *Phys. Rev. D* **85**, p. 064001, doi:10.1103/PhysRevD.85.064001, URL `http://link.aps.org/doi/10.1103/PhysRevD.85.064001`.

[243] Peacock, J. A. (1999). *Cosmological Physics* (Cambridge University press).
[244] Peleg, Y. (1991). On the third quantization of general relativity, *Class.Quant.Grav.* **8**, pp. 827–842, doi:10.1088/0264-9381/8/5/008.
[245] Penzias, A. A. and Wilson, R. W. (1965). A Measurement of excess antenna temperature at 4080-Mc/s, *Astrophys.J.* **142**, pp. 419–421, doi:10.1086/148307.
[246] Peres, A. (1962). On cauchy's problem in general relativity - ii, *Il Nuovo Cimento Series 10* **26**, 1, pp. 53–62, doi:10.1007/BF02754342, URL `http://dx.doi.org/10.1007/BF02754342`.
[247] Perez, A. (2006). Regularization ambiguities in loop quantum gravity, *Phys. Rev. D* **73**, p. 044007, doi:10.1103/PhysRevD.73.044007, URL `http://link.aps.org/doi/10.1103/PhysRevD.73.044007`.
[248] Perez, A. and Rovelli, C. (2006). Physical effects of the immirzi parameter in loop quantum gravity, *Phys. Rev. D* **73**, p. 044013, doi:10.1103/PhysRevD.73.044013, URL `http://link.aps.org/doi/10.1103/PhysRevD.73.044013`.
[249] Randono, A. (2006a). Generalizing the Kodama state. I. Construction, arXiv:gr-qc/0611073.
[250] Randono, A. (2006b). Generalizing the Kodama state. II. Properties and physical interpretation, arXiv:gr-qc/0611074.
[251] Regge, T. (1961). General Relativity without coordinates, *Nuovo Cim.* **19**, pp. 558–571, doi:10.1007/BF02733251.
[252] Robertson, H. (1935). Kinematics and World-Structure, *Astrophys.J.* **82**, pp. 284–301, doi:10.1086/143681.
[253] Rothe, H. (1992). Lattice gauge theories: An Introduction, *World Sci.Lect.Notes Phys.* **43**, pp. 1–381.
[254] Rovelli, C. (1991a). Ashtekar formulation of general relativity and loop space nonperturbative quantum gravity: A Report, *Class.Quant.Grav.* **8**, pp. 1613–1676, doi:10.1088/0264-9381/8/9/002.
[255] Rovelli, C. (1991b). Time in quantum gravity: An hypothesis, *Phys. Rev. D* **43**, pp. 442–456, doi:10.1103/PhysRevD.43.442, URL `http://link.aps.org/doi/10.1103/PhysRevD.43.442`.
[256] Rovelli, C. (1996). Black hole entropy from loop quantum gravity, *Phys.Rev.Lett.* **77**, pp. 3288–3291, doi:10.1103/PhysRevLett.77.3288, arXiv:gr-qc/9603063.
[257] Rovelli, C. (1997). Quantum gravity as a 'sum over surfaces', *Nucl.Phys.Proc.Suppl.* **57**, pp. 28–43, doi:10.1016/S0920-5632(97)00351-4.
[258] Rovelli, C. and Smolin, L. (1990). Loop Space Representation of Quantum General Relativity, *Nucl.Phys.* **B331**, p. 80, doi:10.1016/0550-3213(90)90019-A.
[259] Rovelli, C. and Speziale, S. (2010). On the geometry of loop quantum gravity on a graph, *Phys.Rev.* **D82**, p. 044018, doi:10.1103/PhysRevD.82.044018, arXiv:1005.2927.
[260] Rovelli, C. and Thiemann, T. (1998). The Immirzi parameter in quantum general relativity, *Phys.Rev.* **D57**, pp. 1009–1014, doi:10.1103/PhysRevD.57.1009, arXiv:gr-qc/9705059.

[261] Rovelli, C. and Vidotto, F. (2008). Stepping out of Homogeneity in Loop Quantum Cosmology, *Class.Quant.Grav.* **25**, p. 225024, doi:10.1088/0264-9381/25/22/225024, arXiv:0805.4585.

[262] Sakurai, J. (1985). *Modern quantum mechanics* (Addison-Wesley).

[263] Samuel, J. (2001). Comment on holst's lagrangian formulation, *Phys. Rev. D* **63**, p. 068501, doi:10.1103/PhysRevD.63.068501, URL `http://link.aps.org/doi/10.1103/PhysRevD.63.068501`.

[264] Shestakova, T. and Simeone, C. (2004a). The Problem of time and gauge invariance in the quantization of cosmological models. 1. Canonical quantization methods, *Grav.Cosmol.* **10**, pp. 161–176, arXiv:gr-qc/0409114.

[265] Shestakova, T. and Simeone, C. (2004b). The Problem of time and gauge invariance in the quantization of cosmological models. II. Recent developments in the path integral approach, *Grav.Cosmol.* **10**, pp. 257–268, arXiv:gr-qc/0409119.

[266] Singh, P., Vandersloot, K. and Vereshchagin, G. (2006). Non-singular bouncing universes in loop quantum cosmology, *Phys.Rev.* **D74**, p. 043510, doi:10.1103/PhysRevD.74.043510, arXiv:gr-qc/0606032.

[267] Singh, P. and Wilson-Ewing, E. (2013). Quantization ambiguities and bounds on geometric scalars in anisotropic loop quantum cosmology, arXiv:1310.6728.

[268] Smolin, L. (2002). Quantum gravity with a positive cosmological constant, arXiv:hep-th/0209079.

[269] Sotiriou, T. P. and Faraoni, V. (2010). f(R) Theories Of Gravity, *Rev.Mod.Phys.* **82**, pp. 451–497, doi:10.1103/RevModPhys.82.451, arXiv:0805.1726.

[270] Szulc, L. (2007). Open FRW model in Loop Quantum Cosmology, *Class.Quant.Grav.* **24**, pp. 6191–6200, doi:10.1088/0264-9381/24/24/003, arXiv:0707.1816.

[271] Szulc, L., Kaminski, W. and Lewandowski, J. (2007). Closed FRW model in Loop Quantum Cosmology, *Class.Quant.Grav.* **24**, pp. 2621–2636, doi:10.1088/0264-9381/24/10/008, arXiv:gr-qc/0612101.

[272] Taub, A. H. (1951). Empty space-times admitting a three parameter group of motions, *The Annals of Mathematics* **53**, 3, pp. 472–490.

[273] Taveras, V. and Yunes, N. (2008). The Barbero-Immirzi Parameter as a Scalar Field: K-Inflation from Loop Quantum Gravity? *Phys.Rev.* **D78**, p. 064070, doi:10.1103/PhysRevD.78.064070, arXiv:0807.2652.

[274] Thiemann, T. (1996). Reality conditions inducing transforms for quantum gauge field theory and quantum gravity, *Class.Quant.Grav.* **13**, pp. 1383–1404, doi:10.1088/0264-9381/13/6/012, arXiv:gr-qc/9511057.

[275] Thiemann, T. (1998a). QSD 3: Quantum constraint algebra and physical scalar product in quantum general relativity, *Class.Quant.Grav.* **15**, pp. 1207–1247, doi:10.1088/0264-9381/15/5/010, arXiv:gr-qc/9705017.

[276] Thiemann, T. (1998b). Quantum spin dynamics (QSD), *Class.Quant.Grav.* **15**, pp. 839–873, doi:10.1088/0264-9381/15/4/011, arXiv:gr-qc/9606089.

[277] Thiemann, T. (1998c). Quantum spin dynamics (qsd). 2. *Class.Quant.Grav.* **15**, pp. 875–905, doi:10.1088/0264-9381/15/4/012, arXiv:gr-qc/9606090.

[278] Thiemann, T. (2001). Gauge field theory coherent states (GCS): 1. General properties, *Class.Quant.Grav.* **18**, pp. 2025–2064, doi:10.1088/0264-9381/18/11/304, arXiv:hep-th/0005233.

[279] Thiemann, T. (2006a). Complexifier coherent states for quantum general relativity, *Class.Quant.Grav.* **23**, pp. 2063–2118, doi:10.1088/0264-9381/23/6/013, arXiv:gr-qc/0206037.

[280] Thiemann, T. (2006b). Solving the Problem of Time in General Relativity and Cosmology with Phantoms and k-Essence, arXiv:astro-ph/0607380.

[281] Thiemann, T. (2006c). The Phoenix project: Master constraint program for loop quantum gravity, *Class.Quant.Grav.* **23**, pp. 2211–2248, doi:10.1088/0264-9381/23/7/002, arXiv:gr-qc/0305080.

[282] Thiemann, T. (2007). Loop Quantum Gravity: An Inside View, *Lect.Notes Phys.* **721**, pp. 185–263, doi:10.1007/978-3-540-71117-9$_$10, arXiv:hep-th/0608210.

[283] Thiemann, T. (2008). *Modern canonical quantum general relativity* (Cambridge University press), arXiv:gr-qc/0110034.

[284] Thiemann, T. and Winkler, O. (2001a). Gauge field theory coherent states (GCS). 2. Peakedness properties, *Class.Quant.Grav.* **18**, pp. 2561–2636, doi:10.1088/0264-9381/18/14/301, arXiv:hep-th/0005237.

[285] Thiemann, T. and Winkler, O. (2001b). Gauge field theory coherent states (GCS): 3. Ehrenfest theorems, *Class.Quant.Grav.* **18**, pp. 4629–4682, doi:10.1088/0264-9381/18/21/315, arXiv:hep-th/0005234.

[286] Thiemann, T. and Winkler, O. (2001c). Gauge field theory coherent states (GCS) 4: Infinite tensor product and thermodynamical limit, *Class.Quant.Grav.* **18**, pp. 4997–5054, doi:10.1088/0264-9381/18/23/302, arXiv:hep-th/0005235.

[287] Tomlin, C. and Varadarajan, M. (2013). Towards an Anomaly-Free Quantum Dynamics for a Weak Coupling Limit of Euclidean Gravity, *Phys.Rev.D* **87**, 4, p. 044039, doi:10.1103/PhysRevD.87.044039, arXiv:1210.6869.

[288] Vandersloot, K. (2007). Loop quantum cosmology and the $k = -1$ robertson-walker model, *Phys. Rev. D* **75**, p. 023523, doi:10.1103/PhysRevD.75.023523, URL `http://link.aps.org/doi/10.1103/PhysRevD.75.023523`.

[289] Varadarajan, M. (2008). Towards new background independent representations for loop quantum gravity, *Class.Quant.Grav.* **25**, p. 105011, doi:10.1088/0264-9381/25/10/105011, arXiv:0709.1680.

[290] Varadarajan, M. (2013). Towards an Anomaly-Free Quantum Dynamics for a Weak Coupling Limit of Euclidean Gravity: Diffeomorphism Covariance, *Phys.Rev.* **D87**, 4, p. 044040, doi:10.1103/PhysRevD.87.044040, arXiv:1210.6877.

[291] Vilenkin, A. (1989). Interpretation of the wave function of the universe, *Phys. Rev. D* **39**, pp. 1116–1122, doi:10.1103/PhysRevD.39.1116, URL `http://link.aps.org/doi/10.1103/PhysRevD.39.1116`.

[292] Vollick, D. N. (2003). $1/r$ curvature corrections as the source of the cosmological acceleration, *Phys. Rev. D* **68**, p. 063510, doi:10.1103/PhysRevD.68.063510, URL `http://link.aps.org/doi/10.1103/PhysRevD.68.063510`.

[293] Von der Heyde, P. (1976). The field equations of the Poincaré gauge theory of gravitation, *Phys.Lett.* **A58**, pp. 141–143, doi:10.1016/0375-9601(76) 90266-8.

[294] Wald, R. M. (1984). *General Relativity* (University of Chicago Press).

[295] Weinberg, S. (2000). *The Quantum Theory of Fields: Modern Applications*, Vol. 3 (Cambridge University Press).

[296] Weyl, H. (1950). *The Theory of Groups and Quantum Mechanics* (Dover).

[297] Wilson, K. G. (1974). Confinement of quarks, *Phys. Rev. D* **10**, pp. 2445–2459, doi:10.1103/PhysRevD.10.2445, URL `http://link.aps.org/doi/10.1103/PhysRevD.10.2445`.

[298] Wilson-Ewing, E. (2010). Loop quantum cosmology of bianchi type ix models, *Phys. Rev. D* **82**, p. 043508, doi:10.1103/PhysRevD.82.043508, URL `http://link.aps.org/doi/10.1103/PhysRevD.82.043508`.

[299] York, J., James W. (1972). Role of conformal three geometry in the dynamics of gravitation, *Phys.Rev.Lett.* **28**, pp. 1082–1085, doi:10.1103/PhysRevLett.28.1082.

Index

Printed in the United States
By Bookmasters